Biomining

D.E. Rawlings · B.D. Johnson (Eds.)

Biomining

With 72 Figures, 5 in Color, and 37 Tables

 Springer

Douglas E. Rawlings
Professor and HOD of Microbiology
University of Stellenbosch
Private Bag X1
Matieland 7602
South Africa

D. Barrie Johnson
School of Biological Sciences
University of Wales
Bangor LL57 2UW
United Kingdom

ISBN 978-3-642-07115-7   e-ISBN 978-3-540-34911-2

Springer-Verlag is a part of Springer Science + Business Media

springer.com

© Springer-Verlag Berlin Heidelberg 2010

Editor: Dr. Christina Eckey, Heidelberg
Desk Editor: Anette Lindqvist, Heidelberg

Cover Design: Design & Production, Heidelberg

Preface

Biomining is the generic term that describes the processing of metal-containing ores and concentrates using (micro-) biological technology. This is an area of biotechnology that has seen considerable growth in scale and application since the 1960s, when it was first used, in very basically engineered rock "dumps" to recover copper from ores which contained too little of the metal to be processed by conventional smelting. Refinements in engineering design of commercial biomining operations have paralleled advances in our understanding of the biological agents that drive the process, so biomining is now a multifaceted area of applied science, involving operators and researchers working in seemingly disparate disciplines, including geology, chemical engineering, microbiology and molecular biology. This is reflected in the content of this book, which includes chapters written by persons from industry and academia, all of whom are acknowledged leading practitioners and authorities in their fields.

Biomining has a particular application as an alternative to traditional physical-chemical methods of mineral processing in a variety of niche areas. These include deposits where the metal values are low, where the presence of certain elements (e.g., arsenic) would lead to smelter damage, or where environmental considerations favor biological treatment options. Commercial-scale biomining operations are firmly established in all five continents, with the exception of Europe, though precommercial ("pilot-scale") investigations have recently been set up in Finland to examine the feasibility of extracting nickel and copper from complex metal ores, in engineered heaps. While copper recovery has been, and continues to be, a major metal recovered via biomining, ores and concentrates of other base metals (such as cobalt) and precious metals (chiefly gold) are also processed using this biotechnology.

Developments and refinements of engineering practices in biomining have been important in improving the efficiency of metal recovery. The application of heap leaching to mineral processing continues to expand and, whereas this was once limited to copper processing, considerable experience has been gained in using heaps for gold recovery in the Carlin Trend deposits of the USA. Also, in recent years, there has been industrial-scale application of a radically different approach for heap leaching (the GEOCOAT process), which is described in this book. The other major engineering approach used in biomining – the use of stirred-tank bioreactors – has been established for over 20 years. Over this time, these systems, used mostly for processing refractory gold ores, have been found to be far more robust than was initially envisaged. Huge mineral leaching tanks are in place in various parts of the world, and are described in this book by the commercial operators who have designed and constructed the majority of them. This book also includes a chapter describing how the use of high-temperature stirred-tank bioreactors is being explored as an option to recover copper from chalcopyrite, a mineral (quantitatively the most abundant copper mineral) that has so far proven recalcitrant to biological processing.

Two other important aspects of biomining are covered in this book. One is the nature and diversity of the microorganisms that are central to the core function of bioprocessing of ores, and how these may be monitored in commercial operations. The biophysical strategies used by different microorganisms and microbial consortia for the biodegradation of the ubiquitous mineral pyrite, as well as what is known about the pathways and genetics of the enzymes involved in iron and sulfur oxidation are also described. Significant advances that are being made in what has for long been a black box – the modeling of heap reactors – are also described.

This book follows a previous text entitled *Biomining: Theory, Microbes and Industrial Processes*, also published by Springer (in 1997) and which became out of print a short time after its publication. We believe that, owing to the efforts of colleagues who have contributed to this completely rewritten and updated text, this book is a worthy successor.

Douglas E. Rawlings
Barrie Johnson
May 2006

Contents

List of Contributors

MURRAY BATH
GeoBiotics, LLC, Suite 310, 12345 W. Alameda Parkway, Lakewood, CO 80228, USA

JOHN D. BATTY
Johannesburg Technology Centre, BHP Billiton, Private Bag X10014, Randburg, 2125, South Africa

VIOLAINE BONNEFOY
CNRS, Laboratoire de Chimie Bactérienne, Institut de Biologie Structurale et de Microbiologie, 31 chemin Joseph Aiguier, 13402 Marseille Cedex 20, France

JAMES A. BRIERLEY
Brierley Consultancy LLC, 2074 E. Terrace Drive, Highlands Ranch, CO 80126, USA

DAVID W. DEW
Johannesburg Technology Centre, BHP Billiton, Private Bag X10014, Randburg, 2125, South Africa

DAVID G. DIXON
Department of Materials Engineering, University of British Columbia, 6350 Stores Road, Vancouver, BC, V6T 1Z4, Canada

PATRICK D'HUGUES
BRGM, 3 Avenue Claude Guillemin, 45060 Orléans Cedex 2, France

CHRIS A. DU PLESSIS
Johannesburg Technology Centre, BHP Billiton, Private Bag X10014, Randburg, 2125, South Africa

ESTEBAN M. DOMIC
DOMIC SA, Office 61, Santa Magdalena 10, Providencia, Chile, and Mining Engineering Department, Universidad de Chile, Santiago, Chile

PETER D. FRANZMANN
Centre for Environment and Life Sciences, CSIRO Land and Water, Private Bag No. 5, Wembley, WA 6913, Australia

KEVIN B. HALLBERG
School of Biological Sciences, University of Wales, Bangor LL47 4UF, UK

TODD J. HARVEY
GeoBiotics, LLC, Suite 310, 12345 W. Alameda Parkway, Lakewood, CO 80228, USA

REBECCA B. HAWKES
School of Biological Sciences and Biotechnology, Murdoch University, South Street, Murdoch, WA 6150, Australia

DAVID S. HOLMES
Laboratory of Bioinformatics and Genome Biology, Andrés Bello University and Millennium Institute of Fundamental and Applied Biology, Santiago, Chile

D. BARRIE JOHNSON
School of Biological Sciences, University of Wales, Bangor LL47 4UF, UK

ANNA H. KAKSONEN
Institute of Environmental Engineering and Biotechnology, Tampere University of Technology, P.O. Box 541, 33101 Tampere, Finland

THOMAS C. LOGAN
Newmont Mining Corporation, 10101 E. Dry Creek Road, Englewood, CO 80112, USA

DOMINIQUE HENRI ROGER MORIN
BRGM, 3 Avenue Claude Guillemin, 45060 Orléans Cedex 2, France

PAUL R. NORRIS
Department of Biological Sciences, University of Warwick, Coventry CV4 7AL, UK

WALDEMAR OLIVIER
Goldfields Limited, St. Andrews Road, Parktown, Johannesburg, 2193, South Africa

JOCHEN PETERSEN
Department of Chemical Engineering, University of Cape Town, Private Bag, Rondebosch, 7701, South Africa

JASON J. PLUMB
Centre for Environment and Life Sciences, CSIRO Land and Water, Private Bag No. 5, Wembley, WA 6913, Australia

JAAKKO A. PUHAKKA
Institute of Environmental Engineering and Biotechnology, Tampere University of Technology, P.O. Box 541, 33101 Tampere, Finland

DOUGLAS E. RAWLINGS
Department of Microbiology, University of Stellenbosch, Private Bag X1, Matieland, 7602, South Africa

MARJA RIEKKOLA-VANHANEN
Talvivaara Mining Company Limited, Salmelantie 6, 88600 Sotkamo, Finland

JOSÉ ROJAS-CHAPANA
Nanoparticle Technology Department, Research Center Caesar, 53175 Bonn, Germany

THOM SEAL
Newmont Mining Corporation, Carlin Operations, P.O. Box 669, Carlin, NV 89822, USA

HELMUT TRIBUTSCH
Solare Energetik Department, Hahn–Meitner-Institut Berlin, 14109 Berlin, Germany

PIETER C. VAN ASWEGEN
Goldfields Limited, St. Andrews Road, Parktown, Johannesburg, 2193, South Africa

JAN VAN NIEKERK
Goldfields Limited, St. Andrews Road, Parktown, Johannesburg, 2193, South Africa

1 The BIOX™ Process for the Treatment of Refractory Gold Concentrates

PIETER C. VAN ASWEGEN, JAN VAN NIEKERK, WALDEMAR OLIVIER

1.1 Introduction

Gencor has pioneered the commercialization of bioxidation of refractory gold ores. Development of the BIOX™ process started in the late 1970s at Gencor Process Research, in Johannesburg, South Africa. The successful development of the technology led to the commissioning of a BIOX™ pilot plant in 1984, followed by the first commercial BIOX™ plant at the Fairview mine in 1986 (van Aswegen et al. 1988). The BIOX™ process was fully commercialized in 1991 when the Fairview plant was expanded to treat the total concentrate production of the mine and the Edwards roasters were finally shut down.

Commissioning of a further three BIOX™ plants at Harbour Lights (Barter et al. 1992) in 1992, Wiluna (Stephenson and Kelson 1997) in 1993 and Sansu (Nicholson et al. 1993) in 1994 followed. Toward the end of 1990 a single BIOX™ tank was commissioned at the Saõ Bento mine in Brazil (Slabbert et al. 1992) to operate also series with two pressure oxidation autoclaves. In 1998 the Tamboraque BIOX™ plant (Loayza and Ly 1999) was commissioned in Peru and concluded what could be considered as a first generation of commercial BIOX™ plants. For all these BIOX™ plants the technology was provided under a technology license agreement.

The robustness, simplicity of operation, environmental friendliness and cost-effectiveness of the technology has been demonstrated at all of these operations. The BIOX™ process has been a technical and economic success and offers real advantages over conventional refractory processes, such as roasting and pressure oxidation. Ongoing development work on bench and pilot scales, as well as on operating plants, is aimed at improving the efficiency and cost-effectiveness of the process even further.

When the interests of Gencor and Gold Fields of South Africa were merged in February 1998 to form the new Gold Fields, the BIOX™ process technology and its holding company, Biomin Technologies Limited were transferred to the new company. With the increase in the gold price, there has been a renewed interest in the development of refractory gold ore deposits and the application of the BIOX™ technology to treat such ores. The year 2005 can be considered to mark the development and commissioning of a new generation BIOX™ plants to treat refractory gold ore concentrates. Both the Suzdal

Biomining
(ed. by Douglas E. Rawlings and D. Barrie Johnson)
© Springer-Verlag Berlin Heidelberg 2007

BIOX™ plant in Kazakhstan and the Fosterville plant in Australia were commissioned during May 2005. During 2006 and 2007, BIOX™ plants will be
commissioned at the Jinfeng (China), Bogoso (Ghana) and Kokpatas
(Uzbekistan) projects.

1.2 The BIOX™ Process Flow Sheet

The typical process flow sheet for the BIOX™ process is shown in Fig. 1.1. The
sulfide concentrate from the flotation section of the plant is pumped to the
BIOX™ stock tank. Flotation concentrate is thickened to a density of at least
50% solids to minimize carryover of flotation reagents to the BIOX™ reactors. A minimum sulfide-S concentration of approximately 6% is usually
required to ensure adequate bacterial activity during the biooxidation stage.

A regrind circuit may be included in the circuit before the stock tank, especially when a portion of the concentrate is produced using flash flotation. The
feed concentrate to BIOX™ is typically milled to 80% smaller than 75 μm with
a minimum diameter of more than 150 μm. An increase in the grind size
would reduce of particles with a sulfide oxidation rate and would result in a
lower overall oxidation for similar BIOX™ treatment periods. Fine grinding
to 80% smaller than 20 μm will enhance the sulfide oxidation rate but may
influence the downstream processes negatively, for example to increase the
settling area required or to increase the viscosity of the slurry.

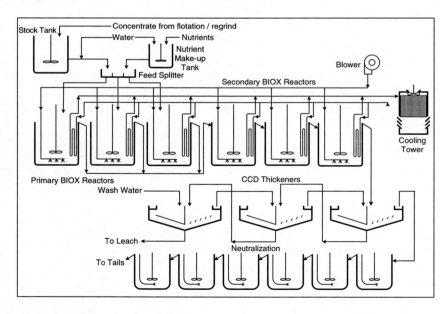

Fig. 1.1. Typical BIOX process flow sheet

A biooxidation plant typically consists of six equidimensional reactors configured as three primary reactors operating in parallel followed by three secondary reactors operating in series. The feed concentrate from the stock tank is diluted to 20% solids by mass before being fed to the primary BIOX™ reactors. The operating slurry solids content is determined mainly by the oxygen mass transfer requirement of the process. In cases of low sulfide-S concentrations, it may be possible to operate the reactors at a higher solids concentration.

The pulp residence time in the biooxidation reactors is typically 4–6 days depending on the oxidation rates achieved, and is a function of the sulfide-S content and mineralogical composition of the concentrate. Generally, half of the retention time is spent in the primary reactors to allow a stable bacterial population to be established and to prevent bacterial washout. Once a stable bacterial population has been established, a shorter retention time can be tolerated in the secondary reactors where sulfide-S oxidation is completed.

Nutrients in the form of nitrogen, phosphorus and potassium salts are also added to the primary reactors to promote bacterial growth. The standard addition rates and nutrient sources specified by Gold Fields are listed in Table 1.1. Low concentrations of nutrients are often present in the concentrate and this creates the opportunity to reduce the nutrient addition rates once stable operation has been achieved at the plant (Olivier et al. 2000).

The mixed culture of mesophilic bacteria used in the BIOX™ process can operate at temperatures ranging from 30 to 45°C. The pulp temperature in commercial reactors is controlled between 40 and 45°C. This temperature allows maximum sulfide oxidation rates to be achieved while minimizing cooling requirements. The oxidation of sulfide minerals is an exothermic process and the reactors must be cooled continuously by circulating cold water through a series of cooling coils installed inside the reactors. Evaporative cooling towers are used to remove heat from the cooling water.

A minimum carbonate content of 2% in the flotation concentrate is usually required to ensure that sufficient CO_2 is available in the concentrate to promote bacterial cell production. If no carbonate is present, limestone or $CO_2(g)$ must be added to the primary reactors as a source of carbon for cell production.

Low-pressure air is injected into the BIOX™ reactors to supply oxygen for the oxidation reactions. It is extremely important that a dissolved oxygen concentration of more than 2 mg L^{-1} be maintained at all times in the slurry.

Table 1.1. Standard nutrient addition rates and sources

Nutrient	Addition (kg t^{-1})	Source
Nitrogen	1.7	Ammonium sulfate, ammonium phosphate salts and urea
Phosphorus	0.9	Ammonium phosphates and phosphoric acid
Potassium	0.3	Potassium sulfate, hydroxide and phosphate salts

The supply and dispersion of the air is one of the main capital and operating cost components for a commercial biooxidation plant. This is discussed in more detail in Sect. 1.5.

The oxidation of pyrite produces acid, while the oxidation of arsenopyrite and pyrrhotite and the dissolution of carbonate minerals consume acid. Limestone and sulfuric acid are used to control the pH in the BIOX™ reactors within the optimum range of pH 1.2–1.8.

The BIOX™ product contains high concentrations of dissolved ions and must be washed in a three-stage countercurrent decantation (CCD) circuit before cyanide leaching. The washed BIOX™ product would normally contain less than 1 g L^{-1} total iron in solution with a pH of 1–3. Iron removal is necessary before cyanide leaching to promote gold recovery and reduce cyanide consumption. The CCD wash thickener overflow liquor is neutralized in a two-stage process to pH 7–8 to produce a stable precipitate containing all the iron and arsenic. The final precipitates are stable and safe for disposal on a tailings dam.

The process requirements, engineering design and operation of the BIOX™ process are described in detail in the following sections of this chapter.

The BIOX™ process can also be integrated with other metallurgical processes to either increase the treatment capacity of an existing plant or to remove certain contaminants from the material being treated.

The Saõ Bento operation in Brazil is a good example where the BIOX™ process was combined with an existing pressure oxidation plant to increase the capacity of the plant (Slabbert et al. 1992). In this application, BIOX™ was used as a preoxidation step to oxidize a portion of the sulfur before the material was fed to the autoclave, thereby reducing the sulfide-S loading on the autoclave. A total of three BIOX™ reactors were installed over a period, operating in parallel. Biooxidation was a quick and low-cost option to increase the capacity of the existing pressure oxidation plant.

The BIOX™ process can also be combined with other unit processes. The BIOX™ process can be used to remove arsenic or base metal contaminants from the concentrate feed to smelter operations. The arsenic can then be precipitated as a stable product suitable for land disposal. The configuration of the BIOX™ plant and the location in the process flow sheet can be selected to fit the specific application.

Recent testwork has also confirmed the ability of the BIOX™ process to treat arsenic trioxide produced during the roasting of arsenopyrite-containing concentrates. Arsenic trioxide is recovered as a dry powder from the roaster off-gas and disposing of it is both difficult and expensive owing to the toxicity of As(III). Pilot plant testwork and commercial scale plant experience indicated that the BIOX™ process can successfully oxidize the As(III) to As(V) in the BIOX™ reactors (Osei-Owusu 2001; van Niekerk 2001). The arsenic can then be precipitated as a stable ferric arsenate during neutralization.

1.3   Current Status of Operating BIOX™ Plants

Full descriptions of the eight BIOX™ plants mentioned in the "Introduction" have been described in a number of papers (Barter et al. 1992; Loayza and Ly 1999; Nicholson et al. 1993; Slabbert et al. 1992; Stephenson and Kelson 1997; van Aswegen et al. 1988). Table 1.2 gives a summary of the commercial BIOX™ plants, previously and currently in operation, as of late 2005. A short summary of the five operations currently in operation is presented in this section.

1.3.1   The Fairview BIOX™ Plant

The BIOX™ process has been in operation for 19 years at the Fairview mine in South Africa. The pilot plant was commissioned in 1986 to treat 10 t day^{-1} in parallel with the aging Edwards roasters. The process proved to be robust and the capacity of the BIOX™ section was increased in 1991 to treat the full 35 t day^{-1} concentrate. The capacity of the plant was again increased in 1994 and 1999 to the current design capacity of 62 t day^{-1}.

The reactor configuration at Fairview is not the standard BIOX™ configuration owing to the addition of new reactors with each expansion phase. The performance of the plant over the years has, however, proven the stability and adaptability of the process to varying concentrate characteristics and operating conditions (Irons 2001). The Fairview BIOX™ plant has played a vital role in the development of the process. The size of the operation and the close

Table 1.2. A summary of the commercial BIOX operations, currently and previously in operation, at the date of publication

Mine	Country	Concentrate treatment capacity [t day^{-1}]	Reactor size [m^3]	Date of commissioning
Fairview	South Africa	62	340[a]	1986
Saõ Bento[b]	Brazil	150	550	1990
Harbour Lights[c]	Australia	40	160	1991
Wiluna	Australia	158	480	1993
Sansu	Ghana	960	900	1994
Tamboraque[d]	Peru	60	262	1998
Fosterville	Australia	211	900	2005
Suzdal	Kazakhstan	196	650	2005

[a]The volume of the two primary reactors at Fairview.
[b]The BIOX reactors are in care and maintenance due to concentrate shortages.
[c]Mining operations were completed in 1999 and the plant was decommissioned.
[d]Operations were ceased in 2002 due to mining and financial difficulties.

proximity to Johannesburg lends itself perfectly to the testing of new equipment, design modifications and process optimization.

1.3.2 The Wiluna BIOX™ Plant

The BIOX™ process for the treatment of the refractory gold concentrate at the Wiluna gold mine in Western Australia was selected after an extensive metallurgical testwork program. The testwork program included whole-ore roasting, two-stage concentrate roasting, biooxidation and pressure oxidation. The BIOX™ process was finally selected on the basis of improved gold recoveries, lower capital and operating costs, a shorter permitting and construction period and environmental compatibility (Stephenson and Kelson 1997).

Batch BIOX™ amenability tests were performed in 1990 and indicated that the concentrate was amenable to biooxidation. The gold recovery was improved from 27% in the untreated concentrate to more than 98% in the BIOX™ product. Continuous pilot plant testwork was performed in 1990 and 1991 to confirm the amenability of the concentrate to BIOX™ and to generate the necessary data for the design of the commercial operation.

The Wiluna BIOX™ plant was initially designed to treat nominally 115 t day^{-1} concentrate with an average sulfide-S grade of 24% and 10% arsenic. The plant consisted of six equidimensional reactors configured in the standard BIOX™ configuration. The reactors have a working volume of 468 m^3 each giving an overall retention time of 5 days at the design feed rate. The plant, was commissioned early in 1993. The performance of the plant exceeded the design sulfide oxidation rate, averaging 96.5% in the 7-day performance guarantee test in December 1993. The capacity of the plant was expanded in 1996 to the current nominal capacity of 158 t day^{-1} with the addition of two primary reactors and one secondary reactor.

1.3.3 The Sansu BIOX™ Plant

The installation of the BIOX™ process for the treatment of the refractory gold concentrate at the Sansu Sulfide Treatment Plant at Obuasi in Ghana was a major breakthrough for the BIOX™ technology. The BIOX™ process was again selected after an extensive metallurgical testwork program and was selected on the basis of reduced capital and operating cost, reduced technical risk, reduced environmental impact and for the simplicity of operation (Nicholson et al. 1993).

The plant was designed to treat nominally 720 t day^{-1} concentrate in three modules of six 900-m^3 reactors, with a concentrate containing 11.4% sulfide-S and 7.7% arsenic. The nominal treatment capacity of the plant was expanded in 1995 to 960 t day^{-1} concentrate with the addition of a fourth reactor module.

The plant was successfully commissioned in February 1994, exceeding the design sulfide oxidation in May 1994. The successful installation and operation of the Sansu BIOX™ plant clearly demonstrates the scale-up potential of the process using the modular design. The simplicity and ease of operation was also demonstrated, enabling the use of the technology in remote locations. Process optimization and innovations have led to significant savings in operating cost while maintaining steady operation of the BIOX™ reactors (Osei-Owusu 2001).

1.3.4 The Fosterville BIOX™ Plant

The Fosterville BIOX™ Plant, situated in Victoria, Australia, is designed to treat 211 t day^{-1} concentrate at a sulfide-S grade of 20.5%. The plant consists of six 900-m^3 reactors in the standard three primary and three secondary configurations, resulting in a 5-day slurry retention time at the design throughput rate. The BIOX™ is followed by a three-stage CCD circuit with two-stage neutralization of the acidic thickener overflow. Construction of the plant was started in March 2004 and commissioning was in March 2005. The first BIOX™ gold was produced at the end of May 2005 and the design concentrate throughput rate was achieved in June 2005.

The Fosterville concentrate has a design pyrite content of 33% with 13% arsenopyrite. The concentrate is extremely refractory, achieving less than 10% gold recovery upon direct cyanidation. The concentrate also contains organic carbon and a carbon-in-leach circuit must be used to limit the effect of preg-robbing on the overall gold recovery during leaching of the BIOX™ product.

1.3.5 The Suzdal BIOX™ Plant

The Suzdal BIOX™ plant is located in north Kazakhstan, close to the city of Semey. Suzdal is the first BIOX™ plant that will operate at subzero temperatures and is also the first BIOX™ plant in Central Asia. The plant is designed to treat flotation concentrate at a feed rate of 192 t day^{-1} at 12% sulfide-S. The plant consists of six 650-m^3 reactors in the standard three primary and three secondary tank configuration, resulting in a 4-day slurry retention time at the design throughput rate. The process is acid-consuming owing to a fairly high carbonate concentration of the ore. The carbonate can, however, be used as a neutralizing agent during the neutralization of BIOX™ product solution.

The emphasis during the detailed design of the BIOX™ section was to ensure robustness, taking into account uncertainties created with the limited testwork performed and the subzero temperatures experienced during winter. Civil work on the concentrator section commenced during January 2004.

The majority of equipment fabrication, construction and the electrical installation were performed by local contractors under supervision of an engineering team from South Africa. The BIOX™ section was cold-commissioned during March 2005 and the inoculation of the first BIOX™ tank took place during April 2005. The bacterial culture multiplied very quickly, and this resulted in the first gold bar being produced on 27 May 2005.

1.3.6 Future BIOX™ Operations

The development of the BIOX™ process and the testing of new concentrate samples for amenability to the BIOX™ process is of primary concern to Gold Fields. Currently, Biomin Technologies is involved in and provides the technology to a number of BIOX™ projects, of which three are in the construction phase at the date of publication.

The Jinfeng BIOX™ project is located in the Guizhou province in China. The plant will have a design capacity of 790 t day^{-1} concentrate at a sulfide grade of 9.0–12.5%. The plant will consist of two modules of eight 1, 000-m^3 reactors configured as four primary and four secondary reactors, giving a 4-day retention time at the design feed rate. The plant is scheduled to be commissioned during the fourth quarter of 2006.

The Bogoso BIOX™ project in Ghana will have a design capacity of 750 t day^{-1} concentrate. The primary BIOX™ reactors will have an operating volume of 1,500 m^3 each, making these the largest BIOX™ tanks when in operation. The plant will consist of two modules of seven reactors, each to give an overall retention time of 5 days. Commissioning of the plant is scheduled for mid-2006.

The first phase of the Kokpatas BIOX™ plant in Uzbekistan will have a design capacity of 1,069 t day^{-1} at a sulfide-S grade of 20%, making it the largest BIOX™ plant in the world. For phase 1 the plant will consist of four modules of six 900-m^3 reactors each. The plant will have a final design capacity of 2,163 t day^{-1} after the completion of the second phase. The commissioning of the first phase is scheduled for mid-2007.

1.4 The BIOX™ Bacterial Culture

The process utilizes a mixed population of *Acidithiobacillus ferrooxidans*, *At. thiooxidans* and *Leptospirillum ferrooxidans* to break down the sulfide mineral matrix, thereby liberating the occluded gold for subsequent cyanidation.

Acidithiobacilli grow as straight (1–3.5-µm-long) rods, while *Leptospirillum* has similar dimensions but occurs as vibroid cells when young and as a highly motile spiralla when mature. The bacteria are believed to attach themselves to the metal sulfide surfaces in the ore, where they cause accelerated oxidation of the sulfides. The composition of the population is influenced by factors such as temperature and pH. *Leptospirillum* numbers are enhanced by a low pH and by a high slurry temperature (Lawson 1991). Shake-flask tests conducted on *At. ferrooxidans* showed that the oxidative activity was inhibited in a pH range from 2 to 3. Tests with *At. thiooxidans* showed very little growth between pH 0.5 and 1.0.

Because *L. ferrooxidans* is only known to oxidize ferrous iron and *At. thiooxidans* can only oxidize sulfur compounds, it is important to control the pH and temperature within narrow ranges to maintain the right balance of bacterial species to optimize the rate of oxidation. The typical operating pH range in the BIOX™ process is 1.2–1.8. Lime, limestone and/or sulfuric acid are used to control the pH in the reactors. The BIOX™ culture operates best at a temperature of 40°C; however, it is possible to run the process at 45°C in the primary stage and even at 50°C in the final secondary reactors. The oxidation reactions of sulfide minerals are exothermic; therefore, it is necessary to cool the process to maintain the slurry temperature within the optimum range.

The bacteria require sufficient carbon dioxide to promote cell growth. Carbon dioxide is obtained from the carbonate minerals in the ore and from the air added to the process. The bacteria also require inorganic nutrients: nitrogen, phosphorus and potassium, added to the primary reactors as a solution of ammonium sulfate and potassium and phosphate salts.

Certain substances are potentially toxic to the bacteria. These include:

- Thiocyanate and cyanide at very low concentrations
- Bactericides, fungicides and descaling reagents that are normally used for water treatment
- Oil, grease and degreasing compounds
- Chloride concentrations above 7 g L^{-1}, which inhibit bacterial activity (there are indications that the chloride causes membrane damage to the bacterial cells) (Lawson et al. 1995)
- Arsenic at high concentrations [although the culture is tolerant to As(V) concentrations as high as 20 g L^{-1}, a high concentration of As(III) can be toxic]

1.5 Engineering Design and Process Requirements

1.5.1 Chemical Reactions and the Influence of Ore Mineralogy

The oxidation reactions of the main sulfide minerals usually present in refractory ores may be summarized as follows:

$$2FeAsS + 7O_2 + H_2SO_4 + 2H_2O \rightarrow 2H_3AsO_4 + Fe_2(SO_4)_3, \qquad (1.1)$$

$$4FeS_2 + 15O_2 + 2H_2O \rightarrow 2Fe_2(SO_4)_3 + 2H_2SO_4, \qquad (1.2)$$

$$4FeS + 9O_2 + 2H_2SO_4 \rightarrow 2Fe_2(SO_4)_3 + 2H_2O. \qquad (1.3)$$

The oxidation reactions indicate the high oxygen demand of sulfide oxidation. Large volumes of air have to be injected into and dispersed in the slurry. This is one of the main engineering challenges in the design of a full-scale bioreactor, as will be described.

Important secondary reactions include precipitation of ferric arsenate ($FeAsO_4$), acid dissolution of carbonates and precipitation of jarosite, according to the following reactions:

$$2H_3AsO_4 + Fe_2(SO_4)_3 \rightarrow 2FeAsO_4 + 3H_2SO_4, \quad (1.4)$$

$$CaMg(CO_3)_2 + 2H_2SO_4 \rightarrow CaSO_4 + MgSO_4 + 2CO_2 + 2H_2O, \quad (1.5)$$

$$3Fe_2(SO_4)_3 + 12H_2O + M_2SO_4 \rightarrow 2MFe_3(SO_4)_2(OH)_6 + 6H_2SO_4, \quad (1.6)$$

where M^+ is K^+, Na^+, NH_4^+ or H_3O^+

The relative proportions of each mineral dictate various process requirements, such as cooling, acid consumption/production, oxygen demand, degree of precipitation and neutralization. The heat of reaction, acid demand and oxygen demand for oxidation and chemical leaching of principal refractory gold ore minerals are presented in Table 1.3. The actual values, for treatment of a particular concentrate, will be dictated by the relative proportions of the major minerals. Typically the overall heat of reaction is about 30 MJ kg^{-1} sulfide with an oxygen demand of 2.2 kg kg^{-1} sulfide oxidized.

Examples of the effects of the major minerals upon the operation of biooxidation and the BIOX™ process follow.

1.5.1.1 Pyrite

Bacterial oxidation of pyrite (FeS_2) is highly acid-producing; therefore, treatment of a concentrate with a high pyrite content will be acid-generating and

Table 1.3. Process data for sulfide mineral oxidation

Mineral	Pyritic sulfur (%)	Heat of reaction		Oxygen demand (kg O_2 kg^{-1} S^{2-})	H_2SO_4 demand (kg kg^{-1} mineral)
		Mineral (kJ kg^{-1})	Sulfide (kJ kg^{-1} S^{2-})		
Pyrrhotite FeS	36.4	−11,373	−31,245	2.25	0.557
Arsenopyrite FeAsS	19.6	−9,415	−48,036	3.51	0.301
Pyrite FeS_2	53.3	−12,884	−24,173	1.88	−0.408
Ankerite $Ca(Fe,Mg)(CO_3)_2$	–	−219.2	–	–	0.979
Siderite $FeCO_3$	–	-326.7	–	(0.069 kg kg^{-1} mineral)[a]	1.267

[a]Oxidation of Fe(II) to Fe(III)

maintenance of the pH within the required operating range requires addition of lime or limestone.

1.5.1.2   Pyrrhotite/Pyrite

Owing to the acid-consuming nature of pyrrhotite (FeS), the relative proportion of pyrite to pyrrhotite is an important factor affecting the overall lime and/or acid requirements, and one that also influences solution redox potential. The acid dissolution of pyrrhotite releases ferrous iron and elemental sulfur. Although the formation of elemental sulfur by this means is reversed by the bacteria present in the culture, excessive elemental sulfur formation, due to an abnormally high pyrrhotite content, cannot be accommodated in the plant and may lead to an increase in cyanide requirements and lower gold recovery.

The higher ferrous level in solution is beneficial in that it promotes a large bacterial population in the liquor phase of the primary reactors, which in turn reduces the possibility of bacterial washout occurring. However, the higher ferrous concentration lowers the redox potential, which can alter the oxidation chemistry of the process. The most serious effect of a low redox potential, combined with a low iron-to-arsenic ratio in solution, is the possibility of As(III) formation. As(III) may precipitate as the less stable calcium arsenite compound; hence, formation of As(III) should be minimized as far as possible. In addition, As(III) has a greater toxicity effect upon the bacteria than As(V).

1.5.1.3   Arsenopyrite

The ratio of arsenopyrite (FeAsS) to pyrite also influences acid consumption, but to a lesser extent than that of the pyrrhotite-to-pyrite ratio. More critical is the ratio of arsenopyrite to pyrrhotite and pyrite as given by the iron-to-arsenic ratio.

The iron-to-arsenic ratio is critical as it dictates the stability of ferric arsenate precipitates formed on neutralization of the BIOX™ waste liquor. The molar ratio of iron to arsenic in a concentrate is generally required to be greater than 3 to achieve stable effluent products, with respect to arsenic solubilization.

1.5.1.4   Carbonate Minerals

Carbonate content has two major effects on the BIOX™ operation. Firstly, a minimum content is required to ensure production of sufficient CO_2 to promote bacterial growth. If no carbonate is present, limestone must be added to the primary vessels, or the CO_2 content of the air injected must be further enriched with CO_2 gas.

The second effect is that of carbonate dissolution on pH. At a low sulfide-to-carbonate ratio, the primary stage becomes acid-consuming. The degree of precipitation increases and results in coating of the sulfide surfaces. Formation of coatings may result in lower oxidation rates, which in turn reduces liberation of gold for dissolution on cyanidation. The presence of carbonate at high sulfide-to-carbonate ratios is beneficial, not only for CO_2 production, but also in reducing lime requirements for pH control during biooxidation.

1.5.2   Effect of Temperature and Cooling Requirements

The BIOX™ bacterial culture is an adapted mixed culture of mesophilic bacteria as described in the previous section. The operating temperature range for mesophilic bacteria is 30–45°C although the reactors can be operated at temperatures up to 50°C for short periods. The BIOX™ process is normally operated within the temperature range 40–43°C, but both pilot plant testwork and plant experience have indicated that operating at a temperature of up to 45°C is not detrimental to the performance of the bacteria.

The oxidation of sulfide minerals is extremely exothermic as shown in Table 1.3. Constant cooling of the BIOX™ reactors is therefore necessary to control the temperature to within the optimum operating temperature range. The possible heat loads and sinks on the system are:

- Heat of reaction of sulfide mineral oxidation
- Heat generation by the absorption of agitator power
- Heat loss from the heating of incoming slurry to the vessel operating temperature
- Heat gain or loss from the adjustment of the incoming air to the reactor operating temperature
- Evaporative cooling provided by the sparged air at slurry temperature
- Heat loss due to air expansion
- Convection and radiation heat loss

The reaction heat is by far the largest contributor to the net heat load on each reactor, with absorbed agitator power also contributing. Convection heat loss is relatively small for atmospheric temperatures above 0°C. Convection heat loss must, however, be taken into consideration when calculating the heat balance of the last secondary reactors when the atmospheric temperature is below 0°C.

Heat is removed from the slurry by passing cooling water through internal coils installed in the reactors. The coils are configured as four to eight baffles of two to four coils each feeding from a header. Evaporative cooling towers are used to remove heat, from the cooling water. The current operating plants all use open-circuit cooling towers, but depending on the quality of the make-up water, closed-circuit towers can also be used. The closed-circuit towers will enable better control over the cooling water quality, thereby reducing scaling in the cooling coils. Closed-circuit towers are, however, also approximately 3 times as expensive to install. The decision to install open-circuit or closed-circuit cooling towers has to be determined for each project.

The climatic conditions, principally the wet bulb temperature, will also have a large influence on the design of the cooling circuit. The efficiency of evaporative cooling towers is poor under humid conditions, resulting in increases in the size of the cooling towers. The higher design maximum wet bulb temperature will also increase the cold sump water temperature, thus influencing the number and the size of the cooling coils in the reactors.

1.5.3 pH Control

pH is an extremely important parameter for the successful operation of a biooxidation plant. The optimum pH range for the process was found to be 1.1–1.5, although the process can operate over a wider pH range of 1.0–1.8. Poor pH control is often found to be the cause of low bacterial activity in the BIOX™ reactors in commercial operations.

The mineralogical composition has a large influence on the acid balance during the biooxidation of the concentrates, as described previously. The limestone or sulfuric acid requirement, to control the pH of the slurry in each reactor to within the optimum range, will be a function of the concentrations of the various minerals in the flotation concentrate and the extent of oxidation of the minerals in the BIOX™ reactors.

A comprehensive investigation into the effect of lowering the pH on the performance of the BIOX™ process was conducted using a test reactor at the Fairview mine (Chetty et al. 2000). The results proved that sulfide oxidation decreased, and that foaming of the reactor became problematic when the pH of the slurry was allowed to decrease to below pH 1.0. A high pH may also reduce the extent of oxidation and can decrease gold recovery owing to metal salt precipitation resulting in occlusion of the gold particles. If the pH is allowed to increase to above 2.0, the risk of killing the bacteria increases significantly, which can result in the total loss of the bacterial culture.

pH control in the BIOX™ reactors can account for a significant portion of the operating cost for the plant. Sourcing of a low-cost limestone supply can reduce the operating cost considerably for an acid-producing concentrate. In the case of an acid-consuming concentrate, the flotation conditions must be optimized to reject acid-consuming carbonate minerals to the flotation tailings. An acid recycle, returning acidic liquor from the CCD circuit, can also be used to reduce the acid consumption.

1.5.4 Oxygen Supply

The supply of oxygen represents the largest consumer of power in biooxidation and is therefore a major part of both the capital and the operating cost for a biooxidation plant (also see Sect. 1.6). The oxygen requirement is driven by the chemical oxygen demand for the oxidation of the sulfide minerals, and typical values for oxygen demand will vary from 1.8 to 2.6 kg oxygen per kilogram sulfide oxidized, depending on the mineralogical composition of the concentrate and the oxidation rates achieved.

The oxygen for commercial biooxidation is normally supplied by sparging compressed air into the reactors. The BIOX™ reactors are designed with a height-to-diameter ratio of close to 1 in order to minimize the static slurry pressure and thus enable the use of low-pressure blowers for aeration. The volume of air sparged into the system must be adequate to meet the process oxygen demand and to maintain a dissolved oxygen concentration in solution of not less than 2 mg L^{-1}. The aeration rate for each reactor is calculated on the basis of the process oxygen demand, the oxygen content of the aeration air and the utilization of oxygen achieved in the reactors. Typical aeration rates per unit volume in biooxidation reactors can vary from 0.05–0.10 per minute in the primary reactors.

The design of the agitator is one of the most important aspects for the design of a commercial operation. The agitator must be able to achieve good dispersion of the sparged air in order to attain the required oxygen transfer rate and oxygen utilization. The process criteria for the agitator specification can be summarized as follows:

1. Sufficient power must be provided to the impeller to prevent flooding.
2. The oxygen mass transfer coefficient ($k_{1.a}$) for the agitator/aeration system must meet or exceed the required $k_{1.a}$, determined by the oxygen demand of the process.
3. The impeller pumping rate must be sufficient to achieve uniform solids suspension and to maintain uniform temperature, pH and concentration profiles through the reactor.

Flooding of an impeller occurs when the air passes through the impeller and is not dispersed by the fluid flow. This is accompanied by a decrease in the agitator power draw, the oxygen transfer rate achieved and a reduction in the solids suspension capability of the impeller. The power input for the design of the agitator is normally determined by either the oxygen mass transfer rate required or the aeration rate, to prevent flooding of the impeller. An axial-flow impeller is currently the preferred impeller for biooxidation applications in mineral slurries. Alternative impellers for high gas dispersion are available, but the axial-flow fluid-foil impellers offer improved efficiency, giving equivalent oxygen transfer rates at reduced power consumption (Fraser et al. 1993; Lally 1987). The fluid flow developed by these impellers is also much higher than that generated by radial-flow turbines, per unit power input, allowing solids suspension at reduced shear rates and lower power levels.

1.5.5   Process Modeling and Effect of Bioreactor Configuration

The development of operating curves to describe biooxidation performance is important during the design of a BIOX™ plant. The logistic model (Miller

1991) is used to develop the operating curves for each project from the continuous BIOX™ pilot plant data. The logistic model describes the lag phase, the exponential growth phase and the declining growth phase for the bacterial culture in the BIOX™ reactors. The operating curves relate the plant retention, concentrate feed rate and sulfur grade to the extent of sulfide oxidation and the mass of sulfur oxidized. The effect of various reactor configurations on the overall oxidation achieved can also be determined using the logistic model.

The operating curves for a BIOX™ project, based on the results from continuous pilot plant testwork, are shown in Fig. 1.2. The concentrate has a sulfide content of 20% and the plant is designed to treat 100 t day^{-1} concentrate at a 4-day retention time. Three curves, sulfide oxidation, sulfide-S oxidized and the corresponding gold recovery, are shown as a function of the concentrate feed rate. The operating curve defines the upper limit of sulfide oxidized at any given concentrate grade and feed rate under optimum operating conditions. The concentrate feed rates of 50–140 t day^{-1} are equal to plant retention times of 8.0–2.9 days. The mass of sulfur oxidized per day increases almost linearly up to the design feed rate, after which the rate of increase slows down. The sulfide oxidation achieved in the overflow product decreases rapidly over the same range. This has a significant influence on the gold recovery achieved as shown on the graph.

The sulfide oxidation achieved will decrease rapidly as the retention time in the plant reduces to below 3.0 days, because the bacterial numbers can no longer be sustained. Operating the plant under these conditions for extended periods may lead to "washout" of the bacterial culture, the condition where the retention time in the primary BIOX™ reactors is not sufficient for the bacteria to sustain their numbers.

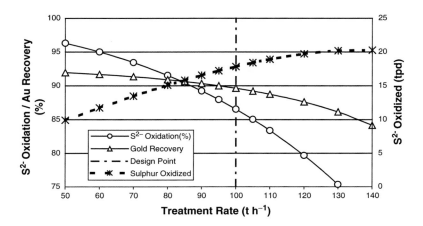

Fig. 1.2. Process operating curves

The optimum reactor configuration for a biooxidation plant is related to the rate of sulfide mineral oxidation and the corresponding rate of bacterial growth achieved. The primary stage is the heart of the BIOX™ process and sufficient residence time must be allowed in the primary stage for a stable bacterial population to develop. If the residence time is too short, bacterial washout will occur and the rate of sulfide oxidation will decrease.

Using the logistic equation, we can demonstrate that the optimum reactor configuration for most applications is three primary reactors operated in parallel followed by three secondary reactors in series as shown in Fig. 1.3. There are, however, specific circumstances where the reactor configuration can be changed to accommodate the specific requirements of the project.

The effect of taking one reactor off-line for maintenance must also be taken into consideration during the design of a BIOX™ plant. Taking one reactor off-line for maintenance will reduce the total retention time in the BIOX™ circuit if the feed rate is kept constant and will change the reactor configuration. The logistic model can be used to predict the effect of taking one or more reactors off-line, with or without a change in the concentrate feed rate. The BIOX™ reactors are usually designed so that any one of the reactors can be taken off-line while maintaining the same number of primary reactors on-line.

1.5.6   Effect of Various Toxins on Bacterial Performance

Sustaining an active bacterial culture is critical for the successful operation of a bioleaching plant. It is therefore important to prevent toxic elements from entering the process or to build up to concentrations that inhibit the mineral-oxidizing bacteria. A number of potentially toxic trace elements may be present in the flotation concentrate. The maximum acceptable levels of these

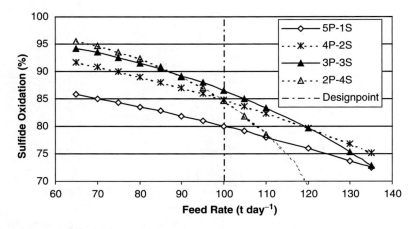

Fig. 1.3. Effect of BIOX reactor configuration on overall oxidation

elements are shown in Table 1.4. Water used for dilution, hosing, gland service, cooling and reagent make-up in the BIOX™ plant should comply with the specifications listed in Table 1.4. All cyanide species are highly toxic, even at very low concentrations.

The BIOX™ bacteria may operate at chloride concentrations up to 5 g L^{-1} but at this concentration jarosite precipitation will be promoted and may result in a decreased gold recovery during cyanidation. The BIOX™ bacteria are tolerant to As(V) concentrations of 15–20 g L^{-1}; however, the bacteria are less tolerant to As(III) and are inhibited at As(III) concentrations of more than 6 g L^{-1}. Lead, which is also regarded as potentially toxic, does not remain in solution after bacterial oxidation of the lead sulfides but precipitates as an insoluble lead sulfate.

Gold Fields has an extensive database on levels of toxicity of certain products to the mineral-oxidizing bacteria. Some reagents commonly used on a plant may be toxic or inhibitory to the BIOX™ culture. Reagents that may typically show toxic effects toward the BIOX™ culture include:

• Cyanide and thiocyanate (at very low concentrations)
• Grease and oil
• Detergents, solvents and degreasing components
• Biocide descaling reagents and other water-treatment reagents
• Certain flocculants, flotation reagents and nutrients

All new reagents to be used on the plant, including flotation reagents, nutrients and flocculants, must undergo toxicity testing prior to implementation.

Table 1.4. Acceptable concentrations of elements in concentrate and water

Element/compound	Maximum acceptable limit (mg L^{-1})
Flotation concentrate	
Sodium	7,000
Mercury	7.5%
Antimony	10%
Solution phase	
SCN	1.0
CNO	2.50
Metal cyanides	5.00
Na	500
Ca	250
Cl	5,000
Total dissolved solids	2,100

1.6   BIOX™ Capital and Operating Cost Breakdown

The BIOX™ process is primarily a sulfide-S treatment process and the capital and operating costs are proportional to the amount of sulfide-S oxidized per unit time. A simple rule of thumb can be used as a first test of the economic viability of applying the BIOX™ process to a refractory gold ore:

Gold grade (grams per tonne)/sulfur grade (percent)>0.7.

The formula can be applied to either whole ore or flotation concentrate. It is site-specific and also dependent on the prevailing gold price.

1.6.1   Capital Cost Breakdown

The capital cost breakdown generated for a recent feasibility study of a potential BIOX™ project is used as an example. The following major design criteria were used for the design:

- Concentrate feed rate 100 t day^{-1}
- Sulfide-S 15%
- Sulfide oxidation 96%
- BIOX™ product settling area requirement 4.5 m^2 t^{-1} h^{-1}
- Overall BIOX™ retention time 4 days

The capital cost distribution is displayed in Table 1.5. The mechanical equipment supply cost is based on budget quotes obtained for all major equipment units and can be classified as a preliminary estimation with a target accuracy of −15 to +25%. The total capital cost estimate is an *order of magnitude* estimation and is based on cost data from previous cost studies. The breakdown of supplied equipment cost per plant section and by equipment type is displayed in Tables 1.6 and 1.7, respectively.

The biooxidation section contributes to more than 50% of the total BIOX™ plant equipment cost, with the BIOX™ reactors and the BIOX™ agitator costs contributing the bulk of this cost. The BIOX™ reactor cost is mainly affected by the retention time required and the materials of construction. BIOX™ reactors, agitator shafts and impellers are usually constructed from either 304L or 316L stainless steel. Duplex stainless steel grades such as 2205 may be required in applications where the chloride levels exceeds 500 mg L^{-1}, because of the increased resistance to pitting corrosion of these steel grades. The BIOX™ agitators are selected and sized specifically to meet the oxygen mass transfer requirements of the process. The supplier and agitator must also have a proven track record of mechanical reliability in this application.

The CCD thickener cost is mainly determined by the settling area requirement, but the process water quality will also affect the materials of construction. A trade-off study between settling area requirement (affecting capital cost) and flocculant consumption (affecting the operating cost) is usually performed for the selection of the thickeners.

Table 1.5. Breakdown of the capital cost estimation for a BIOX™ plant

DESCRIPTION	Factor (% of MES)	(%) of total
DIRECT COST		
Mechanical equipment supply (MES)	100	38.2
Mechanical equipment installation	20	7.6
Earthworks	5	1.9
Platework supply and installation	8	2.9
Concrete supply and installation	10	3.8
Structural steel supply and installation	10	3.8
Piping supply and installation	22	8.6
Buildings supply and installation	5	1.9
Electrical supply and installation	11	4.2
Instrumentation supply and installation	5	1.9
Mob, Demob and Misc	1	0.4
Freight allowance	3	1.1
Total Direct Cost	200	76
INDIRECT COST		
EPCM (% of direct cost)	25	19.1
Contingency (% of direct cost)	2	1.5
Spares (% of MES)	8	3.1
Total Indirect Cost		24
Total Project Cost		100

MES mechanical equipment supply, *EPCM* engineering, procurement and construction management

Table 1.6. Breakdown of the MES cost per plant section

Area	Percentage of total
BIOX reactors	51
Countercurrent decantation	15
Neutralization	8
Water recovery	6
Reagent make-up	6
BIOX™ services	13
Inoculum build-up	1
Total	100

Table 1.7. Breakdown of the mechanical equipment supply MES cost per item type

Equipment		% of Total
Tanks		31
BIOX reactors	21.0	
Neutralisation reactors	5.5	
Other stainless steel tanks	2.2	
Mid steel tanks	2.0	
Agitators		24
BIOX® agitators	21.7	
Other agitators	2.6	
Thickeners		15
CCD thickeners	10.8	
Water-recovery thickeners	4.2	
Blowers		8
Pumps		8
Cooling towers		4
Stainless steel piping		4
BIOX cooling coils		3
Other		1
General		2
Total		104

Blowers are a major capital cost item for any biooxidation plant. The blower discharge pressure and volume are the main parameters determining the cost of these units.

The optimum combination of cooling coil area, approach to wet bulb temperature and cooling water circulation rate must be determined for the design of the BIOX™ cooling circuit. Open-circuit evaporative cooling towers are suitable for most applications, but closed-circuit cooling towers may become viable in the event of poor make-up water quality, subzero operating conditions or dusty environments. These towers allow for better control of the cooling water quality, but are generally up to 3 times the price of open-circuit towers. The cost of pumps is fairly low and the use of reliable pumps from one of the major suppliers is recommended.

1.6.2 Operating Cost Breakdown

The operating cost estimate for the same project is shown in Table 1.8. The operating cost for the project was determined using typical reagent consumptions based on the concentrate characteristics, in conjunction with average reagent cost values for different sites around the world.

Table 1.8. The Operating cost estimation breakdown for a BIOX plant

Description		% of Total
Reagents		49
pH control	29	
Nutrients	16	
Water treatment	4	
Power		27
Labour		9
Maintenance		15
Total		100

Limestone, sulfuric acid and lime, used for pH control in the BIOX™ and neutralization sections, are the major contributors to the BIOX™ operating cost. The use of a cheap, local limestone supply can significantly reduce the operating cost if available. The removal of acid-consuming species during flotation or by utilizing available carbonates in the flotation tailings stream as a neutralizing agent can also significantly reduce the cost. Power is mainly consumed in the generation and dispersion of air required in the BIOX™ process and the power input required per tonne of ore is therefore closely related to the sulfide-S grade of the concentrate. The use of energy-efficient impellers (e.g., Lightnin A315) was a major step forward in reducing the power consumption.

The BIOX™ process is relatively simple, can be operated with a limited number of operators and high skill levels are not required. This is reflected in the fairly low labor cost. It is recommended that maintenance on the BIOX™ reactors be performed every 6 months, with more regular maintenance of pumps, valves and the corrosion protection of structures.

The cyanide consumption of BIOX™ product slurries is relatively high compared with that of the products of alternative technologies such as pressure oxidation or roasting. This can represent a significant portion of the total operating cost for the plant and warrants detailed investigation during the testwork phase. The cost of cyanide can vary between 15 and 20% of the total reagent costs, depending on the location and the concentrate characteristics. The reduction of cyanide consumption during the cyanidation of BIOX™ product solids is an important focus area for Biomin Technologies and is described in more detail in Sect. 1.7.

1.7 New Developments in the BIOX™ Technology

The BIOX™ process has been commercially in operation for nearly 20 years since the commissioning of the first pilot plant at Fairview in 1986. Throughout this period Gold Fields (and formerly Gencor) has maintained a strong focus on research to improve the efficiency of the process and the design of the commercial reactors. Some of the mayor advances included the introduction of energy-efficient impellers (e.g., Lightnin A315), the use of flotation tailings as a neutralizing reagent and changes to the sparge ring design.

There are three major research projects currently under way at Gold Fields. The first project is the evaluation of an alternative mixing system for the BIOX™ reactors. The second project is a general investigation to optimize the cyanidation process by investigating the mechanisms responsible for the consumption of cyanide during leaching of the BIOX™ product, and the third project will test the application of a combination mesophile and thermophile biooxidation process.

1.7.1 Development of an Alternative Impeller

Axial-flow impeller technology in biooxidation reactors is conventionally based on the concept of down pumping. The impeller circulates the slurry in a downward motion with the objective of increasing the gas retention time in the reactor. Alternative agitation systems are being developed in an effort to reduce the overall power input required for aeration and air dispersion. One of the options under investigation is the use of an up-pumping axial-flow impeller in biooxidation reactors.

The reactor is designed with a height-to-diameter ratio of 1:1 and is equipped with two or three A340 impellers. The top impeller is situated just below the surface of the reactor. Air is entrained from the surface by the action of the top impeller, thereby reducing the compressed air requirement. The size of the impeller, the agitation speed and the placement of the top impeller will determine the volume of air that is introduced into the system from the reactor headspace.

The A340 up-pumping impeller has been tested successfully in other applications, including fermentation (Oldshue 1956) and high-pressure oxidation (Adams et al. 1998). The A340 resulted in increased yields from fermentation plants owing to the lower shear rates generated by the A340 impellers while maintaining the mass transfer and blending performance. The A340 also holds several advantages over conventional mixing systems in autoclaves. The A340 does not flood at high gas rates, foaming can largely be eliminated and the sparging of air can, under certain conditions, be reduced or even stopped.

1.7.2 Cyanide Consumption Optimization

During the biooxidation of concentrates, a small percentage of sulfide is not completely oxidized to sulfate, and remains as intermediate sulfur species such as polysulfides. These sulfur species are very reactive with cyanide and form thiocyanate according to the following reactions (Luthy 1979):

$$S_x S^{2-} + CN^- \rightarrow S_{x-1} S^{2-} + SCN^-, \tag{1.7}$$

$$S_2 O_3^{2-} + CN^- \rightarrow SO_3^{2-} + SCN^-. \tag{1.8}$$

Research testwork in the past has been focused on the evaluation of pre-cyanidation treatment processes to deactivate the cyanide-consuming species to reduce cyanide consumption (Broadhurst 1996). Some of these processes include the following:

- *Pre-aeration*: The oxygen demand of washed biooxidation product is usually low since most of the soluble sulfides are already removed during the CCD washing stage. Soluble sulfide is oxidized to thiosulfate and it is preferable to discard the pre-aerated liquor to prevent the formation of thiocyanate.
- *Additional washing*: The addition of an extra washing stage ensures that a minimum concentration of ferrous iron and other soluble cyanicides are present in the feed to the cyanidation section.
- *Treatment with At. thiooxidans*: Oxidation of the reactive sulfur species present in a Fairview BIOX™ product sample by subjecting the slurry to an additional oxidation stage using the sulfur-oxidizing bacterium *At thiooxidans* was attempted. A large percentage of the elemental sulfur (55–74%) and sulfide-S (22–66%) present in the BIOX™ residue was removed, but this did not result in any significant decrease in cyanide consumption.
- *Adjustment of BIOX™ operating parameters*: A combination of feeding the plant with concentrate milled to 80% <20 μm and operating the BIOX™ reactors at a reduced solids concentration will improve bacterial activity and leaching kinetics and may lead to a reduction in the cyanide consumption.
- *Oxidizing agents such as sulfite*: SGS Lakefield Research has developed a process whereby intermediate sulfur species are oxidized with sulfite to form thiosulfate. The pretreatment is typically performed at pH>7 and slurry temperatures above 50°C.
- *Milling and attritioning*: The physical removal of cyanide-consuming species from the particle surface was attempted on various BIOX™ product samples. A recent investigation at the Fairview mine confirmed that not even ultrafine milling (to a grind size of 90% smaller than 13 μm) and preaeration with pure O_2 reduced the cyanide consumption.
- *Acid/base pre-treatment*: The reactions comprise heating the slurry to a temperature range of 50–90°C, with additions of HCl of HNO_3 or NaOH to create oxidizing conditions.
- *Optimization of cyanidation conditions*: Testwork performed by various researchers have confirmed the strong influence that leaching pH and the free cyanide concentration in the slurry have on cyanide consumption (Haque 1992;Kondos et al. 1996).

Significant reduction in cyanide consumption is usually achieved only under aggressive oxidative conditions such as the use of HCl, HNO_3, NaOH or SO_3^{2-} in combination with heating the slurry to temperatures of 50–95°C. This, together with the disposal cost of the oxidizing reagent, makes the majority of the processes not economically viable for implementation in commercial operations.

It was concluded from a recent BIOX™ pilot plant test performed on a bulk concentrate sample that cyanide consumption, and thus the thiocyanate formation, was reduced by up to 67% by controlling the ferric concentration in the primary BIOX™ reactors and by fine-grinding of the concentrate prior to BIOX™ (80% smaller than 20 μm). The cyanidation gold recovery was also increased by 2.1%. The optimization of the cyanidation leaching parameters, especially free cyanide concentration at different stages of the leach, has also proven to be successful in the reduction of cyanide consumption during leaching testwork of various BIOX™ product samples.

Current research performed by Gold Fields is focused on understanding how the intermediate sulfur species are formed and how the formation can be limited during the biooxidation stage. For this investigation, techniques such as X-ray photoelectron spectroscopy to study the relative proportion of intermediate sulfur species and precipitates are used (Parker et al. 2003). Pretreatment procedures identified during previous research testwork showing the most potential will also be investigated in more detail.

1.7.3   Combining Mesophile and Thermophile Biooxidation

Testwork has shown that all the intermediate sulfur species can be fully oxidized by the use of high temperature or thermophilic microorganisms, thereby significantly reducing thiocyanate formation during the cyanidation of the biooxidation product (van Aswegen and van Niekerk 2004). This will result in significantly lower cyanide consumption during leaching of the thermophile product with no loss in gold recovery. The use of thermophilic microorganisms does, however, come at a price.

Increased capital cost: The microorganisms (thermo-acidophilic archaea) operate in a temperature range of 65–80°C, and this increases the corrosive nature of the acidic bioleaching slurry. Standard austenitic stainless steel grades cannot be used for this application and more exotic materials of construction such as duplex stainless steels or acid-lined concrete tanks are required for the manufacturing of the reactors. The process operates at reduced solids concentrations, but this is traded off with faster reaction kinetics.

Increased operating cost: The equilibrium solubility of O_2 in water is very low at these high temperatures and oxygen enrichment of the air introduced into the slurry is required to achieve the required mass transfer rates in the reactors. Water consumption is also increased owing to increased evaporation rates at the higher operating temperatures and O_2 enrichment may be required to reduce the volume of air flowing through the reactors.

A combination mesophile and thermophile process was proposed to make the best use of the advantages of the different microbial strains. In this process the mesophilic BIOX™ bacteria are used for the primary oxidation stage to achieve approximately 70% sulfide oxidation. The BIOX™ process is a robust, relatively cheap and well-established technology. This is then followed by a thermophile oxidation stage to complete the oxidation, targeting the intermediate sulfur species to reduce cyanide consumption.

A simplified flow diagram for the process is shown in Fig. 1.4. The BIOX™ feed and primary reactor configuration is conventional as described in Sect.

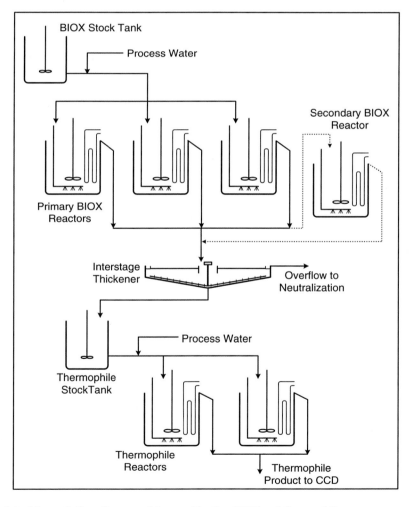

Fig. 1.4. Schematic flow diagram of the combination BIOX and thermophile process

1.2. A secondary reactor may be required depending on the sulfur grade of the concentrate and the oxidation achieved in the primary stage. Semioxidized material is fed to an interstage thickening stage to remove the bulk of the dissolved iron and sulfide species, in order to reduce unwanted iron precipitation reactions during the thermophile stage. The thickener overflow is pumped to the neutralization section. The thickener underflow slurry is pumped to a stock tank before the thermophile stage. The slurry is diluted to 12% solids before being fed to the thermophilic reactors.

Detailed testwork was performed on a concentrate sample from the Fairview BIOX™ plant. The testwork comprised a batch mesophile oxidation stage for a period of 7 days, followed by a batch thermophile stage for a period of 11 days, although complete oxidation was already achieved after 8 days of thermophile treatment. A combined oxidation of 99.5% was achieved with a product elemental sulfur concentration of less than 0.1%. A mineralogical investigation on the combination biooxidation product indicated the presence of a large percentage of gypsum and lower concentrations of iron oxides and hydroxides in the product sample as a result of precipitation reactions.

The focus during the cyanidation optimization stage was to ensure that gold recovery was maintained above 97% while reducing cyanide addition to below the 28 kg t^{-1} required for the Fairview BIOX™ product. A cyanidation test performed on a Fairview BIOX™ product sample indicated that 22.4 kg t^{-1} was consumed through the formation of thiocyanate and that ferrocyanide species consumed 5.6 kg t^{-1} cyanide. Experience over the years has shown that, as a rule of thumb, a minimum free cyanide concentration of 500 mg L^{-1} is required in the leach solution to achieve acceptable gold recoveries. The cyanide consumption, however, increases as the free cyanide concentration increases and an almost linear correlation between cyanide consumption and cyanide addition was found during the testwork.

After the optimum-stage cyanide addition strategy was determined, a series of tests was performed at progressively lower cyanide additions. The best result indicated that 97.1% gold recovery could be achieved with a cyanide addition of only 6 kg t^{-1}. Cyanide consumption that can be contributed to thiocyanate formation was drastically reduced to only 0.4 kg t^{-1}. Leaching testwork performed on samples pretreated using a hot lime process to break down jarosite precipitates showed increased gold recoveries of between 0.5 and 1%, indicating the importance of reducing the formation of precipitates during the biooxidation stage to a minimum.

The final reactor configuration and process design criteria will be determined during combination mesophile and thermopile pilot plant testwork on each concentrate sample individually.

1.8   BIOX™ Liquor Neutralization and Arsenic Disposal

1.8.1   Background

The bacterial oxidation of conventional pyrite and arsenopyrite concentrates produces a liquor phase containing As(V) and ferric sulfate. Arsenic compounds are toxic to all life and strict regulations for the control of arsenic-bearing wastes exist throughout the world. These regulations apply to liquid wastes as well as to any leachate that may be formed by the dissolution of solid wastes owing to exposure to air and/or water.

The conventional method for the fixation of arsenic from solution is by lime neutralization, preferably in the presence of excess iron, to produce ferric arsenate ($FeAsO_4zH_2O$) precipitates. The United States Environmental Protection Agency (US-EPA) considers chemical precipitation of ferric arsenate to be the best demonstrated available technology for the treatment of arsenic-bearing wastes (Rosengrant and Fargo 1990).

Numerous investigations have been carried out at Gold Fields (and previously Gencor Process Research) in order to establish the optimum conditions for the neutralization of BIOX™ liquors, such that the process produces environmentally acceptable effluent liquors and stable ferric arsenate precipitates (Broadhurst 1992; Schaekers 1998).

1.8.2   Development of the Two-Stage BIOX™ Neutralization Process

A two-stage neutralization process, using limestone and/or lime, was developed for the treatment of As(V)-bearing BIOX™ liquors. In the first stage, As(V) is precipitated as stable ferric arsenate by adjusting the pH to 4–5. The pH of the slurry is then increased to an environmentally acceptable level (pH 6–8) in the second neutralizing stage. The use of limestone allows for better pH control, particularly in the first neutralization stage, and is generally more economic than lime. The overall chemistry of the two-stage BIOX™ neutralization process is represented by the following equations:

Stage 1: Neutralization to pH 4–5

$$Fe_2(SO_4)_3 + H_3AsO_4 + CaCO_3 + 2H_2O \rightarrow Fe(OH)_3(s) + CaSO_4(s) + FeAsO_4(s) + 2H_2SO_4 + CO_2. \tag{1.9}$$

Stage 2: Neutralization to pH 6–8

$$H_2SO_4 + CaCO_3 \rightarrow CaSO_4(s) + CO_2 + H_2O \tag{1.10}$$

or

$$H_2SO_4 + Ca(OH)_2 \rightarrow CaSO_4(s) + 2H_2O. \tag{1.11}$$

The effects of neutralization operating parameters have been investigated on a batch laboratory scale using synthetic solutions and on a continuous pilot plant scale using liquors derived from the continuous bacterial oxidation of pyrite/arsenopyrite concentrates (Broadhurst 1992). Batch and/or continuous laboratory neutralization testwork forms part of the testwork program for every new concentrate sample tested by Gold Fields.

The results of the continuous pilot plant testwork indicated that continuous neutralization of BIOX™ liquors, with iron-to-arsenic molar ratios of 3:1 or more, results in environmentally acceptable effluent arsenic concentrations, and produces stable ferric arsenate precipitates over a relatively wide range of operating pH values. The iron-to-arsenic molar ratio has a strong effect on the stability of ferric arsenate precipitates, with the stability increasing as the iron-to-arsenic molar ratio increases from 1 to 16. It is, however, generally accepted that ferric arsenates produced from liquors containing iron-to-arsenic molar ratios of 3–4 or more are sufficiently stable for land disposal. The coprecipitation of gypsum and the presence of base metals (Zn, Cu, Cd) also increases the stability of the ferric arsenate precipitates.

The formation of calcium arsenate, $Ca_3(AsO_4)_2$, precipitates must be avoided during the neutralization of arsenic-bearing solutions. Calcium arsenates are more soluble than ferric arsenate precipitates and are generally not considered to be sufficiently stable for long-term disposal as they decompose to form carbonates under the influence of CO_2 and air. The formation of calcium arsenates can be avoided by maintaining the iron-to-arsenic molar ratio above 3:1 and by increasing the pH in the first stage to not higher than 4–5.

As(III) may be formed to some extent during the oxidation of arsenopyrite concentrates. Testwork at Gencor Process Research, however, indicated that the As(III) is rapidly oxidized to As(V) by the Fe(III) in the biooxidation solution for concentrates containing only pyrite and arsenopyrite as sulfide minerals (Broadhurst 1993). An investigation into 19 arsenopyrite-bearing concentrate samples indicated that the presence of pyrrhotite increased the concentration of As(III) in the BIOX™ product solution, especially for concentrates with pyrite-to-arsenopyrite mass ratios of less than 1:1.

The presence of As(III) in the BIOX™ product solution is undesirable as it is not only toxic to the bacteria at relatively low concentrations, but it also prevents the complete removal of dissolved arsenic from solution and the formation of stable precipitates during neutralization of the liquor if the liquor has an As(III) content of more than 3 g L^{-1} and an iron-to-arsenic molar ratio of less than 6:1. In such cases, the adverse effect of As(III) on the BIOX™ neutralization process can be avoided by the addition of H_2O_2 before or during neutralization (Broadhurst 1994). Testwork indicated that although Fe(III) is capable of oxidizing As(III) to As(V) based on thermodynamics, the kinetics of the process is very slow in practice if no catalyst is present.

1.8.3 BIOX™ Neutralization Process Design and Performance

The standard BIOX™ neutralization flow sheet is shown in Fig. 1.5. The plant consists of six reactors with a retention time of 1 h per tank, to give an overall retention time of 6 h. Lime and/or limestone is used for pH control and the pH of the slurry is increased in two stages. Limestone is generally used for the first stage of neutralization to achieve a pH of 4–5, but lime can also be used if a cheap source of limestone is not readily available. Lime is used for the second stage to increase the pH to 7–8. Flotation tailings can also be used as a neutralizing reagent if it contains sufficient available carbonate minerals, as described later.

The acid solution from the CCD section is fed to the first tank with pH control in tank 2 (stage 1) and tank 5 (stage 2). A recycle stream from tank 3 to tank 1 is incorporated into the circuit. The recycle stream provides seeding for the precipitation reactions in tank 2 and also improves reagent consumption as the unreacted limestone is brought into contact with the acidic solution. The recycle rate is usually 100–200% of the feed rate of acid solution. Air is sparged into all the reactors to maintain dissolved oxygen levels of 6–8 mg L^{-1}.

The results of continuous neutralization testwork performed on BIOX™ liquors derived from the continuous biooxidation of various sulfide concentrates are shown in Table 1.9. The testwork results confirm that the neutralized effluents conform to US-EPA standards with regard to dissolved arsenic in the effluent (EPA standard below 0.5 mg L^{-1}) and stability of the neutralized product solids (EPA standard below 5.0 mg L^{-1} in toxicity characteristic leaching procedure, TCLP, extract) over a wide range of iron-to-arsenic molar ratios (5.0–70) and feed dissolved arsenic concentrations (0.4–9.5 g L^{-1}).

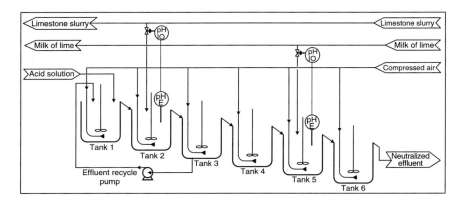

Fig. 1.5. Typical two-stage BIOX neutralization process flow sheet

Table 1.9. Summary of continuous neutralization testwork results

Description of neutralization feed liquor				As (mg L^{-1})	
Fe (g L^{-1})	As (g L^{-1})	Fe-to-As molar ratio	pH	In effluent liquor	In TCLP extract from product solids
21.8	3.45	8.5	1.63	<0.25	1.4
21.9	0.42	70	1.45	<0.02	<0.02
37.9	9.51	5.3	2.04	0.12	0.07
9.48	1.17	10.9	1.59	<0.25	<0.25
12.9	1.6	10.8	1.48	<0.34	<0.34
10.7	2.5	5.7	1.57	<0.34	<0.34

TCLP toxicity characteristic leaching procedure

The effect of ageing on the arsenic-containing compounds was investigated at the Wiluna BIOX™ operation. The tailings facilities were sampled at different levels and the solubility of the arsenic was tested using the standard TCLP test procedure (Raynes and Bird 2001). The results are shown in Table 1.10 and clearly indicate that the leachability of arsenic under the test conditions was very low at all depths in the dam. The Wiluna BIOX™ plant had been in operation for approximately 7 years at the time of the investigation and although no real correlation can be drawn between the depth and the age of a sample, the results confirms that no deterioration in the stability of the arsenic compounds was observed at any depth.

Table 1.10. Stability of arsenic-bearing residues from the Wiluna BIOX plant (Raynes and Bird 2001)

Source/depth (m)	As in TCLP extract (mg L^{-1})	
Tailings dam "C"	Hole 1	Hole 5
0.5–1.0	0.056	0.11
17.5–18.0	<0.02	0.035
30.5	0.023	–
37.0	–	0.068
Tailings dam "Western Cell"	Sample site A	Sample site B
0–1	0.08	0.16
1–2	0.60	0.40
2–3	0.18	0.06
3–4	0.20	0.12
4–5	0.20	0.22

1.8.4 The Use of Flotation Tailings in the Neutralization Circuit

The operating cost for the neutralization of the acidic BIOX™ liquor can be significant if a cheap source of limestone and/or lime is not readily available. For this reason it is important to reduce the consumption rate of the neutralizing reagents to the minimum. Testwork has been conducted by Gold Fields to determine the suitability of using flotation tailings for the first stage of neutralization (Olivier 2001). The success of the flotation tailings will depend on the concentration of carbonate minerals in the flotation tailings as well as the type of carbonate minerals present.

The results of continuous neutralization testwork, using flotation tailings as the stage 1 neutralization reagent, performed on BIOX™ liquors derived from the continuous biooxidation of various sulfide concentrates are shown in Table 1.11. The results confirmed that the neutralization products are acceptable in terms of the dissolved arsenic in the effluent and the solubility of arsenic in the neutralization product solids.

The flow sheet for the neutralization circuit using flotation tailings differs slightly from the standard neutralization flow sheet. The flotation tailings are fed with the acid solution to the first neutralization tank. The on-line pH monitoring and limestone addition to the second tank is therefore not required. The lime addition in tank 5 is the same to ensure that the pH of the final effluent is in the range 6–8. The recycle stream is also not required, as the solids in the flotation tailings stream will supply sufficient sites for seeding.

The retention time required for neutralization must also be confirmed during batch and pilot plant testwork. The reaction rates of certain carbonate minerals are relatively slow and the retention time in the neutralization circuit may need to be increased to allow sufficient time for the carbonate minerals to react. The increased retention time, combined with the additional flow from the flotation tailings, can result in significantly larger neutralization reactors. Testwork was therefore conducted to reduce the volume of flotation tailings fed to the neutralization reactors. The results indicated that there is a definite advantage in removing the finer fractions from the flotation tailings stream and feeding only this stream to the neutralization circuit.

Table 1.11. Summary of continuous neutralization testwork results using flotation tailings as stage 1 neutralizing reagent

Description of neutralization feed liquor				As (mg L^{-1})	
Fe (g L^{-1})	As (g L^{-1})	Fe-to-As molar ratio	pH	In effluent liquor	In TCLP extract from product solids
21.8	3.45	8.5	1.63	<0.25	0.59
9.48	1.17	10.9	1.59	<0.28	<0.25
12.9	1.6	10.8	1.48	<0.34	<0.34
10.7	2.5	5.7	1.57	<0.34	<0.34

Feeding only the finer fraction of the flotation tailings not only reduced the volume of material needed for neutralization, but the testwork also indicated that the finer fraction is more reactive than the courser fraction, thereby reducing the retention time required for the tailings to react.

The use of flotation tailings in neutralization will not be applicable to all ore samples tested, but can result in a significant decrease in the neutralizing reagent cost if the tailings contains sufficient carbonate minerals.

1.9 Conclusion

After nearly 20 years of commercial operation, the BIOX™ process has been established as a viable process route for the treatment of refractory gold ores and concentrates. The process has been proven to offer plants lower capital and operating costs compared with the costs of competing technologies. The neutralized effluent produced from the plants has confirmed that the arsenic complexes produced are stable with reference to redissolution of arsenic. It is believed that with the increase in the price of gold, the BIOX™ process will continue to be the preferred process route for the treatment of refractory gold ores and concentrates.

References

Adams D, Preston M, Post T (1998) Mixing optimisation of high pressure oxidation of gold ore slurries. In: Proceedings Randol 1998, pp 217–221

Barter I, Carter AJ, Holder NHM, Miller DM, Van Aswegen PC (1992) Design and commissioning of a 40 tonne per day concentrate bioxidation treatment plant at the Harbour Lights Mine. In: Proceedings of extractive metallurgy of gold and base metals conference, Kalgoorlie, pp 113–119

Broadhurst JL (1992) The nature and stability of arsenic bearing tailings arising from the BIOX™ gold recovery process. Genmin Process Research, report no PR 92/20F

Broadhurst JL (1993) Formation and behavior of As(III) during BIOX™ process operations: Preliminary investigations. Gencor Process Research, report no PR 93/120C

Broadhurst JL (1994) Neutralisation of arsenic(III) bearing BIOX™ liquors: Laboratory testwork. Gencor Process Research, report no PR 94/26C

Broadhurst JL (1996) Cyanidation of BIOX™ products arising from pyrite/arsenopyrite/pyrrhotite concentrates: a compilation of testwork results relating to cyanide consumption. Gencor Process Research, report no PR96/53

Chetty KR, Marais HJ, Kruger MJ (2000) The importance of pH control in the biooxidation process. In: Proceedings of the colloquium bacterial oxidation for the recovery of metals, Johannesburg, pp 1–12

Fraser GM, Kubera PM, Schutte MD, Weetman RJ (1993) Process/mechanical design aspects for Lightnin A315 agitators in mineral oxidation. In: Proceedings Randol 1993, Beaver Creek, pp 247–253

Haque KE (1992) Role of oxygen in the cyanide leaching of gold ore. CIM Bull 85:31–38

Irons J (2001) Bacterial oxidation at the Afgold Fairview plant – South Africa. In: Proceedings of the 4th BIOX™ users group meeting, Wiluna, pp 4–16

Kondos PD, Griffith WF, Jara JO (1996) Use of oxygen in gold cyanidation. Cana Metall Q 35(4):39–45

Lally KS (1987) A315 – Axial flow impeller for gas dispersion. Lightnin technical article, report number A315-2

Lawson EN (1991) Environmental influences on BIOX™. Paper presented at the 1991 bacterial oxidation colloquium, SAIMM, Johannesburg, pp 1–11

Lawson EN, Nicolas CS, Pellat H (1995) The toxic effect of chloride ions on Thiobacillus ferrooxidans. In: Vargas T et al (eds) Biohydrometallurgical processing, vol 1. University of Chile Press, Santiago

Loayza C, Ly EM (1999) Biooxidation of arsenopyrite concentrate for industrial plant Tamboraque using acid mine drainage. In: Proceedings of the Biomine conference, Perth, pp 162–167

Luthy RG (1979) Kinetics of reactions of cyanide and reduced sulphur species to form thiocyanate. Forsyth, London

Miller DM (1991) Use of the logistic model for the interpretation of bacterial sulfide oxidation. Report no PR91/51F (note for the record) internal report Gencor Process Research, 25 March 1991

Nicholson HM, Lunt DJ, Ritchie IC, Marais HJ (1993) The design of the Sansu concentrator and BIOX™ facility. Biomine '93 conference, Adelaide, pp 138–145

Oldshue JY (1956) Role of turbine impellers in aeration of activated sludge. Ind Eng Chem 48:2194–2198

Olivier W (2001) Flotation tailings as neutralising agent: Bogoso pilot run. In: Proceedings of the 4th BIOX™ users group meeting, Wiluna, pp 30–38

Olivier JW, van Niekerk JA, Chetty KR, Ahern N (2000) BIOX™ plant nutrient optimisation. In: Proceedings of the colloquium bacterial oxidation for the recovery of metals, Johannesburg, pp 1–6

Osei-Owusu J (2001) Operating cost reduction initiatives at the Ashanti BIOX™ plant as at July 2001. In: Proceedings of the 4th BIOX™ users group meeting, Wiluna, pp 39–43

Parker A, Klauber C, Stott M, Watling HR, van Bronswijk W (2003) An X-ray photoelectron spectroscopy study of the mechanism of microbially assisted dissolution of chalcopyrite. In: Proceedings of the 15th international biohydrometallurgy symposium, Athens, pp 111–123

Raynes J, Bird R (2001) The stability of arsenic compounds generated by the neutralisation of BIOX™ liquors at Normandy Wiluna operation. Proceedings of the 4th BIOX™ users group meeting, Wiluna, pp 52–54

Rosengrant L, Fargo L (1990) Final best demonstrated available technology (BDAT) background document for K031, K084, K101, K102, characteristic arsenic wastes (D004), characteristic selenium wastes (D010), and P and U wastes containing arsenic and selenium listing constituents. United States Environmental Protection Agency, report no PB90-234014

Schaekers DM (1998) Evaluation of the stability of residues from biological oxidation of sulfidic ores. Billiton Process Research, report no PR 98/45

Slabbert W, Dew DW, Godfrey MW, Miller DM, Van Aswegen PC (1992) Commissioning of a BIOX™ module at Saõ Bento Mineracao. Randol Gold Forum, Vancouver, pp 447–452

Stephenson D, Kelson R (1997) Wiluna BIOX™ plant – expansion and new developments. In: Proceedings of the Biomine '97 conference, Sydney, pp M4.1.1–8

van Aswegen PC, Marais HJ, Haines AK (1988) Design and operation of a commercial bacterial oxidation plant at Fairview. In: Proceedings of the Perth, international gold conference, Randol, Perth, pp 144–147

van Aswegen PC, van Niekerk J (2004), New developments in the bacterial oxidation technology to enhance the efficiency of the BIOX™ process. In: Proceedings of the BacMin conference, Bendigo, November 2004, pp 181–190

Van Niekerk J (2001) Pilot sale testing of the disposal of arsenic trioxide using the BIOX™ process at Fairview BIOX™ plant. In: Proceedings of the 4th BIOX™ users group meeting, Wiluna, pp 17–29

2   Bioleaching of a Cobalt-Containing Pyrite in Stirred Reactors: a Case Study from Laboratory Scale to Industrial Application

DOMINIQUE HENRI ROGER MORIN, PATRICK D'HUGUES

2.1   Introduction

The Kilembe sulfide deposit in the district of Kasese (located 420 km west of Kampala, Uganda), which was mined for 26 years (1956–1982), produced 16 million tons of copper ore and a cobalt-rich pyrite concentrate that had been stockpiled at the Kasese railway terminal, 12 km downhill from Kilembe Mine. The amount of concentrate available on site was approximately 900,000 t and contained approximately 80% pyrite and 1.38% cobalt, (an estimated cobalt metal reserve of some 11,300 t). Most of the cobalt is disseminated in ionic form within the pyrite lattice.

Heavy rainfalls dispersed the pyrite concentrate in large trails over several kilometers downhill from the stockpile, causing the release of sulfuric acid and heavy metals and consequently the deterioration of the environment in this area (Fig. 2.1).

The "Kasese project" started in the BRGM facilities at Orléans in 1988 with the first laboratory-scale testwork. The objective was to demonstrate that the recovery of cobalt from Kasese pyrite using (bio)hydrometallurgy was technically and economically feasible. In 1992, the Kasese Cobalt Company (KCC) was created to establish the techno-economic viability of the project, and then to find investors for the industrial project. Ten years after the first tests in the BRGM laboratory, in 1998, the project became a reality with the inoculation of the bioleach tanks on-site.

Step by step, the study of the applicability of bioleaching to recover cobalt from the cobaltiferous pyrite concentrate passed all the required feasibility studies. From the beginning, a large number of positive features appeared to give the project a real chance to succeed:

- The pyrite concentrate did not require any mining investment.
- Water was potentially abundant at the Kasese site.
- Limestone is abundant as it is a major component of the subsurface of the Rift Valley near the site.
- The region is served by reasonable infrastructure, including electricity supply.
- There was a strong political motivation to eradicate the environmental black spot represented by the stockpile at the border of a natural reserve

Biomining
(ed. by Douglas E. Rawlings and D. Barrie Johnson)
© Springer-Verlag Berlin Heidelberg 2007

Fig. 2.1. Aerial view of the pyrite concentrate stockpile before the implementation of the bioleaching plant. In the background, devegetated areas downhill from the stockpile due to pyrite run-off

(Elizabeth National Park) that contaminated several square kilometers of land and a lake (Lake George).

- The site topology below the stockpile is characterized by a gentle slope, appropriate for the hydraulic transport of the slurries and solutions.
- There was virtually no activity on the land around the site.
- Natural biodegradation of the pyrite could be observed on-site, and was confirmed by initial laboratory tests.

Notwithstanding the aforementioned points, the feasibility of the project could only be established after a long series of amenability assays and optimization works, the main results of which are described in this chapter. This involved demonstrating the "bioleach-ability" of the pyrite concentrate, and that cobalt could be efficiently solubilized using an acceptable level of energy consumption. This result was obtained after optimizing the operating conditions of bioleaching and by determining specific design criteria for the equipment and the power requirement. A major and unique specificity of the process used in the KCC plant is the management of the pregnant solution to extract four metals, cobalt, zinc, copper and nickel, after the removal of iron. Only cobalt is recovered in economic quantities, the other metals having marginal grades in the concentrate. This makes a quite complex hydrometal-lurgical flowsheet that has, however, been shown to be efficient, robust and flexible.

Beyond the description of the case study, this chapter highlights the spin-offs of the optimization work required to ensure a viable application of the bioleaching technique to the cobaltiferous pyrite of Kasese for the benefit of the development of biohydrometallurgy in general.

2.2 Feasibility and Pilot-Scale Studies

2.2.1 Characteristics of the Pyrite Concentrate

The stockpiled cobaltiferous pyrite is a flotation concentrate, the typical com-position of which is shown in Table 2.1. The cobalt is mainly disseminated in pyrite in an unidentified form, but is also present in very small amounts as siegenite $[(Co,Ni)_3S_4]$ and in a pentlandite $[(Ni,Fe,Co)_9S_8]$ – bravoite $[(Fe,Ni,Co)S_2]$ association. Microprobe analysis showed that the cobalt pres-ent is in ionic form in the pyrite crystal lattice with an average grade of 1.7% and, in this connection, it was observed that the dissolution of cobalt was strictly proportional to the decomposition of pyrite up to at least 85% pyrite degradation. The gangue minerals are mostly quartz, silicates and neogene gypsum; there is no significant amount of carbonates present. The particle size distribution of the concentrate is such that more than 95% of the particles by weight are less than 150 μm in size with 80% of particles (P80) smaller than 90 μm. The average specific gravity of the concentrate is about 4,000 kg m^{-3}.

2.2.2 Bioleaching of the Cobaltiferous Pyrite

Cobalt sulfides are known to be leachable under bacterial activity and cobalt in solution can reach concentrations as high as 30 g L^{-1} (Torma 1988). The

Table 2.1. Average composition of Kasese pyrite flotation concentrate limestone

Fe (%)	Sulfide (%)	Sulfate-S (%)	Sulfur (%)	Co (%)	Ni (%)	Cu (g t^{-1})	Zn (g t^{-1})	Pb (g t^{-1})
38.3	40.9	0.73	0.3	1.38	0.12	2 005	167	24
Cd (g t^{-1})	As (g t^{-1})	Sn (g t^{-1})	Sb (g t^{-1})	Mn (%)	Al (%)	Si (%)	Ti (%)	Ca (%)
<2	303	<10	<10	0.03	1.16	3.5	0.05	1.4

solubilization of cobalt from pure cobaltite, or CoS, or from other sulfide minerals has been reported elsewhere (Thompson et al. 1993). In the case where cobalt is finely disseminated in a sulfide matrix, like pyrite, the bioleach processing is quite similar to the bioleaching of a refractory gold ore (Chap. 1), with the aim being to oxidize and dissolve the pyrite in order to release the metal (in gold processing the metal is exposed but remains in the mineral and is solublized by cyanide).

2.2.3 Inoculation and Microbial Populations

The original culture used to inoculate the first amenability tests of the KCC project originated from acid mine drainage waters sampled by BRGM. In order to adapt the inoculum to the cobaltiferous pyrite, the microorganisms were subcultured several times on this substrate. Originally, Collinet-Latil and Morin (1990) isolated strains of *Acidithiobacillus ferrooxidans* and *At. thiooxidans* from this population. During the first continuous bioleaching testworks, *Leptospirillum*-like bacteria, associated with the rod-shaped bacterial population, were identified (Battaglia et al. 1994b; Battaglia-Brunet et al. 1998).

Since 2000, single-strand conformation polymorphism (SSCP; Chap. 12) has been used to investigate and compare the diversity of the bacterial culture in different operating conditions. This includes at laboratory and industrial scales, in batch and continuous modes, and in air-lift reactors and mechanically agitated reactors (Battaglia-Brunet et al. 2002; d'Hugues et al. 2003; Foucher et al. 2003). The predominant organisms (five strains out of seven) were studied using 16S ribosomal RNA gene sequencing. The results revealed the presence of bacteria affiliated to the genus *Leptospirillum* (two different strains), *At. thiooxidans*, and also an *At. caldus* like organism and a *Sulfobacillus thermosulfidooxidans* like organism. Whichever operating conditions were used, organisms related to *Leptospirillum* spp. (probably *L. ferriphilum*) and an *At. thiooxidans* like organism were always present, within a community of two to six different bacterial species. The occurrences of *At. caldus* like and *S. thermosulfidooxidans* like bacteria were more variable (d'Hugues et al. 2003). Attempts were made to determine the proportions that were attached to solid particles or freely suspended in the medium using a combination of PCR–SSCP and microscopic techniques (Battaglia-Brunet et al. 2002; d'Hugues et al. 2003; Foucher et al. 2003). In the liquid phase, *At. thiooxidans* like bacteria were dominant during the early phase of batch, but were later supplanted by *Leptospirillum*-like bacteria. *Leptospirillum* spp. were always in the majority on the solids. The growth of *S. thermosulfidooxidans* like bacteria seemed to be favored by less intensive agitation–aeration operating conditions. In laboratory batch tests with pyrite, *At. thiooxidans* like bacteria were always present in significant numbers at the beginning of the tests; however, *Leptospirillum* spp. always appeared as the major contributor to bioleaching, especially at industrial scale in continuous conditions. The predominance of *Leptospirillum* organisms over both *At. thiooxidans* and *At. caldus* has been observed in other industrial cultures (Rawlings et al. 1999). *Leptospirillum* spp. have been found by other authors to be the major solid colonizer on pyrite, whereas sulfur-oxidizing *Acidithiobacillus*-like organisms are less represented on the solids fraction (Norris et al. 1988). From these various observations, it was assumed that pyrite dissolution mainly resulted from oxidation by ferric iron located at the interface between the pyrite surface and the attached *Leptospirillum* organisms. The critical role of *Leptospirillum* spp. would be the subsequent reoxidation of ferrous iron produced during pyrite oxidation. The development of *At. thiooxidans* or *At. caldus* would be of less direct importance to the overall bioleaching efficiency.

From the tests carried out on the KCC industrial operation, it was shown that the industrial population was mainly composed of microorganisms that were also present in laboratory-scale cultures growing on the same concentrate; however, it was not determined whether the main industrial strains were indigenous to the Kasese concentrate or originated from the inoculum used at laboratory scale. Appropriate studies of the biodiversity in the Kasese stockpile and in the Kilembe deposit where the pyrite was extracted might be a valuable contribution to the understanding of the role of microbial ecology in the development of new mineral technologies (Johnson 2001). Even though the first tests using different molecular biology approaches led to quite promising, reproducible and coherent results, the work to be carried out on the ecology of iron-oxidizers and sulfur-oxidizers remains considerable. Future studies will have to be cross-checked using different techniques and will have to go beyond merely identifying the microorganisms present, and focus also on the microbial population dynamics of these bioleach systems.

2.2.4 Optimizing the Efficiency of Bioleaching

Preliminary tests showed that the cobaltiferous pyrite could be satisfactorily bioleached to release cobalt in solution (Morin et al. 1993). Extensive studies on the influence of some key operating parameters were carried out in batch tests using 200-mL air-lift tubes (Battaglia et al. 1994a), such as pH, influence of dissolved iron and cobalt, nutrient medium optimization, solids concentration, particle size distribution, inoculum size and temperature. The resulting optimized operating conditions found were:

- Optimum growth temperature: 35°C with thermotolerance up to 46°C
- Solids concentration: 10–15% (by weight)
- Particle size: all particles ground to smaller than 63 μm
- pH in the tanks: between 1.3 and 2.0

In addition, the bioleaching population displayed relatively high tolerance to soluble ferric iron (more than 35 g L^{-1}) and cobalt (more than 5 g L^{-1}). The basic nutrient medium was initially adapted for bioleaching arsenopyrite concentrate, using a modified 9K medium. This 0Km medium (9K without iron, "m" indicating modification of the basal salts) has the following composition: $(NH_4)_2SO_4$, 3.7 g L^{-1}; H_3PO_4, 0.8 g L^{-1}; $MgSO_4 \cdot 7H_2O$, 0.52 g L^{-1}; KOH, 0.48 g L^{-1}. The same medium was tested on the cobaltiferous pyrite using different nutrient concentrations. The best bioleaching rate was obtained when using the classic concentrations of the 0Km medium. Therefore, it seems that the substrate characteristic (pyrite vs. arsenopyrite) did not influence the optimal nutrient requirements of the bacterial population.

After the screening, the most critical operating parameters in batch cultures were further investigated in more detail in continuous operations with agitated tank reactors at laboratory (80 L), pilot (4 m³) and semi-industrial (65 m³) scales (Morin et al. 1995; d'Hugues et al. 1997). It was found that the mineral oxidation rate was at least 30% faster in continuous mode than in batch conditions, showing the importance of running continuous bioleaching tests when the objective is to evaluate performances with a view to application at an industrial scale.

The following operating conditions were investigated in detail:

- Solids concentration
- Residence time
- Oxygen and carbon dioxide: gas–liquid transfers, consumption and limiting concentrations in solution
- Nutrient concentrations
- Recycling of cobalt solutions

The solids concentration for the operation constitutes one of the most critical parameters of the bioleaching process in terms of impact on the size of the equipment. Since the industrial bioleach plant was planned to operate at a pulp density of 20% solids, several factors, such as oxygen requirement, nutrient availability, physical effect of mechanical shearing and turbulence on the bacterial cells generated by agitation, were suspected to limit the bioleaching efficiency at such a high solids concentration. A test was carried out at 20% solids in a continuous laboratory-scale unit composed of four stirred reactors arranged in a cascade (d'Hugues et al. 1997). The gas balance method was used in order to follow O_2 and CO_2 consumption throughout the experiment. Using on-line gas analyzers, it was possible to simultaneously monitor the physiological behavior of the microorganisms and the pyrite oxidation kinetics. The best pyrite oxidation kinetics obtained was 80% cobalt extraction in 4.5-days' residence time. It was later demonstrated that even more efficient kinetics could be achieved at 20% solids, and that the minimum residence time before washing out the reactors was 1 day.

The dissolved oxygen concentration below which oxygen availability was limiting was 1.2 mg L^{-1}. This limiting dissolved oxygen concentration was slightly greater than the values stated in the literature at that time (Pinches et al. 1988; Chapman et al. 1993) and was generally in the range 0.1–1.1 mg L^{-1}. The oxygen requirement for pyrite oxidation was estimated to be on average

0.89 kg of O_2 per kilogram of pyrite oxidized (11% below the theoretical value resulting from the stoichiometry of reaction of 1 kg kg^{-1} to produce sulfur as sulfate and iron as ferric iron).

In the absence of oxygen limiting conditions, the role of the bacteria in the oxidation of the pyritic substrate did not seem to be directly related to their growth. Similar observations had already been made by various researchers (Nagpal et al. 1993; Mandl 1984). However, the quantity of biomass formed in the first steps of bioleaching influenced the overall efficiency of cobalt recovery.

Carbon dioxide uptake rate measurements also allowed it to be demonstrated that high agitation rates could affect bacterial productivity. More than high shearing effects induced by the agitation system, the excessive hydrodynamic turbulence caused by aeration and agitation was suspected to be the main physiological stress factor for the microbial population. It was suspected, as already observed in similar cases (Bailey and Hansford 1993), that the mechanical and hydrodynamic forces limited bacterial productivity by inhibiting bacterial contact with the solid substrate.

Preliminary laboratory- and pilot-scale studies carried out at 10% solids indicated that reagent-grade chemical compounds (orthophosphoric acid and ammonium sulfate) could be successfully replaced by fertilizer-grade chemicals such as diammonium phosphate and urea, which would be, in economic terms, more suitable for industrial-scale bioleaching; however, as limiting phenomena were expected at higher solids concentration, the test of replacing urea by ammonium was also carried at 20% solids (Fig. 2.2).

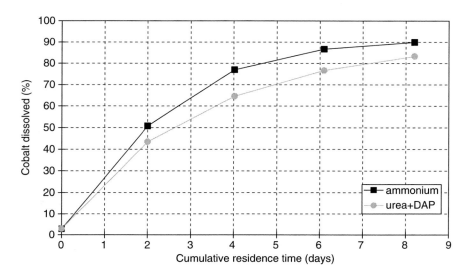

Fig. 2.2. Influence of the nitrogen source on the kinetics of cobalt dissolution. *DAP* diammonium phosphate

When nitrogen was provided as ammonium, improvements in bacterial growth and pyrite oxidation were observed (d'Hugues et al. 1997). These results were correlated with a more efficient attachment of the bacteria to the solid substrate, and also to an increase in the formation of iron precipitates. It was concluded that the use of ammonium rather than urea improved the bioleaching efficiency by favoring bacterial attachment to the solid substrate.

Other authors have also reported a positive influence of precipitates on the oxidation of ores, for example, Southam and Beveridge (1993) showed that bacterial cells were cemented to mineral surfaces by iron precipitates. This phenomenon could play an important role in sulfide mineral oxidation by creating an acidic oxidative microenvironment which would favor chemical and biological substrate degradation.

The influence of variations of solution composition as a result of bioleaching on pyrite degradation was also evaluated. It was shown, for example, that the dissolved cobalt concentration had only a limited inhibitory effect at more than 5 g L^{-1}. This tolerance of the bioleaching microorganisms to high concentrations of soluble cobalt was essential for the industrial application as it meant that pregnant bioleach solution could be recycled to the feed after iron removal to upgrade the concentration of cobalt in solution and consequently minimize water consumption.

The addition of CO_2 (at 1% of the volume of air) injected in the series of continuous stirred bioreactors improved microbial growth, though 2% CO_2 did not bring further improvement. This effect of CO_2 was only observed under optimal conditions of oxygen supply, indicating that control of the kinetics by CO_2 transfer only occurred when the mineral oxidation was no longer limited by the transfer of O_2. In terms of operability, the combination of the effects of the dissolved CO_2, excess acidity and high ferric iron concentration justified having a continuous addition of limestone to the bioleach tanks. As described later, the control of pH in the tanks by limestone addition was a critical operating parameter in the full-scale operation.

Some main operating parameters to be also determined for the engineering of the industrial plant were the power consumption required for dispersing air (to promote oxygen transfer), and the elimination of excess heat. In 1-m³ reactors it was demonstrated that the agitation system could ensure an oxygen transfer efficiency up to 40%. Together with oxygen transfer, power consumption and heat transfer were studied and optimized in a 65-m³ tank equipped with the same agitation system as the laboratory-scale unit. The BROGIM agitation system was designed in a collaborative project between Robin Industries (France, now Milton Roy Mixing) and BRGM (Bouquet and Morin 2005). The agitation system was designed so that the air at the bottom of the tanks is dispersed by a disk turbine, whereas the pumping effect in the tank is ensured by one propeller (or two propellers) placed above the turbine. The pilot testwork in the 65-m³ tank confirmed the efficiency of the agitation system to disperse air and provided data required to scale up the system to industrial size.

2.2.5   Solution Treatment and Cobalt Recovery

2.2.5.1   *Neutralization of the Bioleach Slurry*

As a result of pyrite oxidation, iron and sulfate were the main chemical species in the bioleach liquor. The first step of the treatment process involved neutralizing the solution at a pH close to 3 (Fig. 2.3). Adding a limestone slurry caused the iron(III) to precipitate as jarosites and ferric hydroxide, accompanied by the formation of gypsum.

The solids were then separated from the liquid by thickening and filtration, using a nonionic or cationic flocculant. The filterability, evaluated by means of a 0.5-m^2 pilot belt filter, was in the range 0.32–0.48 t (dry weight) solids h^{-1} m^{-2}. The efficiency and the stability of the filtration could only be ensured if gypsum crystals reached a significant size (more than approximately 100 μm). This was readily achieved by recycling a fraction of the neutralized slurry to the feed of the neutralization stage, thereby seeding with gypsum crystals and improving crystal growth. The recycling of neutralized slurry facilitated minimizing the size of the equipment required for neutralization and filtration at the industrial-scale operation.

A typical solid residue composition is shown in Table 2.2. The cobalt and sulfur contents in the residue (0.1 and 2.4%, respectively) were actually lower

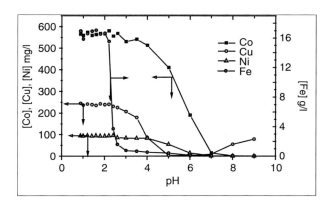

Fig. 2.3. Solubilities of Co, Cu, Ni and Fe versus pH in a typical bioleach solution neutralized with lime slurry

Table 2.2. Composition of the solids residue obtained after neutralizing the bioleach pulp with the substances listed

Co (%)	Fe (%)	Cu (g t^{-1})	Ni (g t^{-1})	Sulfate-S (%)	S total (%)	Sulfide (%)	CO$_3$ (%)	C$_{org}$ (%)	Si (%)	Ca (%)	Moisture (% H$_2$O)
0.1	13.2	470	145	13.9	16.4	2.4	0.1	0.15	1.7	13.6	29.5

after recycling of the unleached pyrite as described in Sect. 2.3.4; gypsum represented about 60% of the solids by weight. Limestone consumption was about 1.1 t per tonne of oxidized pyrite concentrate. Beside the calcium carbonate grade (94% on average), a critical characteristic of the limestone was its magnesium content, as this element was likely to build up along with the water recycling and consequently to interfere with the purification treatment and the extraction of the valuable metals. Moreover, it is costly to eradicate magnesium from the final effluent. It was established that its grade had to be less than 2% to minimize its impact on the process performance.

In order to minimize the volume of organic phases involved in the solvent extraction stages of the hydrometallurgical plant, the pregnant solution from the filtration stage was recycled to the bioleach section, thereby increasing the cobalt concentration to approximately 5 g L^{-1}, the apparent threshold of inhibition for the bacterial activity. A very small amount of cobalt was lost during filtration owing to coprecipitation. However, acid washing of the filter cake resulted in a recovery of more than 99% cobalt in solution in the pilot-scale operation, and this was also confirmed at the industrial scale.

2.2.5.2 Removal of Iron from the Pregnant Solution

The primary solution coming out of the belt filter had the average composition shown in Table 2.3. The final iron-removal step aimed (1) to oxidize the ferrous iron present (approximately 0.5 g L^{-1}) to ferric iron and (2) to complete the precipitation of iron as hydroxide and jarosites in order to achieve a final concentration in the liquor of less than 1–5 mg L^{-1} Fe. This treatment was required to avoid iron contamination and deleterious precipitation of ferric compounds during the further purification and extraction stages. Both air and hydrogen peroxide were used for the oxidation of the ferrous iron. For neutralization, the pH was adjusted to 4 by the addition of limestone; no detectable cobalt losses were observed during this step.

2.2.5.3 Zinc Removal

Despite the low content of zinc in solution, a zinc solvent extraction unit was required to avoid poisoning of the downstream cobalt extraction. The zinc extraction used a mixture of 5% di(2-ethylhexyl)phosphoric acid (D2EHPA) diluted in kerosene. The optimal operating parameters were established by means of continuous bench-scale testwork that demonstrated the efficiency of

Table 2.3. Typical average composition (mg L^{-1}) of the primary biosolution from the belt filter

Co	Cu	Ni	Fe	Zn	Mg	Sn	Sb	Pb	Cd	Al	Mn
4,663	313	406	3,000	46	900	<0.1	4	3.6	2.7	90	12.4

a typical three-stage treatment: extraction/scrubbing (to remove the traces of cobalt extracted by the solvent)/stripping. As was expected, zinc was removed from the aqueous feed below 1 mg L^{-1} and was recovered completely in the stripped liquor. Good recovery of the small amount of cobalt loaded or carried along with the D2EHPA was obtained by scrubbing with a dilute sulfuric acid solution. All scrubbed organic-phase analyses have shown that the cobalt concentration in the solvent, before stripping, was below 1 mg L^{-1}. Nickel was not extracted, but copper (and manganese) was slightly extracted and partially scrubbed. About 1.9% of the feed copper was lost in the stripping solution. Owing to the low level of iron in the feed solution, the balance for this metal was difficult to assess. Iron was extracted and appeared to be stripped. Organic-phase analysis showed the presence of an iron-circulating load, but the fact that no build-up was observed indicated that an acceptable equilibrium could be reached. All lead was extracted and stripped but with time, and owing to recycling of the stripping phase, precipitation of lead sulfate was observed. Only 20% of the feed aluminum was extracted, despite the half-extraction pH (pH value at which the extracted metal is distributed in equal quantities in the aqueous and organic phases) for this metal being in the range 1–2. This was explained by the fact that aluminum loading was kinetically inhibited.

2.2.5.4   Copper Removal

A number of processes such as cementation and sulfidization were considered for the removal of copper, but it was concluded that these were inadequate and/or too expensive. Solvent extraction by LIX 622 combined with electrowinning was tested, but although this gives a satisfactory metallurgical solution, the investment payback was too low for the metal production involved. Finally, it was decided to precipitate copper hydroxide by neutralizing the zinc raffinate with limestone and sodium hydroxide. A two-step precipitation was designed in order to minimize cobalt loss. In the first step the zinc raffinate (pH 2.5) is neutralized in a series of three tanks by addition of limestone in the first two tanks and of caustic soda in the last one. After thickening, the overflow is routed to the second stage, whereas the underflow is filtered on a filter press. The solid residue is collected and sold as a by-product. In the second stage, the pH of the solution is increased to 7 using sodium hydroxide. After thickening and clarification, the liquor is sent to the cobalt solvent extraction. Since coprecipitation of cobalt begins to be significant at this pH, the thickened underflow is recycled back to the process in order to releach the solids that have been precipitated.

2.2.5.5   Cobalt Solvent Extraction and Electrowinning

Final hydrometallurgical purification of the cobalt and electrowinning were first studied using laboratory-scale equipment to determine the general operating parameters. These parameters were then tested in a continuous circuit to optimize (1) the solvent extraction at pilot scale (10 L h^{-1} of aqueous solution) and (2) the cobalt electrolysis in a 2-m^3 cell, with 1-m^2 electrodes.

CYANEX 272 (dissolved in kerosene at 15% v/v) was selected for extracting and purifying the cobalt from the decopperized leachate. The cobalt extraction yield was around 99%, leaving a raffinate with less than 5 mg L^{-1} cobalt. The pH in the mixer-settlers used for loading was adjusted continuously by addition of caustic soda containing 140 g L^{-1} NaOH. Nickel, magnesium and calcium are rejected from the solvent during scrubbing. The scrub solution was prepared from strong electrolyte after (1) dilution in order to achieve a cobalt concentration of about 10 g L^{-1} and (2) adjustment to pH 3 by addition of NaOH.

Results from the laboratory and pilot tests of electrowinning led to the following basic design characteristics:

- Temperature: 70–75°C
- Spent electrolyte: 40 g L^{-1} Co, 10–13 g L^{-1} H$_2$SO$_4$
- Strong electrolyte: 45 g L^{-1} Co, 2 g L^{-1} H$_2$SO$_4$

The reactions involved are as follows:

- Cathodic reactions:
  $Co^{2+} + 2e^- \rightarrow Co$ metal
  $2H^+ + 2e^- \rightarrow H_2(g)$
- Anodic reaction:
  $H_2O \rightarrow 0.5O_2(g) + 2H^+ + 2e^-$

Cathode: SS 316L	Anode: Pb–Sb alloy
Cell voltage: 5 V	Current density: 300 A m^{-2}
Current efficiency: 60–70%	Pulling cycle: 4–5 days

With the earlier electrowinning tests, performed at laboratory scale and without final iron removal or zinc solvent extraction, the chemical grade of the cobalt deposit was 99.73% Co. At the industrial scale, the efficiency of the purification stages results in a cobalt grade often in excess of 99.9%.

2.3   Full-Scale Operation: the Kasese Plant

Construction of the full-scale bioleaching operation began in 1998 and was completed in 1999. Bioleach tanks at the KCC plant are shown in Fig. 2.4.

2.3.1   General Description of the Process Flowsheet

A simplified block diagram of the process flowsheet is shown in Fig. 2.5.

Fig. 2.4. Bioleach tanks (Bioco reactors) at the Kasese Cobalt Company (*KCC*) plant

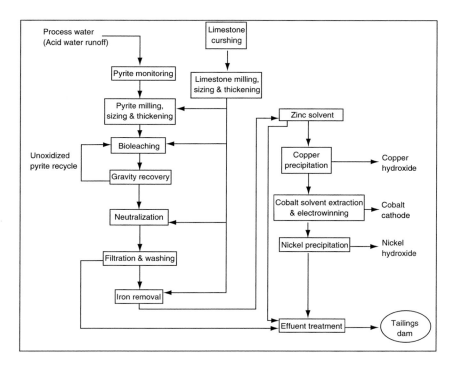

Fig. 2.5. General process flowsheet of the KCC installation

The seven main steps of the process are as follows:

1. Pyrite concentrate reclamation by hydraulic monitoring
2. Physical preparation of the pyrite concentrate and limestone
3. Bioleaching
4. Iron removal
5. Solution purification and solvent extraction
6. Cobalt electrowinning and conditioning
7. Effluent treatment and waste management

The general characteristics of these steps are outlined in the following sections and the bioleaching treatment is described in more detail.

2.3.2 Pyrite Reclamation and Physical Preparations

Pyrite concentrate is reclaimed from the stockpiles at a nominal rate of 245 t day^{-1} using high-pressure monitor guns and is then pumped to the process plant. Reclaimed slurry is thickened and the thickened underflow is pumped to the regrind circuit, where the pyrite particle size is reduced to P80<35 μm ("passing 80%, i.e., size under which 80% by weight of the mineral sample is smaller than"). Reground pyrite, classified by hydrocyclones is thickened to 50% solids prior to bacterial oxidation. After recycling of a fraction of the pregnant solution from filtration of the bioleach slurry to build up cobalt and the addition of a concentrated solution of nutrients, the pulp density of the feed to the bioleach circuit is reduced to 20%.

Limestone is crushed to smaller than 90 mm at the quarry prior to transport to the plant site at Kasese, where it is subjected to secondary crushing to smaller than 9 mm by a energy impact crusher and then ball milling to 80% smaller than 44 μm. Ground limestone is stored in a tank as a slurry at 50% solids in weight and is distributed to the bioleach and neutralization circuits by means of a ring at a steady flowrate in order to avoid plugging of the pipes.

2.3.3 Bioleach Circuit

The bioleach circuit includes a primary stage with three tanks in parallel, and secondary and tertiary stages of one tank each. Theoretically, this configuration is optimized to ensure a safe retention time for the bacterial growth in the primary stage and sufficient mineral biooxidation in the later bioreactors. The design is a compromise between the following constraints: (1) to minimize the overall unit volume; (2) to minimize the number of tanks (and thus the number of mixing systems) and (3) to ensure a maximum height of slurry owing to the admissible head pressure of the blowers (about 150 kPa). Every tank [constructed of stainless steel, 304L (BS) grade] has the same total operating volume (1,380 m^3), and gravity transfer through gutters ensures the bioleach slurry flows from one stage to the next.

Air is supplied to the bioleach tanks by five blowers (HV Turbo), which can feed each tank up to the equivalent value of 20,000 m^3 h^{-1} air in normal conditions (0°C, 100 kPa). The operating airflow rates in the tanks are in the range 10,000–15,000 m^3 h^{-1} in the primary-stage reactors and between 5,000 and 10,000 m^3 h^{-1} in the secondary- and tertiary-stage reactors. Flowmeters on every air feed pipe monitor airflow rates in cubic meters per hour for normal conditions (i.e., they are converted into values at 0°C and 100 kPa and are expressed as normal cubic meters per hour). Aeration and mixing in the tanks are ensured by air injection and the BROGIM system designed according to the size of the tanks and to the oxygen transfer and mixing requirements (Fig. 2.6).

The nominal average temperature in all tanks is 42°C. Heat control is ensured by internal stainless steel cooling coils connected to a cooling tower. Each primary tank contains more than 1.5 km of coils for heat transfer. Heat transfer efficiency is an important design criterion for the mixing system. The pH is kept as constant as possible in every tank by the continuous addition of limestone slurry at a controlled rate. The nominal values are 1.4–1.5 and 1.5–1.7 in the primary and secondary/tertiary tanks, respectively. Limestone addition functions not only to neutralize the acidity produced by the oxidation of the sulfide compounds, but also to generate CO_2 in situ for the autotrophic mineral-oxidizing bacteria.

Fig. 2.6. Bottom part of the agitation system (BROGIM) in the bioleach tank

2.3.4   Recycling of Sulfide in the Bioleach Process

Very early in the study of optimization of the bioleach process, it was thought that the recycling of coarse sulfide solid particles would ensure a longer residence time of this relatively refractory fraction of the concentrate while minimizing the total residence time. Recycling that fraction of the pyrite by gravity separation, using differences in specific gravity between the different compounds, was technically conceivable as it was already used at an industrial scale in other operations, such as processing iron, tin and gold ores. In theory, such a recycling had several other advantages. A fraction of the biomass is recycled, which stabilizes the bacterial growth and the more refractory sulfides bearing copper and nickel are better dissolved, which reduces the grades of these metals in the final residue. On the other hand, improved dissolution of these metals increases their concentrations in the pregnant solution and therefore increases the operating cost of their extraction for removal from the final effluent. The recycling of refractory material may also have the drawback that, under steady-state conditions, the solids concentration in the bioleach tanks is higher than in the feed, or even constantly increases. The innovative idea of recycling the sulfide particles from a bioleach slurry to the feed of the bioleaching treatment is patented by BRGM under the name BIOGRAV.

A simulation exercise showed that pyrite recycling could significantly improve cobalt recovery, and so it was decided to implement a gravity circuit, which had, however, to be simple and low in cost. The equipment used for the recovery of the heaviest fraction of the slurry flowing out from the last bioleach tank includes a hydrocyclone aimed at removing the finest particles of iron hydroxide and a series of gravimetric spirals that produce a concentrate, which is recycled into the feed of the bioleach circuit.

2.3.5   Monitoring of the Bioleach Process Performance: Some Practical Results

Measuring O_2 and CO_2 in the off-gas flows of the tanks is a reliable method of monitoring the bioleaching performance. This method allowed taking measurements required for the determination of the optimal air flow rates. Typical results of oxygen uptake rate for a primary tank are shown in Fig. 2.7.

It appeared that an air flow rate between 11,000 and 12,000 N m^3 h^{-1} was the minimum value to obtain a maximum oxygen uptake rate level. The maximum oxygen uptake rate values obtained were between 1,350 and 1,400 mg L^{-1} h^{-1}, correlating with an oxygen transfer rate of about 1,750 kg O_2 h^{-1} in a primary bioleach tank and an oxygen transfer efficiency of approximately 50%. Such a high oxygen transfer efficiency value was above that originally predicted. Measurements of dissolved oxygen at the top of the bioleach tank showed that the oxygen concentration was in the range 1.5–2.0 mg L^{-1} for

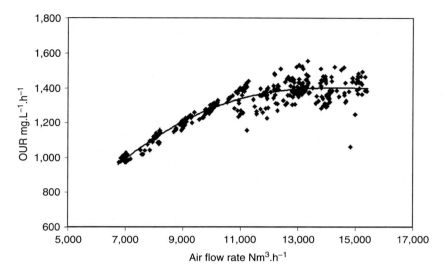

Fig. 2.7. Oxygen uptake rate (*OUR*) vs. air flow rate in a primary bioleach tank of the KCC plant

flow rates from 10,000 to 12,000 N m^3 h^{-1}. In this range of flow rates, the oxygen concentration was above the theoretical limiting value of 1.5 mg L^{-1}.

One critical operating aspect for the aeration was the difficulty in balancing the air pressure in all the bioleach reactors. Differences in pressure load from one reactor to another frequently resulted in blower trips.

Using the gas analysis system, the CO_2 availability in the bioreactor was identified as a critical issue for the bacterial activity as it is the source of carbon for the autotrophic iron-oxidizers and sulfur-oxidizers present. In other industrial bioleach reactors, mainly dedicated to oxidation of gold-containing minerals, CO_2 is usually supplied through the dissolution of carbonates present in the concentrates. As no carbonates occur in the Kasese pyrite concentrate, the CO_2 required for optimal growth is generated by dissolution of the limestone added for pH control. Therefore, it was important that limestone addition was as constant as possible so that CO_2 could be provided to the culture continuously. This change in the limestone addition procedure helped to improve bacterial growth and consequently the bacterial density of the culture.

When the bioleach reactor is affected by technical problems, the drop of pyrite oxidation leads to a "vicious cycle" of reduced CO_2 availability, as the regulation of the CO_2 level is directly related to the oxidation efficiency (acid production) through the pH regulation.

The gas analysis system has proven to be very accurate and reliable for monitoring the bacterial oxidation of the cobaltiferous pyrite. This tool was very helpful to follow the bacterial culture and to assess the consequences of any process deviation. This system could also be used to optimize the process and evaluate the effect of any changes in the operating conditions.

2.3.6   Bioleaching Performance

Approximately 60% of the total pyrite is oxidized in the primary bioleach stage and the rest of the oxidation (20–30%) occurs in the two other stages. The other stages can be limited to one tank only, which allows the shutdown of a reactor for maintenance and the reconfiguration of the remainder to retain three primary bioreactors. The total residence time, accounting for the addition of limestone slurry and the gas hold-up, is about 6 days with the full circuit running, and the cobalt dissolution yield generally approaches 80%.

2.3.7   Processing of the Pregnant Liquor

2.3.7.1   Iron Removal

Iron is removed in two operations. The first treatment is carried out in a series of three agitated tanks of 300 m^3 in series where limestone slurry is added to increase the pH of the bioleach slurry stepwise to 3. Then the neutralized slurry is filtered on a belt filter of 73 m^3. The primary filtrate is pumped to the next iron removal area and the wash filtrate is used to dilute the bioreactor feed to the required density. The solid residues are repulped with raw water and transferred to the effluent treatment area.

Another neutralization of the primary filtrate to pH 5.7 removes the remaining traces of iron with the aim of achieving a final concentration of less than 1 mg L^{-1} to prevent iron poisoning of the zinc extractant.

2.3.7.2   Solution Purification and Solvent Extraction

Zinc is extracted at pH 3 with D2EHPA diluted in high flash point kerosene. Cobalt, copper and nickel coextracted into the organic phase with the zinc are removed by scrubbing and then zinc is stripped from the scrubbed organic phase using 50 g L^{-1} sulfuric acid. Copper is removed by neutralization in two stages from the zinc raffinate using caustic soda solution to prevent copper poisoning of the extractant in the downstream cobalt solvent extraction plant at pH 5–7. Cobalt is extracted by CYANEX 272, diluted in high flash point kerosene. Nickel coextracted onto the organic phase (with the cobalt) is scrubbed and then the cobalt from the scrubbed organic phase is stripped using 130 g L^{-1} sulfuric acid.

Finally, the precipitation of nickel from the cobalt raffinate at pH 9–10 using caustic soda solution produces a saleable nickel hydroxide product.

2.3.7.3 Cobalt Electrowinning and Conditioning

Cobalt is recovered by electrolysis from the pure solution generated by solvent extraction (which contains 40–50 g L^{-1} cobalt), and results in a 99.9% pure product. The cobalt cathodes are crushed, burnished and drummed for shipment.

2.3.7.4 Effluent Treatment and Waste Management

All plant effluent is treated in a single operation. This includes repulped cake from the belt filter, strip solution from zinc solvent extraction and the overflow solution from nickel precipitation. The effluent is mixed with slaked lime slurry to a pH of 9 before being hydraulically transported to the tailings dam. In the tailings dam, the final solid waste products settle and the water is pumped at a controlled rate to the Rukoki river. As a pilot-scale processing of liquid wastes, a fraction of the outflow goes to a reed bed for further polishing. Up to 10% of the effluent is treated in constructed wetlands using locally available reeds (*Phragmites mauritianus*). Results show a significant amount of heavy metals and other chemical components is retained by the reed bed (up to 90% at low flow rates).

The effluent from the wetlands has been observed to encourage vegetation growth in the 150-ha arid trail area that resulted from the earlier runoffs of pyrite before the implementation of the metal processing on the site.

2.4 Conclusion

The optimization of the bioleaching treatment in stirred tanks applied to the cobaltiferous pyrite was a remarkable opportunity to focus attention on the most critical operating parameters of this process. The feedback from the industrial operation has shown that the performances of the plant are quite consistent with the predictions. Nevertheless, the answer to the question of whether chemical engineering or microbial physiology and ecology limits the efficiency of stirred-reactor technology remains uncertain.

The feasibility study estimated the lowest cobalt price for ensuring the economic viability of the project at US $12/lb of metal. The cobalt price was between US $6 and US $8/lb of cobalt when the plant was put on care-and-maintenance status in September 2002. In 2003, and more so in 2004, global demand for cobalt increased to such an extent that the price of the metal reached more than US $20/lb and the production at Kasese could be revived. The plant was restarted within a couple of months of this decision, demonstrating the flexibility and robustness of this technology.

The KCC plant employs 250 workers, among which 96% are Ugandans. Many local contractors also benefit from this industrial activity by providing services to the plant.

After the founding step of the use of bioleaching to recover copper and uranium by heap and in situ leaching, and later the application to refractory gold concentrates in stirred tanks, the KCC plant is the first industrial installation incorporating bioleaching into a sophisticated hydrometallurgical flowsheet allowing the selective extractions of various metals. As such, in the history of biohydrometallurgy, this operation can be considered as the milestone opening an era of the complex application of bioleaching.

Acknowledgements. KCC is gratefully acknowledged for the supportive collaboration with BRGM maintained after the start-up and commissioning of the plant. The authors are quite aware that the paper describes the main features of the KCC installation without essential improvements that have been made by the operators since the commissioning. Hopefully, another paper written with the operators will show how with time the solutions to unpredictable technical issues have been found. This paper was published with the authorization and support of the Research Direction of BRGM under reference number 04001.

References

Bailey AD, Hansford GS (1993) Factors affecting bio-oxidation of minerals at high concentrations of solids: a review. Biotechnol Bioeng 42:1164–1174

Battaglia F, Morin D, Ollivier P (1994a) Dissolution of cobaltiferous pyrite by *Thiobacillus ferrooxidans* and *Thiobacillus thiooxidans*: factors influencing bacterial leaching efficiency. J Biotechnol 32:11–16

Battaglia F, Morin D, Garcia JL, Ollivier P (1994b) Isolation and study of two strains of *Leptospirillum*-like bacteria from a natural mixed population cultured on a cobaltiferous pyrite substrate. Antonie van Leeuwenhoek 66:295–302

Battaglia-Brunet F, d'Hugues P, Cabral T, Cézac P, Garcia JL, Morin D (1998) The mutual effect of mixed *thiobacilli* and *leptospirilli* populations on pyrite bioleaching. Miner Eng 11:195–205

Battaglia-Brunet F, Clarens M, d'Hugues P, Godon JJ, Foucher S, Morin D (2002) Dynamic study of a pyrite-oxidising bacterial population using the single-strand conformation polymorphism technique. Appl Microbiol Biotechnol 60:206–211

Bouquet F, Morin D (2005) BROGIM®: a new three-phase mixing system – testwork and scale-up. In: Harrison STL, Rawlings D, Petersen J (eds) International biohydrometallurgy symposium, IBS 2005, Cape Town 25–29 September 2005, pp 173–182

Chapman JT, Marchant PB, Lawrence RW, Knopp R (1993) Bio-oxidation of a refractory gold bearing high arsenic sulfide concentrate: a pilot study. FEMS Microbiol Rev 11:243–252

Collinet-Latil MN, Morin D (1990) Characterization of arsenopyrite oxidizing *Thiobacillus*. Tolerance to arsenite, arsenate, ferrous and ferric iron. Antonie van Leeuwenhoek 57:237–244

d'Hugues P, Cézac P, Cabral T, Battaglia F, Truong-Meyer XM, Morin D (1997) Bioleaching of a cobaltiferous pyrite: a continuous laboratory-scale study at high solids concentration. Miner Eng 10:507–527

d'Hugues P, Battaglia-Brunet F, Clarens M, Morin D (2003) Microbial diversity of various metal-sulfidesulfides bioleaching cultures grown under different operating conditions using 16S-rDNA analysis. In: Tsezos M Remoudaki E, Hatzikioseyian A (eds) International biohydrometallurgy symposium, IBS 2003, Athens, paper no 145

Foucher S, Battaglia-Brunet F, d'Hugues P, Clarens M, Godon JJ, Morin D (2003) Evolution of the bacterial population during the batch bioleaching of a cobaltiferous pyrite in a suspended-solids bubble column and comparison with a mechanically agitated reactor. Hydrometallurgy 71:5–12

Johnson DB (2001) Importance of microbial ecology in the development of new mineral technologies. Hydrometallurgy 59:147–157

Mandl M (1984) Growth and respiration kinetics of *Thiobacillus ferrooxidans* limited by CO_2 and O_2. Biologia (Bratislava) 39 (4):429–434

Morin D, Battaglia F, Ollivier P (1993) Study of the bioleaching of a cobaltiferous pyritic concentrate. In: Torma AE, Wey JE, Lakshaman VL (eds) Biohydrometallurgical technologies, vol 1. The Minerals, Metals and Materials Society, Warrendale, pp 147–155

Morin D, Ollivier P, Hau JM (1995) Treatment of a cobaltiferous pyritic waste by bioleaching process. In: Rao SR, Amaratunga LM, Richards GG, Koudos PD (eds) Waste processing and recycling in mineral and metallurgical industries II. The Canadian Institute of Mining, Metallurgy and Petroleum, Montreal, pp 22–23

Nagpal S, Dahlstrom D, Oolman T (1993) Effect of carbon dioxide concentration on the bioleaching of a pyrite-arseno-pyrite ore concentrate. Biotechnol Bioeng 41:459–464

Norris PR, Barr DW, Hinson D (1988) Iron and mineral oxidation by acidophilic bacteria: affinities for iron and attachment to pyrite, In: Norris PR, Kelly DP (eds) Biohydrometallurgy. Proceedings of the international symposium, Warwick, 1987. Science and Technology Letters, Kew, pp 43–61

Pinches A., Chapman JT. Te Riele WAM, Van Staden M (1988) The performance of bacterial leach reactors for the preoxidation of refractory gold-bearing concentrates. In: Norris PR, Kelly DP (eds) Biohydrometallurgy. Proceedings of the international symposium, Warwick, 1987. Science and Technology Letters, Kew, pp 329–344

Rawlings DE, Tributsh H, Hansford GS (1999) Reason why '*Leptospirillum*'-like species rather than *Thiobacillus ferrooxidans* are the dominant iron-oxidizing bacteria in many commercial processes for the biooxidation of pyrite related ores. Microbiology 145:5–13

Southam G, Beveridge TJ (1993) Examination of Lipopolysaccharide (O-antigen) populations of *Thiobacillus ferrooxidans* from two mine tailings. Appl Environ Microbiol 59:1283–1288

Thompson DL, Noah KS, Wichlacz PL, Torma AE (1993) Bioextraction of cobalt from complex metal sulfidesulfides. In: Torma AE, Wey JE, Lakshaman VL (eds) Biohydrometallurgical technologies, vol 1. The Minerals, Metals and Materials Society, Warrendale, pp 653–666

Torma AE (1988) Leaching of metals. Biotechnology 6B:367–399

3 Commercial Applications of Thermophile Bioleaching

CHRIS A. DU PLESSIS, JOHN D. BATTY, DAVID W. DEW

3.1 Introduction

The focus of this chapter is the commercial applications of thermophilic bioleaching technologies. Because of the prominent role of these applications, particularly in processing copper ores and concentrates, the commercial exploitation and technical features of the thermophilic technologies will be discussed in the context of this metal.

3.2 Commercial Context of Copper Processing Technologies

Before discussing the application of thermophile technologies, a brief overview of the competitive technical and commercial landscape is warranted. There are a number of processing options that could be considered when evaluating a particular copper mining project. A diagrammatic overview of the main processing options is given in Fig. 3.1. These processing options have important commercial implications and provide insights into the role of various technologies in selecting the most economically favorable processing route. The four main copper processing technology options are smelting, concentrate leaching, heap leaching and in situ leaching. A brief description of each technology is provided, followed by a delineation of the commercial niche for thermophile bioleaching technologies compared with other available processing technologies.

3.2.1 In Situ Leaching

In situ leaching is the only processing option that does not require the metal-containing material to be removed from the ground (Bosecker 1997; Liu and Brady 1999). In such cases a network of drill-holes is typically used to inject acidic solutions directly into the subsurface ore body where it percolates until reaching an impermeable layer. A secondary, deeper set of drill-holes is used for extraction of the pregnant liquor solution (PLS) to the surface for further processing. An important requirement for in situ leaching is an impermeable hydrogeological barrier layer immediately below the leaching

Biomining
(ed. by Douglas E. Rawlings and D. Barrie Johnson)
© Springer-Verlag Berlin Heidelberg 2007

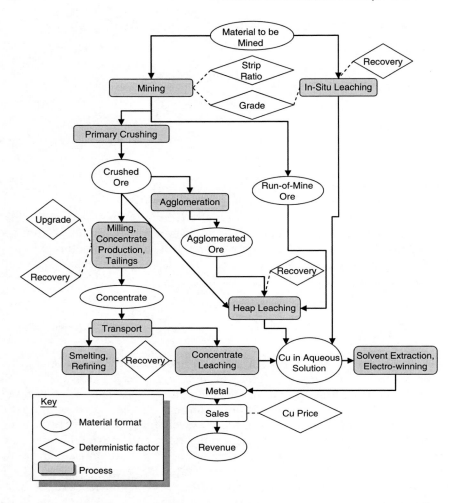

Fig. 3.1. Overview of copper processing technology routes

zone in combination with other containment measures, for obvious environmental reasons. The PLS is further processed by conventional solvent extraction and electrowinning methods to produce cathode metal of the desired grade (Kordosky 2002). Copper recoveries are relatively low, with leaching periods typically measured in years. This processing method is generally only a viable technology option for the recovery of copper from acid-soluble minerals, mainly copper oxides, with lesser success in recovery of secondary copper sulfides (chalcocite, Cu_2S and covellite, CuS), and is unsuitable for recovery of primary copper sulfides such as chalcopyrite ($CuFeS_2$). The economics of the process is governed principally by the copper grade, the size of the ore body, the recovery of the copper from the leaching material, and environmental risk factors. Although capital and operating costs for this processing route are relatively low, the environmental risks, liabilities, regulatory approvals, and closure costs may be substantial. The main environmental risk is associated with the failure to adequately capture and contain the metal-bearing acid solutions that may impact the groundwater resources. This technology may be difficult to implement for high production rates (i.e., more than 50,000 t year^{-1}). For these reasons, this technology has declined in popularity during the last few decades.

3.2.2 Smelting

Smelting is the most common processing option for producing copper metal from ore. One of the main reasons for the popularity of this process stems from its relative insensitivity to the mineralogy in which the copper occurs. In the case of smelting, as is the case with all processing routes except for in situ leaching, the copper-containing material has to be removed from the subsurface prior to further processing. The magnitude of the mining and associated costs is influenced by a number of factors (Sect. 3.3). Once mined, the ore is put through a sequence of crushing, milling, and flotation processes in order to produce a copper concentrate. The concentrate represents an upgraded version of the ore material, with a copper concentration as high as 35%. The processes involved in producing a concentrate, by definition, reject a substantial amount of gangue material which has to be disposed of in tailings containment facilities. Some losses of copper also occur to tailings, thus making copper recovery during concentrate production an important factor. Smelting is essentially a pyrometallurgical processing route where the concentrate is fed into a smelter process and subsequently refined into the metal grade of choice (Moskalyk and Alfantazi 2003). The overall smelting process typically results in virtually a total conversion of copper contained in the concentrate into metal, so that recovery percentages are near 100%.

3.2.3 Concentrate Leaching

Concentrate leaching follows the same pretreatment route as smelting, but makes use of either hydrometallurgical or biohydrometallurgical methods of extracting metal from the concentrate. The process efficiency and costs associated with concentrate leaching are governed by the copper mineralogy, with primary copper sulfide minerals such as chalcopyrite being the most refractory. A vast array of concentrate leaching processes are now available (Table 3.1), and can be broadly grouped into biohydrometallurgical (e.g., BioCop™ and GEOCOAT®) and hydrometallurgical processes (Carranza et al. 2004; Wang 2005). With most of these concentrate leaching technologies the copper is converted from a mineral phase to a solution phase in acidic sulfate

Table 3.1. Summary of concentrate leaching technologies

Name	Description
BioCop™	Microorganisms used in atmospheric tank leaching at temperatures below 85°C. Grinding to 80% passing 38 µm. PLS to SX/EW
GEOCOAT®	Concentrate material coated onto inert particles, where aerobic microbial bioleaching occurs. Heap construction similar to conventional heap leaching. PLS to SX/EW
Cuprochlor heap leaching	Crushed ore agglomerated using calcium chloride and sulfuric acid. PLS to SX/EW
Intec	Circumvents SX by electrolytic deposition at the cathode of copper from a purified sodium chloride–sodium bromide electrolyte
Outokumpu Hydrocopper	Atmospheric leaching with a cupric chloride–brine solution. Caustic soda is used to precipitate copper from cuprous oxide. Reagents are regenerated using chloralkali cell technology. Copper, powder, rods or dendrites are produced
CESL	Fine grinding to reduce the feed size to about 16–18 µm and low-pressure oxidation with chloride followed by atmospheric leaching and SX/EW circuit
Activox	Ultrafine grinding to reduce the feed size to below 10 µm and low-pressure oxidation. PLS to SX/EW
Phelps Dodge pressure leach	Leaching in a pressure vessel with oxygen followed by blending with lower-grade stockpile leach solutions, SX/EW, and/or direct EW
Anglo American Corporation/University of British Columbia (AAC/UBC) process	Medium-pressure oxidation process operating at 150°C, to which surfactant is added to disperse the molten sulfur. The feed is finely ground to 10–20 µm. PLS to SX/EW
MIM/Highlands Albion (Nenatech) process	Fine grinding to reduce the feed size to about 16–18 µm, then ferric sulfate leaching at about 80°C and atmospheric pressure, with oxygen or air sparging. PLS to SX/EW
Dynatec process	Pressure oxidation process operating at 150°C. Low-grade coal is added as molten sulfur dispersant. The feed is ground to 30–40 µm, and unleached copper is floated and recycled to maximize copper extraction. PLS to SX/EW
NSC process	Moderate pressure oxidation at 125–155°C, catalyzed with nitrogen species supplied from sodium nitrite, preceded by an ultrafine grind to 10 µm. PLS to SX/EW

PLS pregnant liquor solution, *SX* solvent extraction, *EW* electrowinning

solution. The PLS is processed by conventional solvent extraction and electrowinning methods. However, Outokumpo's HydroCopper process allows for interesting circumvention of both conventional solvent extraction and electrowinning processes (Hyvärinen and Hämäläinen 2005) by following a chloride leaching approach.

3.2.4 Heap and Dump Leaching

Dump and heap leaching are generic processes whereby either run-of-mine (dump) or crushed and agglomerated (heap) ore is typically placed either by dumping or stacking at heights ranging from 4 to 10 m for heap leaching and of 18 m or higher for dump leaching. Heaps are irrigated with acid solutions and may be aerated from the bottom via a manifold air-distribution system. The PLS solution is collected in a drainage system embedded immediately above an impermeable layer or membrane, and copper is again recovered by solvent extraction and electrowinning. Dump and heap leaching are usually only applied to marginal ores that cannot be economically processed via a concentrate production route. While copper oxide minerals readily leach in the presence of acid (typically sulfuric acid), secondary copper sulfide minerals are more effectively leached by the combination of acid and ferric iron. The role of bacteria in such systems is to convert ferrous iron to ferric iron at a rate that exceeds the rate at which ferric iron is consumed in the leaching reaction. However, some primary copper sulfides, such as chalcopyrite, are not chemically leached in the presence of acid and ferric iron (Antonijević and Bogdanović 2004) under mesophilic (below 40°C) bacterial bioleaching conditions. Effective chalcopyrite leaching is only achieved under bioleaching conditions at elevated temperatures (above 45°C, preferably above 55°C) (Clark and Norris 1996). To achieve effective recovery of copper in the case of chalcopyrite, microbes are required to not only convert ferrous iron to ferric iron, but also to efficiently generate heat by sulfur oxidation (van Staden et al. 2005).

3.3 Key Factors Influencing Commercial Decisions for Copper Projects

The choice of processing technology depends on a complex set of parameters that include operating and capital costs, long-term copper price, total size of the project, geographic location, transport costs, as well as ore grade and associated mineralogy and other factors. An optimum technology choice can only be made upon detailed consideration of the relevant facts for each specific project. There are, however, a number of important factors that influence the processing technology selection decision. While not exhaustive, the following paragraphs and Fig. 3.2 highlight some of the most prominent of these.

3.3.1 Operating Costs

The flow diagram in Fig. 3.3 illustrates approximate operating costs expressed as US cents per pound of final copper metal or cathode produced, for a reasonably sized copper producing facility (approximately 100,000 t year^{-1}). Typical (operating costs) for the three main processing routes are:

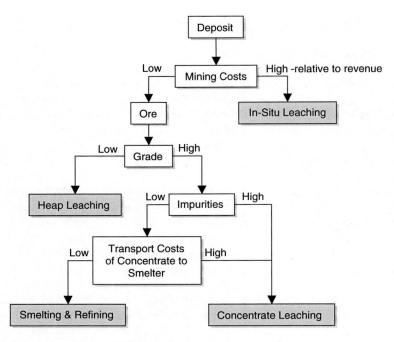

Fig. 3.2. Summary of decision flow diagram of the main factors governing the choice of copper processing technologies

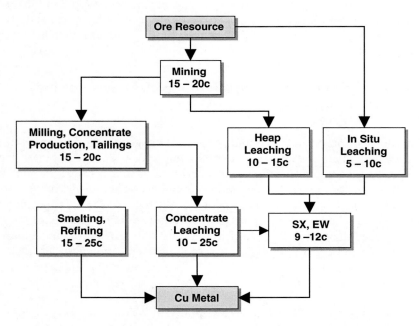

Fig. 3.3. Summary of operating costs for copper processing technologies (costs in US cents per pound of final copper product). Heap leaching costs include primary crushing costs. Transport costs assumed to be minimal (i.e., below 5 ¢ and are not included). *SX* solvent extraction, *EW* electrowinning

- Smelting: 45–65 ¢ lb^{-1}
- Concentrate leaching (including solvent extraction/electrowinning): 49–77 ¢ lb^{-1}
- Heap leaching: 34–47 ¢ lb^{-1}

The values listed in Fig. 3.3 are typical ranges for the year 2003, and are by no means fully inclusive. These costs should be considered only as indicative of how most copper operations behave, and should not be considered in isolation. For example, while heap leaching operating costs seems attractive, it should be taken into account that 15–60% of the total copper may not be recoverable via this processing route. A significant portion of unrecoverable copper is generally economically unacceptable, except in certain marginal-grade project scenarios where the copper would otherwise be unrecoverable.

3.3.2   Capital Costs

In addition to operating costs imperatives, the capital costs associated with a particular project also play a critical role in influencing the choice of processing technology. In many instances the capital costs plays an overriding role in the choice of processing technology. An important influence on capital costs considerations is the fact that the world has a current oversupply of smelting and refining facilities. This implies that the capital costs of smelting and refining facilities are considered to be sunk costs and are, therefore, not taken into account when considering this option. Because concentrate transport costs to smelting facilities are generally modest (typically less than 5 ¢ lb^{-1}), concentrates can be readily sold to any of a large number of smelting and refining facilities around the world. This phenomenon currently represents a powerful advantage of the smelting processing out over on-site concentrate leaching. In remote undeveloped areas, with a lack of existing typical mining infrastructure where concentrate transport costs are high, on-site concentrate leaching with subsequent cathode production becomes economically attractive. The main reason for this is because of the mass loss incurred in the production of metal cathode. The establishment of capital costs infrastructure, to facilitate concentrate handling and transport, may also be significant in areas without such existing facilities. This may further add to the attractiveness of concentrate leaching processes in such cases.

3.3.3   Mining Costs

The most important factors governing the mining costs of an operation are the volume of material to be mined and the so-called strip ratio. The strip ratio is the ratio of waste material relative to ore that has to be removed. Strip ratios typically range from 1 to 10, with high strip ratio projects (ratio above 5) usually only being feasible if the copper grade is relatively high (above 1.5%). For example the strip ratio of Escondida is 7, Tintaya 10, and most deposits in the Zambian Copper Belt have a strip ratio larger than 10. Strip ratios only apply to open-cast or pit mining. In cases where the strip ratio is excessive, underground mining may be considered, although at a cost premium. Mining costs comprise a significant portion of the operating costs for copper projects (Fig. 3.3). Because these costs are significant, the strip ratio is also of particular importance. The influence of mining cost is most prominently highlighted in the sulfide heap leaching project at the Escondida mine in Chile, the world's largest copper mine. This mine primarily produces concentrate which is directly sold to smelters. Marginal sulfide copper ore with a copper content between 0.7 and 1.0%, previously mined to access high-grade ores, was stockpiled in waste dumps. The fact that the ore was marginal, according to the established cut-off grade, precluded smelting and concentrate leaching as processing options. With the advent of heap bioleaching technology, processing of this material and even lower grade (0.4–0.7%) material became viable, and developed into a significant heap leaching project of its own. Testing confirmed copper heap bioleaching recoveries to be in the range of at least 35%. A critically important factor that enabled the project to be viable was that mining costs were already covered by mining of the higher-grade ore for concentrator feed. All future marginal ore will be mined as waste and the only additional cost is the differential costs in the distance hauled between the waste dump and the leach dump. These unique sets of circumstances culminated in the construction of one of the world's largest heap bioleaching operations commencing in 2006. This heap bioleaching operation is anticipated to treat up to 20 Mt of marginal ore per annum for a period of approximately 20 years and produce up to 200,000 t of copper cathode per annum.

3.3.4 Impurities

Impurities contained in the ore, and subsequently the concentrate, pose a number of problems when processed via the smelting route. These problems range from hazardous emissions to the quality of the final metal product. Typical impurities that negatively impact on the smelting process are arsenic, antimony, and bismuth (Villarroel 1999). Arsenic is the most commonly encountered impurity in copper ore, with enargite (Cu_3AsS_4) often a significant mineral contained in such ore bodies (Riveros et al. 2001). Smelters start imposing penalties on the feed intake concentrate when the arsenic concentration exceeds 0.15%, with penalties escalating at higher levels. At concentrations of less than 1%, smelters may, in some instances, mitigate the effect of arsenic content in the smelter feed by dilution and other measures; however, at arsenic concentrations exceeding 1% most smelters will not accept the concentrate. The arsenic concentration at these levels thus provides strong impetus for a move towards concentrate leaching methods that are either unaffected by the arsenic or convert arsenic to arsenate. Microbial strains used in biohydrometallurgy applications are, in general, remarkably resistant to a wide range of metals and impurities (Dopson et al. 2003), thus making tank bioleaching particularly effective in this niche. An example of this technology is BHP Billiton's BioCop™ technology (US Patent 5,919,674) which was demonstrated in a joint venture (Alliance Copper) to treat arsenic-containing ores owned by Coldelco.

3.3.5   Level of Sulfur Oxidation Required for Disposal

The presence of high levels of impurities in high-grade concentrate material presents a niche opportunity for concentrate leaching in general. An important technical consideration in comparing concentrate leaching technologies, apart from the obvious cost comparison, is the level of sulfur oxidation required, with particular reference to oxygen and disposal costs. For example, in certain concentrate leaching processes, such as tank bioleaching, the sulfur is fully oxidized to sulfate (Gericke et al. 2001). This requires a higher level of oxygen supply than some autoclave processes where the sulfur is only partially oxidized. If oxygen is supplied in the form of enriched air, the complete oxidation of sulfur to sulfate may represent a significant operating cost. Such operating costs should, however, be balanced against disposal costs of residue where the sulfur is not fully oxidized. While partial sulfur oxidation during the concentrate leaching process may represent an operational saving in oxygen, it results in a residue with remaining oxidizable sulfur and its associated disposal costs and liabilities. Disposal risks are not only limited to the remaining sulfur content of the residue, but because of its acid-generating capacity sulfur is one of the main considerations in tailings disposal. Disposal risks, costs, and legal liabilities vary between climatic regions, countries and geographical locations and need to be explicitly considered in selecting an appropriate copper leaching technology.

3.3.6   Alternative Acid Use

In smelting processes, the sulfur content of the concentrate feed material represents an important energy source. The reduced sulfur in the feed concentrate is essentially converted to sulfur dioxide during the process. In the case of concentrate hydro/biohydrometallurgical processes the sulfur can be converted to sulfuric acid, which may be beneficially applied elsewhere on a particular site operation. In addition, a large amount of acid can be gained from the exchange of Cu^{2+} ions for H^+ ions in the solvent extraction stage of the downstream processing of the biohydro/hydrometallurgical process (i.e. 2 mol H^+ for each mol Cu^{2+}, equivalent to approximately 1.5 ton sulfuric acid for every tonne of Cu extracted). This acid, which exits the solvent extraction plant as raffinate, may be circulated through an adjacent heap leach, thus diminishing the amount of make-up acid required. Because of the relatively high cost of sulfuric acid, the acid gain from such processes may be a significant advantage over smelting processes that do not generate acid in this manner. One such example, applied by Phelps Dodge at their Baghdad operation, is that acid generated by the concentrate pressure leach process is subsequently utilized in the (net acid consuming) heap leaching process (Brewer 2004).

3.4   Techno-commercial Niche for Thermophilic Bioleaching

The techno-commercial niches for heap leaching and concentrate leaching, and swing factors that influence these niches, were described in the preceding sections (Fig. 3.2). While a relatively large number of biohydrometallurgical leaching processes are industrially and commercially applied (Rawlings et al. 2003), the niche applications of two thermophilic bioleaching processes, namely, thermophilic tank bioleaching and thermophilic heap bioleaching, will be explored in more detail. Both of these technologies represent a further specialized application within the context of concentrate leaching and heap leaching, respectively. These two processes will be discussed mainly in the context of copper leaching, but we will also include a description of application opportunities for other metals.

3.4.1   Thermophilic Tank Bioleaching Features

3.4.1.1   Requirement for Thermophilic Conditions

The technical and industrial application of mesophilic tank bioleaching, using bacteria, is well known in the industry and has been thoroughly described elsewhere (Dew et al. 1997; Rawlings 2002). The differentiating factor in tank bioleaching that requires the use of thermophilic conditions (typically 60–80°C), as opposed to mesophilic conditions (below 40°C), is the presence of a significant portion of the copper as refractory primary copper sulfides such as chalcopyrite (Norris and Owen 1993; Stott et al. 2003). The high-temperature requirement, in turn, has important implications for the microbiology of the process. While bacteria are typically used at mesophilic conditions, thermophilic conditions require the use of archaea (Norris et al. 2000). It has already been established that the presence of high concentrations of impurities (i.e., arsenic) favors concentrate leaching of high-grade material over smelting. The ability to process impurities, in addition to the requirement for high temperatures, favors the use of a thermophilic tank bioleaching process. Examples of impurities are high concentrations of arsenic-containing sulfides (e.g., arsenopyrite, FeAsS) or copper sulfides (e.g., enargite, Cu_3AsS_4) that occur together with the refractory primary copper sulfide (e.g., chalcopyrite).

3.4.1.2 Microbial-Catalyzed Reactions

The BioCop™ process has been developed and applied successfully at proto-type scale (Batty and Rorke 2005). The operating conditions of a typical ther-mophilic BioCop™ bioleaching tank reactor are provided in Table 3.2. A complex set of microbial-catalyzed aerobic reactions influence the electro-chemistry, and facilitate metal dissolution in a number of ways (Schippers and Sand 1999; Crundwell 2003). The most important of these reactions in most biohydrometallurgy applications can be summarized by the following interdependent steps (Suzuki 2001):

- In an acid solution, initial proton attack on the ferrous sulfide containing minerals results in an initial solubilization of ferrous iron.
- Soluble ferrous iron is microbially oxidized to ferric iron.
- Soluble ferric iron, together with acid, facilitates the initial chemical attack and oxidation of the mineral sulfide, while sulfur-oxidizing microorgan-isms further oxidize the reduced sulfur species to sulfate.
- The microbial-mediated bioleaching process generates acid, to counter acid-consuming reactions (Stott et al. 2003); it also generates heat (mainly via sulfur oxidation), which is important in initiating the leaching process, main-taining the copper in solution, and expediting the leaching reaction.

Table 3.2. Typical BioCop™ thermophilic tank bioleaching operating conditions

Operating condition	Unit values and description
Temperature	78°C
Pulp grind size	d_{80} 33 μm
Pulp density of concentrate feed	12% (w/w)
Primary reactor residence time	48 h
Overall reactor-train residence time	96 h
pH	1.5
Dissolved oxygen concentration	1–4 mg L^{-1}
Microbial cell concentration in solution	~109 cells mL^{-1}
Carbon dioxide supplementation	1% (v/v) of total gas inflow
Copper recovery	>98%
Microbial population	*Sulfolobus*-like spp., *Metallosphaera*-like spp., *Acidianus*-like spp.
Gas mass transfer, agitation and solid suspension	By agitation (Lightning A315)
Redox potential	~700 mV (Ag/AgCl in 3 M KCl)

3.4.1.3 Reactor Configuration

In the BioCOP™ process, the reactions described earlier occur continuously in a series of six equal-size continuous culture reactors. The reactors are arranged as primary and secondary reactors with three reactors in parallel comprising the primary reactors and three reactors in series making up the secondary reactors. This configuration of reactors is utilized in order to approximate a single large continuously stirred tank reactor followed by a plug flow reactor in order to increase overall process efficiency (Konishi et al. 2001; Gonzalez et al. 2004). The bulk of the microbial process, and about 50–70% of the metal dissolution, occurs in the primary reactors, with the subsequent reactors acting as polishing steps in which the remaining leaching action occurs as a result of chemical ferric iron attack. Downstream processing of acid-soluble metals such as copper is achieved by gravitational settling and thickening of the residue sludge, with the supernatant PLS reporting to conventional solvent extraction and electrowinning. The high rate of iron oxidation in these reactions results in an overall high redox potential leaching environment. While high redox potentials are generally considered to cause passivation of chalcopyrite (Third et al. 2000; Hiroyoshi et al. 2004), the high temperature and microbial action on the mineral surfaces result in metal leaching rates that exceed the rate of passivation, thus facilitating an effective leaching process (Tshilombo et al. 2002; Sandström et al. 2005). Each of the continuous culture reactors is supplied with aeration and agitation. Agitation is critically important owing to the required mass transfer of oxygen and carbon dioxide to the liquid phase, in addition to the solid suspension requirement.

3.4.1.4 Oxygen Supply

The high temperatures at which these reactors operate not only impose the requirement for special microbial cultures but also impose some constraints in terms of gas mass transfer. The various microbial and chemical oxidation reactions involved in the oxidation and dissolution of minerals impose an overall oxygen demand in the solution phase. This demand can only be satisfied if the rate of transfer from the gas phase equals or exceeds the rate of demand within the solution phase. The increased demand for oxygen under these conditions is generally facilitated by increased aeration rates, higher impeller agitation rates, and improved agitator designs. In the case of mesophiles, sufficient oxygen may be supplied to the reactor by simply sparging with air underneath a suitable agitator. The volumes of air required are substantial as the oxygen utilization from air is limited to about 40%. The large air requirement implies high power consumption and the use of multistage blowers is preferred over higher-pressure compressors. Agitation speeds and power inputs cannot be indefinitely increased to improve mass transfer limitations as cell shear stress damage to bioleaching microbes becomes a limiting factor at high agitation speeds and power inputs increase disproportionately in the presence of high pulp densities (Chong et al. 2002). Under thermophilic bioleaching conditions, mass transfer limitation due to the reduced solubility of oxygen is a process challenge that cannot be overcome by simply increasing agitation speeds and aeration rates. In the thermophile situation, the use of oxygen or oxygen-enriched air is required owing to higher demand and also the reduced overall mass transfer driving force (d'Hugues et al. 2002). The use of oxygen-enriched air in such reactors at large scale does, however, pose its own set of process risks. For example, the use of oxygen-enriched air (as high as 90%) could potentially result in the increase of dissolved oxygen concentrations as high as 15 mg L^{-1} and this could have catastrophic effects on archaea as their capacity to dissipate oxygen-derived free radicals may become overloaded. The optimal dissolved oxygen concentration has been determined to be 1.5–4.1 mg L^{-1} (de Kock et al. 2003). Addition and maintenance of dissolved oxygen at a concentration within this narrow range is the subject of key patents governing the operation of thermophilic tank bioleaching (US Patents 6,883,020 B1 and 6,860,919 B1).

3.4.1.5 Oxygen Production

On an industrial scale, oxygen is either generated through a cryogenic process whereby air is liquefied and then separated into its different components through distillation columns, or through the use of vacuum pressure swing adsorption (VPSA) plants. In the VPSA system, air is pumped under pressure into zeolite adsorption columns where the oxygen is adsorbed onto the zeolite. When the pressure is reduced in these columns, the oxygen desorbs and is then compressed to the required pressure. VPSA plants are typically only specified where the total daily oxygen requirement is less than 300 t per day and above this cryogenic plants are used. Cryogenic plants deliver higher-purity gas and can produce essentially pure oxygen, whereas the VPSA plants can produce a maximum purity of about 94%. High-purity oxygen is not a requirement and 94% oxygen would be ideal if the economic scale of base metal bioleach plants did not effectively preclude the use of these plants.

3.4.1.6 Carbon Dioxide

Because of the autotrophic nature of the primary mineral-oxidizing thermophilic archaea, CO_2 is required as a carbon source. Air contains 0.033% CO_2 by volume and this is generally insufficient for sustaining microbial cultures in tank leach systems. Many mineral concentrates contain carbonates which react quickly in the highly acidic bioleach liquors, releasing CO_2. In such cases, it may not be necessary to supplement air with CO_2. Some concentrates are highly acid producing when leached, and require addition of an alkaline reagent to prevent the pH falling too low. In such cases, the use of limestone ($CaCO_3$) is recommended as this supplements the CO_2 supply as well as controlling the pH. Where a concentrate is neither highly acid producing nor does it contain natural carbonate components it is necessary to enrich the (oxygen-enriched) air supply with CO_2 at a level between 1 and 2% by volume. CO_2 is expensive, as there are no cost-effective means of production in small quantities.

3.4.1.7   Agitation

It is most cost-effective to introduce gas into the bioleach reactor at the minimum possible pressure, but as a consequence the gas requires assisted mechanical agitation to effectively disperse. In the first tank bioleach plant (the Fairview BIOX™ Plant) a Rushton turbine was used to agitate the tanks. Rushton turbines are high power consumers and have an additional drawback that they provide essentially no solid suspension capabilities. This limitation is usually overcome by having an axial flow impeller higher up on the same agitator shaft. In contrast, high-solidity axial-flow impellers have very high gas handling capability, and provide sufficient suspension of solids. Impellors of this type are now widely used in tank bioleach plants.

3.4.1.8   Pulp Density

Contrary to some reports on batch studies (Gómez et al. 1999; Rubio et al. 2002), the limitation on the pulp density usage in continuous thermophilic tank bioleaching is governed by the copper tenor in solution rather than shear stress effects. The upper limit for effective operation is usually reached at approximately 36 g L^{-1}. Copper inhibition effects, on microbial growth, are already evident at copper concentrations of more than 7 g L^{-1}. The approximate cell-doubling times of the archaea used in the thermophilic process are typically in the range 11–17 h, while the hydraulic retention time in the primary reactors is typically 48 h. The relatively large excess of hydraulic retention time compared with cell-doubling time provides process robustness and the ability to absorb growth-inhibition effects from factors such as inhibitory copper concentrations (Crundwell 2001). Alternatively, methods may be employed to continuously remove copper from solution, thus limiting its potential inhibitory effects on the bioleaching archaea.

3.4.1.9   Arsenic Conversion to Arsenate

One of the main advantages that bioleaching has over smelting is that it can be applied to cases where there are very high concentrations of arsenic in the concentrate (Dopson and Lindström 1999; Casiot et al. 2003). Arsenic impurity is most pronounced in concentrates containing either arsenopyrite (FeAsS) or enargite (Cu_3AsS_4). In the case of smelting or any pyrometallurgical process, the arsenic is oxidized to As_2O_3, which will remain in a gaseous form until the temperature drops to below 193°C. Consequently, arsenic is released directly to the atmosphere or, in the case where smelters have suitable off-gas handling, the arsenic will accumulate in the flue dust. As_2O_3 is readily soluble in water, and the handling of flue dusts is a major problem for smelting and roasting operations. Since the world market for arsenic is extremely small, and is currently contracting, smelters and roasters are forced to create and store large stockpiles of the highly toxic arsenic by-product. In bioleaching and pressure leaching of arsenic-bearing concentrates, the arsenic(III) is oxidized to arsenate(V). The arsenate-bearing solutions, which also contain a high ferric iron content, are then usually neutralized using lime or limestone, or a combination thereof. The product of this neutralization step is ferric arsenate, which is environmentally very stable and suitable for disposal on conventional tailings dams.

3.4.1.10 BioCyn™

In the case of gold recovery from the residue via cyanidation, the mass loss and sulfide reduction is beneficial as it significantly reduces the amount of cyanide required for subsequent gold recovery. Cyanide reacts with sulfides, so the sulfide content of the concentrate effectively reduces the amount of cyanide available for reacting with the gold. The benefit of thermophilic conditions in this application is the higher rate of oxidation resulting in smaller reactor size and favorable reactor operating costs. This BHP Billiton thermophilic tank bioleaching process (BioCyn™) has been licensed to Gold Fields for application at the Fairview mine in Barberton, South Africa.

3.4.1.11 Cost Factors

The capital costs of a bioleaching operation are typically less than for a smelting operation but the operating costs can be higher and are dependent on plant location and the costs of services at the particular site. In a tank bioleach plant the capital costs are mainly influenced by the materials of construction and the cost of delivering oxygen to the microbes. The provision of sufficient oxygen to the microbes consists of blowers or compressors or an oxygen plant and high power agitators to disperse this gas into the reactor.

3.4.1.12 Materials of Construction

The bioleach environment is extremely aggressive. The choice of materials used to construct the reactors and associated equipment is, therefore, an important consideration. The bioleach process is usually conducted in the pH range 1–2 at temperatures ranging from 40°C (mesophile) to about 78°C (thermophile), with ferric ion concentrations from about 5 to over 50 g L^{-1}. For operations where there is a low chloride concentration in the process water, materials of construction can be relatively inexpensive, such as rubber-lined mild steel or stainless steel grades such as 304L or 316L. With increased chloride concentrations, it is necessary to use more exotic and costly stainless steel alloys such as SAF2205, SAF2507, or 904L. In the case of BHP Billiton's prototype plant, which operates at 78°C and with a chloride concentration of about 300 mg L^{-1}, it was decided to construct the reactor vessels using a proprietary lining system consisting of ceramic tiles and bricks which sandwich a polyurethane membrane. This was considered the most cost-effective option at the time. Stainless steel prices are very dependent on the nickel price and, during periods where nickel is very costly, alternatives such as the ceramic/membrane system should also be considered.

3.4.2 Thermophilic Tank Bioleaching Application Options and Opportunities

3.4.2.1 Copper–Gold Applications

There are numerous ores where copper and gold are found together in sufficient concentrations for the exploitation of these deposits to be warranted. In some cases, either the copper or the gold is of sufficient value that exploitation of the deposit for one component alone can be justified. Examples of significant copper gold deposits are Yanacocha (Peru), Salobo (Brazil), and Olympic Dam (Australia).

3.4.2.2 Expansion Applications

Recently, various existing operations that use on-site smelters have considered plant expansions. Owing to the nature of the smelting operations it is not possible to make small incremental increases in smelter capacity simply by adding additional equipment. Smelters that increase capacity will typically do so in integer multiples of the original capacity. Bioleaching has now provided an alternative for smelters wanting a partial increase in capacity. This is achieved by either adding a complete bioleach plant to handle the additional tonnage or, more preferably, by treating the concentrate partially in the bioleach plant, with the aim of reducing the mass into the smelter. In concentrates containing high concentrations of magnesium, smelting operations are often seriously affected in that the slag becomes too viscous. Pretreatment through a bioleach stage will remove the readily soluble magnesium and hence make the concentrate more suitable for smelting.

3.5   Thermophilic Heap Bioleaching of Marginal Ores

Heap leaching technology has been applied in the copper and gold industries for a number of years (Brierley and Brierley 2001). Interesting hybrid processes have also been developed in which concentrate material is coated, by agglomeration, onto inert support material with the leaching process operated as a heap bioleaching process (Sampson et al. 2005). Applications of these processes typically operate with ore or concentrate material of relatively high sulfur content, i.e., more than 2%, with the majority of the sulfur typically contributed by either pyrite (FeS_2) or pyrrhotite ($Fe_{x-1}S_x$). When a significant proportion of sulfur occurs in an amorphous form, it greatly facilitates rapid oxidation and thus heat generation. Microbial oxidation of sulfur to sulfate, rather than iron oxidation, is generally the most important-heat-generating reaction in heap bioleaching, as illustrated by the following reactions:

$$Fe^{2+} + 0.25O_2 + H^+ \rightarrow Fe^{3+} + 0.5H_2O \ (\Delta H^0_j = -97.3 \text{ kJ mol}^{-1}),$$
$$S^0 + 1.5O_2 + H_2O \rightarrow SO_4^{2-} + 2H^+ \ (\Delta H^0_j = -606 \text{ kJ mol}^{-1}).$$

A high starting concentration of sulfur is, therefore, an important parameter in considering the heat-generation capacity of a heap. This section focuses on thermophilic heap bioleaching of marginal ores, i.e., with relatively low sulfur content, as opposed to the high sulfur content processes operated by Geobiotics and Newmont (US Patents 6,277,341 B1 and 5,914,441, and US Patent application 2005/0044990 A). The processing of such marginal ores pertains particularly to the copper industry. A stark reality of exploration geology is that large high-grade readily and commercially accessible copper deposits are increasingly difficult to find. An increasing proportion of copper reserves are found in marginal ores (less than 1% Cu), of which an increasing proportion occurs as refractory primary sulfide copper, mainly chalcopyrite. The marginal and refractory nature of these reserves precludes conventional smelting, concentrate leaching, and in situ and acid heap leaching processing technologies. The commercial importance of these marginal reserves has, not surprisingly, sparked considerable efforts to enable the effective economic recovery of copper from such material. Techniques such as chloride leaching, redox-potential-controlled, or alternating microbial oxidation/reduction processes have been proposed (Lu et al. 2000; Third et al. 2000); however, the most comprehensively demonstrated technology for accessing chalcopyrite from marginal ore has been through thermophilic heap bioleaching. The microbial reactions that facilitate the dissolution of copper during the heap leaching of marginal ore or during concentrate leaching are similar, as summarized earlier in this chapter.

3.5.1 Basic Heap Design and the Importance of Heat Generation

Heap bioleaching technology relies on very similar operational heap configurations, and civil and geotechnical engineering as for conventional acid
heap leaching. The basic configuration comprises an impermeable leaching
pad upon which the material to be leached is stacked. The PLS is collected via
a system of drainage pipes contained in a 1–2-m inert overburden material at
the bottom of the heap. An aeration distribution pipe system is used to supply air into the heap. The PLS typically reports to intermediate ponds and
eventually to conventional solvent extraction and electrowinning, while the
raffinate is added to the top of the heap via a network of irrigation pipelines.
The leach material can either be run-of-mine or crushed and agglomerated
ore. The microbial reactions described previously occur at the ore–liquid
interface and in the solution and result in the release of copper from the mineral to the solution phase. Typical heap operational parameters are provided
in Table 3.3. The key distinguishing process feature of marginal copper sulfide heaps is the achievement of elevated heap temperature (above 55°C) in
order to facilitate effective recovery of copper from chalcopyrite. The attain-
able temperature is the net result of heat-generation and heat-loss/heat-
retention factors. Heat generation is mainly the result of microbial oxidation
of sulfur to sulfate (Petersen and Dixon 2002). The two main factors that gov-
ern the overall capacity for heat generation are therefore the available sulfur
content of the ore and microbial activity.

3.5.2 Sulfur Availability

The available sulfur content of a deposit depends upon the ore mineralogy
and the physical characteristics and deportment of the mineral particles.
Usually, the most abundant sulfur-containing mineral found associated with
copper minerals of interest is pyrite. In general, ores with a pyrite (or equivalent sulfur-containing minerals) content of more than 2% are most
amenable to thermophilic heap bioleaching. The sulfur content of the ore

Table 3.3. Typical heap bioleaching operational parameters

Parameter	Unit values and description
Height	4–10 m
Acid in raffinate	6–8 g L^{-1}
Acid during agglomeration	2–8 g L^{-1}
Leaching period	300–450 days
Air-flow rate	0.02–0.08 N m^3 t^{-1} h^{-1}
Raffinate irrigation flow rate	4–18 L m^2 h^{-1}

should be considered with regard to sulfur availability for microbial oxidation, rather than total content, and thus has to be determined on a case-by-case basis. Geological formation of porphyry intrusions most typically results in declining pyrite content with an increase in chalcopyrite content. For this reason ore blending, to obtain the optimal pyrite content, is important for the implementation of this technology.

3.5.3 Microbial activity, CO_2, and O_2

Microbial activity is not only governed by the availability of reduced sulfur and iron, from which they obtain their energy, but also by a wide range of other factors. Air supplied from the bottom of the heap provides oxygen as an electron acceptor and carbon dioxide as a carbon source to the autotrophic bioleaching microbes (Lizama 2001). An important consideration is that air is supplied mainly from one direction, and in a plug-flow manner. The consumption of these two gases as a function of heap height, and the kinetic response of microbial growth towards decreased concentrations of CO_2 and O_2, is an important process consideration (Nemati et al. 1998). Typical CO_2 and O_2 consumption rates are in the range 0.05–0.20 and 5–10 g t^{-1} h^{-1} respectively. CO_2 availability may also be significantly influenced by the carbonate content of the gangue material. Although the ore usually contains a sufficient range of micronutrients to support microbial growth, macronutrients such as ammonium and phosphate have to be supplied via the raffinate solution at concentrations of approximately 50 mg L^{-1}.

3.5.4 Inoculation

Optimal heat generation also depends on uniform distribution of microbial activity across the entire heap height. This concept was first patented by Newmont (US Patent 5,246,486) and has since been the subject of numerous other patent applications (PCT applications WO 2004-027099 A1 and WO 2004-027100 A1). Also important is the presence of complementary (Johnson 2001) and temperature-sequential microbial populations that can facilitate the heating of the heap from mesophilic (ambient to 45°C), through moderately thermophilic (45–60°C) to thermophilic (60–80°C) temperatures (Franzmann et al. 2005; Brierley 2001). While such strains are typically well known to exist, the process management of the heap at 52–55°C is particularly important owing to the fact that the biodiversity and activity of acidophilic prokaryotes, and particularly sulfur-oxidizers, is apparently less in this temperature range than at both lower and higher temperatures. Heap inoculation is achieved during agglomeration, or stacking of the heap, or via the raffinate solution (Gericke et al. 2005). These inoculation methods have both advantages and disadvantages. Inoculation during agglomeration or stacking has the advantage of facilitating a uniform distribution of microorganisms. However, the placement of moderate thermophiles and thermophilic microbes several months before optimal conditions for their growth are developed in the heap is a disadvantage as during this waiting period adverse conditions negatively affect the survival rate of the moderate thermophile and thermophile inoculum. Introducing microbial inocula via the raffinate solution has the advantage of allowing the introduction of specific microbial cultures at the precise times when they are required. However, the strong tendency of acidophiles, particularly sulfur-oxidizing prokaryotes, to attach to minerals (Bouffard and Dixon 2004) restricts the migration of such cells via the percolating raffinate solution. This phenomena is more pronounced in tall heaps. Techniques for achieving effective inoculation, therefore, represent important proprietary information for companies that implement heap bioleaching technologies.

3.5.5   pH

Obtaining a uniform pH distribution is as vital as obtaining a uniform microbial distribution throughout the heap. A solution pH value of less than 2 is important from a hydrometallurgy as well as from a microbial point of view. Microbial activity in general, and particularly that of thermophilic archaea, is negatively affected at solution pH values above 2. Achieving a solution pH of less than 2 is a function of the acid used during agglomeration, the acid content of the raffinate, the irrigation flow rate, and the acid-generating/acid-consumption properties of the sulfide and gangue minerals, which are site-specific. The acid consumption of the gangue minerals is a key technical and a key economic factor (Dixon 2004). There are also important boundary limitations to operational parameters that can be used to achieve the desired solution pH. The amount of acid used during agglomeration, and in the raffinate solution, cannot be increased to a point where damage occurs to the microbial inoculum. In addition, the use of increased acid concentrations typically also increases the reactivity of the gangue minerals, resulting in an increase in gangue mineral dissolution. This, in turn, has two potential impacts. The first is that the concentration of total dissolved salts in solution is increased, which may result in detrimental ionic strength effects on microbial activity (Blight and Ralph 2004; Shiers et al. 2005). Secondly, increased acid concentrations may compromise the structural integrity of the heap owing to the dissolution, particularly of clay minerals, possibly causing compromised hydraulic properties of the heap. This problem is particularly prevalent in heap leaching technologies when applied to nickel laterites.

3.5.6   Inhibitory Factors

Accumulation in heaps of compounds that can inhibit microbial growth needs to be avoided. These include chloride, nitrate, and fluoride, with solution concentrations greater than 0.5 g L^{-1} (F^-) or 1.5 g L^{-1} (NO_3^- and Cl^-) being problematic (Gómez et al. 1999). These anions may be present in on-site process waters, or may be derived from minerals present in the target ore. Aluminum and sulfate concentrations exceeding 10 and 100 g L^{-1}, respectively, also have a negative impact on microbial activity, as do high osmotic potentials in general (Suzuki et al. 1999; Harahuc et al. 2000). An important, and often overlooked, source of inhibitory compounds is the solvent extraction process used to recover copper from the PLS (Bosecker 1997). While the solvents used in this process are highly insoluble in water, a small fraction of the solvents may end up in the water phase. Water-solubility of solvent extraction organic compounds may be dramatically increased owing to fungal growth at the air–solvent interface. Such organic molecules may be carried into the heap irrigation solution either as discrete droplets (micelles) or as water-soluble compounds, and may result in inhibition of microbial bioleaching. Management and control of the solvent extraction plant should, therefore, be considered an integral component of an effective heap bioleaching process.

3.5.7 Heat Retention, Air-Flow Rate, and Irrigation Rate

Retention and loss of thermal energy in a heap is mainly a function of climatic conditions, altitude, heap height, irrigation rate, and air-flow rate. The last three are the main practical operational parameters available to control the heat loss (Dixon 2000). In order to increase insulation, large heaps measuring several kilometers in length and width and with heights of up to 18 m may be constructed.

3.5.7.1 Heap Height

The heap height, in particular, is important in cold climatic conditions where maximum insulation from cold atmospheric temperatures is required. This height, however, also exaggerates some of the gradient effects discussed earlier with respect to pH profiles, microbial inoculation migration, and colonization, as well as carbon dioxide and oxygen via the air supply. The benefits of increased heap height have to be carefully weighed up against the potentially detrimental gradient effects on factors that influence microbial activity. Heat retention may also be assisted by the use of insulating materials as well as solar heating mats to elevate the temperature of the raffinate.

3.5.7.2 Irrigation and Air-Flow Rates

Apart from heap height and insulation, the two main operational controls that facilitate the management of heat retention in a heap are the air-flow and irrigation rates (Bouffard and Dixon 2001). In general, increased air-flow and irrigation rates have direct or indirect beneficial effects on microbial activity, and thus on heat generation. Reduced air-flow and irrigation rates promote heat conservation, though they negatively impact heat generation. Reduced irrigation rates cause increased gradient effects and may result in the accumulation of soluble salts and the development of high osmotic potentials, both of which could have a detrimental impact on microbial activity. This dichotomy in control philosophy of irrigation and air-flow rates represents an important tension between heat generation and heat conservation and has to be managed in a balanced manner in order to achieve elevated heap temperatures for effective thermophilic heap bioleaching.

3.6　Summary

The advent of thermophilic bioleaching has opened up exciting technological and commercialization opportunities, both in tank and in heap bioleaching. The key feature of thermophilic tank bioleaching is its ability to deal with impurities, while the key feature of thermophilic heap bioleaching is the ability to recovery copper from chalcopyrite through the combination of heat generation and iron oxidation. These present useful technology processing routes, particularly for the copper industry, but the commercial feasibility of each project has to be carefully assessed, taking full cognizance of the often subtle factors that influence its success.

References

Antonijević MM, Bogdanović GD (2004) Investigation of the leaching of chalcopyrite ore in acidic solutions. Hydrometallurgy 73:245–256

Batty JD, Rorke GV (2005) Development and commercial demonstration of the BioCop™ thermophile process. In: Harrison STL, Rawlings DE, Petersen J (eds) Proceedings of the 16th international biohydrometallurgy symposium, 25–29 September 2005, pp 153–161

Blight KR, Ralph DE (2004) Effect of ionic strength on iron oxidation with batch cultures of chemolithotrophic bacteria. Hydrometallurgy 73:325–334

Bosecker K (1997) Bioleaching: metal solubilization by microorganisms. FEMS Microbiol Rev 20:591–604

Bouffard SC, Dixon DG (2001) Investigative study into the hydrodynamics of heap leaching processes. Metall Trans B 32B:763–776

Bouffard SC, Dixon DG (2004) Evolution of bacterial community in a pyritic refractory gold ore column leaching environment. Miner Process Extr Metall Rev 25:313–319

Brewer RE (2004) Copper concentrate pressure leaching – plant scale-up from continuous laboratory testing. Miner Metall Process 21:202–208

Brierley CL (2001) Bacterial succession in bioheap leaching. Hydrometall 59:249–255

Brierley JA, Brierley CL (2001) Present and future commercial applications of biohydrometallurgy. Hydrometallurgy 59:233–239

Carranza F, Iglesias N, Mazuelos A, Palencia I, Romero R (2004) Treatment of copper concentrates containing chalcopyrite and non-ferrous sulfides by the BRISA process. Hydrometallurgy 71:413–420

Casiot C, Morin G, Juillot F, Bruneel O, Personn J-C, Leblanc M, Duquesne K, Bonnefoy V, Elbaz-Poulichet F (2003) Bacterial immobilization and oxidation of arsenic in acid mine drainage (Carnoulès creek, France). Water Res 37:2929–2936

Chong N, Karamanev DG, Margaritis A (2002) Effect of particle–particle shearing on the bioleaching of sulfide minerals. Biotechnol Bioeng 80:349–357

Clark DA, Norris PR (1996) Oxidation of mineral sulfides by thermophilic microorganisms. Miner Eng 9:1119–1125

Crundwell FK (2001) Modeling, simulation, and optimization of bacterial leaching reactors. Biotechnol Bioeng 71:255–265

Crundwell FK (2003) How do bacteria interact with minerals? Hydrometallurgy 71:75–81

d'Hugues P, Foucher S, Gallé-Cavalloni P, Morin D (2002) Continuous bioleaching of chalcopyrite using a novel extremely thermophilic mixed culture. Int J Miner Process 66:107–119

de Kock SH, Barnard P, du Plessis CA (2003) Oxygen and carbon dioxide kinetic challenges for thermophilic mineral bioleaching processes. Biochem Soc Trans 32:273–275

Dew DW, Lawson EN, Broadhurst JL (1997) The BIOX® Process for biooxidation of gold-bearing ores or concentrates. In: Rawlings DE (ed) Biomining: theory, microbes and industrial processes. Springer, Berlin Heidelberg New York, pp 45–80

Dixon DG (2000) Analysis of heat conservation during copper sulfide heap leaching. Hydrometallurgy 58:27–41

Dixon S (2004) Definition of economic optimum for the leaching of high acid-consuming copper ores. Miner Metall Process 21:198–201

Dopson M, Lindström EB (1999) Potential role of Thiobacillus caldus in arsenopyrite bioleaching. Appl Environ Microbiol 65:36–40

Dopson M, Baker-Austin C, Koppineedi PR, Bond PL (2003) Growth in sulfidic mineral environments: metal resistance mechanisms in acidophilic micro-organisms. Microbiology 149:1959–1970

Franzmann PD, Haddad CM, Hawkes RB, Robertson WJ, Plumb JJ (2005) Effects of temperature on the rates of iron and sulfur oxidation by selected bioleaching Bacteria and Archaea: application of the Ratkowsky equation. Min Eng 18:1304–1314

Gericke M, Muller HH, Neale JW, Norton AE, Crundwell FK (2005) Inoculation of heap-leaching operations. In: Harrison STL, Rawlings DE, Petersen J (eds) Proceedings of the 16th international biohydrometallurgy symposium, 25–29 September 2005, pp 255–264

Gericke M, Pinches A, van Rooyen JV (2001) Bioleaching of a chalcopyrite concentrate using an extremely thermophilic culture. Int J Miner Process 62:243–255

Gómez E, Ballester A, González F, Blázquez ML (1999) Leaching capacity of a new extremely thermophilic microorganism, Sulfolobus rivotincti. Hydrometallurgy 52:349–366

Gonzalez R, Gentina JC, Acevedo F (2004) Biooxidation of a gold concentrate in a continuous stirred tank reactor: mathematical model and optimal configuration. Biochem Eng J 19:33–42

Harahuc L, Lizama HM, Suzuki I (2000) Selective inhibition of the oxidation of ferrous iron or sulfur in Thiobacillus ferrooxidans. Appl Environ Microbiol 66:1031–1037

Hiroyoshi N, Kuroiwa S, Miki H, Tsunekawa M, Hirajima T (2004) Synergistic effect of cupric and ferrous ions on active-passive behaviour in anodic dissolution of chalcopyrite in sulfuric acid solutions. Hydrometallurgy 74:103–116

Hyvärinen O, Hämäläinen M (2005) HydroCopper – a new technology producing copper directly from concentrate. Hydrometallurgy 77:61–65

Johnson DB (2001) Importance of microbial ecology in the development of new mineral technologies. Hydrometallurgy 59:147–157

Konishi Y, Tokushige M, Asai S, Suzuki T (2001) Copper recovery from chalcopyrite concentrate by acidophilic thermophile Acidianus brierleyi in batch and continuous-flow stirred tank reactors. Hydrometallurgy 59:271–282

Kordosky GA (2002) Copper recovery using leach/solvent extraction/electrowinning technology: Forty years of innovation, 2.2 million tonnes of copper annually. S Afr J Min Metall Nov– Dec:445–450

80   Chris A. du Plessis, John D. Batty, David W. Dew

Liu J, Brady BH (1999) Evaluation of one-dimensional *in-situ* leaching process. Int J Numer Anal Methods Geomech 23:1857–1872
Lizama H (2001) Copper bioleaching behaviour in an aerated heap. Int J Miner Process 62:257–269
Lu ZY, Jeffrey MI, Lawson F (2000) An electrochemical study of the effect of chloride ions on the dissolution of chalcopyrite in acidic solutions. Hydrometallurgy 56:145–155
Moskalyk RR, Alfantazi AM (2003) Review of copper pyrometallurgical practice: today and tomorrow. Miner Eng 16:893–919
Nemati M, Harrison STL, Hansford GS, Webb C (1998) Biological oxidation of ferrous sulfate by *Thiobacillus ferrooxidans*: a review on the kinetic aspects. Biochem Eng J 1:171–190
Norris PR, Burton NP, Foulis NAM (2000) Acidophiles in bioreactor mineral processing. Extremophiles 4:71–76
Norris PR, Owen JP (1993) Mineral sulfide oxidation by enrichment cultures of novel thermoacidophilic bacteria. FEMS Microbiol Rev 11:51–56
Petersen J, Dixon DG (2002) Thermophilic heap leaching of a chalcopyrite concentrate. Miner Eng 15:777–785
Rawlings DE (2002) Heavy metal mining using microbes. Annu Rev Microbiol 56:65–91
Rawlings DE, Dew DW, du Plessis CA (2003) Biomineralization of metal-containing ores and concentrates. Trends Biotechnol 21:38–44
Riveros PA, Dutrizac JE, Spencer P (2001) Arsenic disposal practices in the metallurgical industry. Can Metall Q 40:395–420
Rubio A, Garcia Frutos FJ (2002) Bioleaching capacity of an extremely thermophilic culture for chalcopyritic materials. Miner Eng 15:689–694
Sampson MI, Van der Merwe JW, Harvey TJ, Bath MD (2005) Testing the ability of a low grade sphalerite concentrate to achieve autothermality during biooxidation heap leaching. Miner Eng 18:427–437
Sandström A. ShchukarevA, Paul J (2005) XPS characterisation of chalcopyrite chemically and bio-leached at high and low redox potential. Miner Eng 18:505–515
Schippers A, Sand W (1999) Bacterial leaching of metal sulfides proceeds by two indirect mechanisms via thiosulfate or via polysulfides and sulfur. Appl Environ Microbiol 65:319–321
Shiers DW, Blight KR, Ralph DE (2005) Sodium sulfate and sodium chloride effects on batch culture of iron oxidising bacteria. Hydrometallurgy 80:75–82
Stott MB, Sutton DC, Watling HR, Franzmann PD (2003) Comparitive leaching of chalcopyrite by selected acidiphilic bacteria and archaea. Geomicrobiol J 20:215–230
Suzuki I (2001) Microbial leaching of metals from sulfide minerals. Biotechnol Adv 19:119–132
Suzuki I, Lee D, Mackay B, Harahuc L, Oh JK (1999) Effect of various ions, pH, and osmotic pressure on oxidation of elemental sulfur by *Thiobacillus thiooxidans*. Appl Environ Microbiol 65:5163–5168
Third KA, Cord-Ruwisch R, Watling HR (2000) The role of iron-oxidizing bacteria in stimulation or inhibition of chalcopyrite bioleaching. Hydrometallurgy 57:225–233
Tshilombo AF, Petersen J, Dixon DG (2002) The influence of applied potentials and temperature on the electrochemical response of chalcopyrite during bacterial leaching. Miner Eng 15:809–813
van Staden PJ, Shaidaee B, Yazdani M (2005) A collaborative plan towards the heap bioleaching of low grade chalcopyritic ore from a new Iranian mine. In: Harrison STL, Rawlings DE, Petersen J (eds) Proceedings of the 16th international biohydrometallurgy symposium, 25–29 September 2005, pp 115–123
Villarroel D (1999) Process for refining copper in solid state. Miner Eng 12:405–414
Wang S (2005) Copper leaching from chalcopyrite concentrates. J Met 57:48–51

4 A Review of the Development and Current Status of Copper Bioleaching Operations in Chile: 25 Years of Successful Commercial Implementation

Esteban M. Domic

4.1 Historical Background and Development of Copper Hydrometallurgy in Chile

The technologies of leaching, solvent extraction and electrowinning, although adopted separately and at different times throughout the history of metallurgy in Chile, have today achieved an advanced stage of development and maturity, which has resulted in their broad-scale implementation. Bioleaching has been an integral part of this advance and has also reached a significant level of maturity, as described in this chapter.

A great deal of progress has been made since the introduction of vat leaching percolation followed by direct electrowinning at Chuquicamata in 1915 and at Potrerillos in 1928. Initially, a vat leaching system followed by scrap iron precipitation was widely used in a variety of small and mid-sized mining operations, primarily at state-owned plants through the Empresa Nacional de Minería (ENAMI). Another application, using higher-grade ore, was agitation leaching followed by countercurrent decantation and washing prior to scrap iron precipitation, also known as "cementation". Even though the product was high-grade copper precipitate or "copper cement" (75–85% copper), it nonetheless required subsequent smelting and refining treatment, which increased costs, and the high levels of water pollution produced by the residual solutions discarded from the precipitate circuit were an additional negative side effect. Examples of this process include the plants at Michilla, Ojancos, Taltal and Vallenar.

Subsequently, the advent of the process of agglomeration and acid curing followed by nonflooding or "trickle" leaching, known as the TL Leaching process, permitted mines such as the Sociedad Minera Pudahuel Lo Aguirre facility (1980) to be brought into commercial operation. This initiated a period during the 1980s that saw an intensification in the use of heap leaching and its variations. These techniques made it possible to successfully process increasingly lower grades of ore.

The experimental use of solvent extraction began early on (1970) at the pilot plant at Chuquicamata. Although followed by several other similar endeavors, only the tests run at Lo Aguirre in the late 1980s were successful and Lo Aguirre became Chile's first commercial solvent-extraction-based operation. During the 1980s, the only modern leaching projects for new

Biomining
(ed. by Douglas E. Rawlings and D. Barrie Johnson)
© Springer-Verlag Berlin Heidelberg 2007

minerals to be implemented were the solvent extraction–electrowinning plant at El Teniente to treat the mine's acidic waters (1984), the solvent extraction–electrowinning plant at the Ripios Project (involving the retreatment of the old vat leach tailings) at Chuquicamata (1987) and the secondary thin-layer heap bioleaching (TL) solvent extraction–electrowinning operation at Lo Aguirre (1985).

At El Teniente, the former operation consisted of natural in-place bioleaching as winter melt-off water ran through the broken ores of the block-caving operation, producing diluted solutions containing some 1–1.5 g Cu L^{-1}. The Ripios plant was a reprocessing facility of the old oxide copper dumps used to dispose of the early vat leach tailings at Chuquicamata. When coupled with other strong solutions from the new vats, this provided a solvent extraction feed of 10–12 g Cu L^{-1}. The Lo Aguirre operation was one of retreatment of the residual copper sulfide in the tailings from the initial TL leaching of the oxides that characterized the early years of operations at the Lo Aguirre mine. Those copper sulfide ores continued to grow as a proportion of the feed, and rapidly became the only copper source in the feed.

In the 1990s, a combination of rising copper prices, the depletion of higher-grade ore, increased costs associated with traditional processes, a growing concern for the environment and the arrival of foreign investors to Chile who were interested in exploiting new technologies all led to changes in methods for copper recovery. As a result, a variety of projects began, many of which used new leaching techniques, over 90% of them based on the principles of the TL leaching process followed by solvent extraction and electrowinning, for the clean production of premium copper cathodes of "higher grade," as measured by current BSI and ASTM standards.

Since then, numerous innovations in leaching have been studied, including on-site pilot plant scale trials. These include (1) the use of seawater to leach copper oxide ores, (2) the use of a strictly copper sulfide ore feed (3) mining at high elevations (up to 4,000 m above sea level) and (4) the presence of exotic and refractory copper oxide minerals in the feed, while maintaining the essential features inherent to the TL process. These have low operating costs, good environmental compatibility and generate a high-quality product.

The first of these new projects was the Lince Project, borne of an association between the Finnish and Chilean mining companies Outokumpu and Minera Michilla. This effort was unique in that it was the first operation in the world to use seawater as the sole water source in a controlled heap leach operation that followed all the TL leaching concepts to successfully produce high-grade cathodes in spite of the high chloride content (up to 50–60 g L^{-1}) of the leach solutions using leaching–solvent extraction–electrowinning.

In the recent textbook *Hidrometalurgia: Fundamentos, Procesos y Aplicaciones* (Domic 2001), a separate chapter (Chap. 19) is devoted to describing the initial developments leading to the industrial implementation of TL leaching in most of the current operating projects. The principles of this technology are also reviewed. Additionally, this and other chapters of that text (Chaps. 7, 19–21) assess in detail most of the subsequent developments that have constituted state-of-the-art developments of that technology up to the present day. By 2001, after 20 years of successful operation, the Lo Aguirre mine shut down because of the depletion of ore reserves in the nearby mining district. The processing plant was eventually sold, disassembled and rebuilt at La Cascada mine, near Iquique, in 2004, where it is again fully operative at its original design capacity of 15,000 t year^{-1} of copper cathode.

This chapter will review the level of industrial implementation and the current status of copper bioleaching in Chile, 25 years after the commissioning of the Lo Aguirre mine, the first commercial operation using bioleaching for a controlled bioprocessing of copper sulfide ores (Editor 2005). The most prominent events that occurred in Chile, in relation to the implementation of the leaching, solvent extraction and electrowinning technology applied to copper, are presented by year in Table 4.1 (Domic 1998, updated).

4.2 Technical Developments in Chile in the Direct Leaching of Ores

The leaching techniques used in Chile throughout the 1970s consisted almost exclusively of developments attained in the early years of this century when large US mining operations installed flooded percolation leaching in vats, primarily at Chuquicamata (1915) and Potrerillos (1928). In both cases, copper recovery was achieved through direct electrowinning. The technique produced cathodes of acceptable quality for the standards of the time; however, the copper cathode was not always of uniform quality and would be inadequate by current standards. At other vat operations installed subsequently, such as Mantos Blancos (1961) and Sagasca (1972), copper recovery was secured via precipitation, with SO_2 in the first case and scrap iron in the second.

During this period (the 1960s and 1970s) some noteworthy progress was made in agitation leaching followed by countercurrent decantation and washing, along with copper recovery using scrap iron. The plants at Ojancos, Michilla, Exótica, and the ENAMI state-owned stations at Vallenar and Taltal are included in this group. In this context, heap leaching emerged as an innovation. The success of the technique was strengthened by the broad-scale introduction of plastic piping and liners. The new approach provided an elegant, inexpensive solution to the problems of assuring impervious surfaces for leaching. This problem had previously been confronted, without much success, by using asphalt derivatives as at the Peruvian Cerro Verde pad leaching operation in 1974.

The introduction of this new technology in Chile came with the development of Pudahuel's Lo Aguirre mine. Specifically, in 1975 the company opted

Table 4.1. Prominent events in the implementation of the leaching–solvent extraction–electrowinning technology in Chile (Domic 1998, updated)

1969–1970	*First solvent extraction-electrowinning pilot trials in Chile*, at Chuquicamata, testing vat leach solutions, from the Exótica (currently Mina Sur) mine
1980	Pudahuel's Lo Aguirre mine startup: *first commercial application of the TL leaching system and of solvent extraction-electrowinning in Chile*. Bioleaching pilot plant studies were successful and it was decided to implement a full-scale operation as soon as the presence of sulfide ore in the feed justified it
1981	*Pudahuel was granted a 15-year patent for the TL Leaching process*; in essence: agglomeration and acid curing followed by nonflooded heap leaching, with abilities for heap leaching copper oxide and copper secondary sulfide ores
1984	First commercial plant using solvent extraction–electrowinning with *diluted leach solutions* was started: in situ crater bioleaching at the El Teniente block-caving mine; solutions were 1–1.5 g Cu L^{-1}
1985	Commercial TL bioleaching commenced at Pudahuel's Lo Aguirre facility with a second solvent extraction circuit entirely engineered in Chile. This was the first *controlled fully engineered bioleaching of copper sulfide ores to be commercially implemented* in Chile
1987	First commercial plant using solvent extraction–electrowinning with *highly concentrated solutions* started; using high tenor solutions from the leach of *ripios* (old vat tailings) retreatment plus current vat-leach of the Mina Sur oxide ores at Chuquicamata; leach solutions were 10–12 g Cu L^{-1}
1991	First leaching–solvent extraction–electrowinning plant using Pudahuel's TL leaching technology and license *using seawater for leaching* started at Lince (today Mina Michilla)
1993–1994	Two leaching-solvent extraction–electrowinning plants using Pudahuel's bacterial TL technology and license, *on exclusively copper sulfide ore feed for leaching*, were operated at the Cerro Colorado and Quebrada Blanca mines (at elevations of 3,300 and 4,200 m, respectively)
1994	First *flotation concentrates partial ammonia chemical leach–solvent extraction –electrowinning*, at the Coloso plant of La Escondida, following the principles of the Arbiter process commenced. Shut down in 1998, owing to failures in the construction materials that negatively influenced overall economics
2004	First *thermophilic bacterial agitation leaching of mixed copper sulfide/ arsenide flotation concentrates*, at semicommercial scale in a prototype plant for 20,000 t year^{-1} of copper cathode; using BHP Billiton technology in a joint venture with Codelco, at Alliance Copper, for treating the high arsenic copper concentrates from Mansa Mina, near Chuquicamata

for the TL heap leaching process that had been patented earlier that year by Holmes & Narver (H&N) of California. Unfortunately, the principles of the technique were not fully understood at the time and the conceptual engineering for the process implementation, initially offered by H&N, proved to be ineffective. Hence, by following efforts to resolve the original deficiencies, Pudahuel obtained a second patent for this process in 1981. The modified and improved process – which continued to be called TL Leaching – was commercially implemented at Lo Aguirre, in September 1980, with great success.

For some time, and through 1985, Pudahuel and H&N joined forces in the marketing and promotion of both TL process patents.

The most significant improvement brought about under the Pudahuel patent was that not only copper oxide ores could be leached, but the bacterial action required for proper leaching of the secondary copper sulfide present in the ore, such as chalcocite, covellite and some of the bornite, was now also possible, as noted by Beckel (2001) and recently commented on by the journal *Minería Chilena* (Editor 2005).

The TL process consists of the combination of two primary stages with a number of variables. These two stages are:

1. *Curing:* consisting of adding sulfuric acid, preferably concentrated, to finely crushed (typically 80% smaller than 10 mm), and water-dampened ore, followed by a period of settling or "curing". In this way the ore is agglomerated, forming a highly permeable substrate.
2. *Leaching:* irrigating, by spraying or drip-feed, diluted acid solutions onto the "cured" ore; care is taken to not flood the orebed.

The variables applicable to these basic stages include:

- Ore grain size
- Dose and addition sequence of water and sulfuric acid
- Agglomeration of the fines, which is obtained during the curing
- The porosity and permeability of the orebed
- The structural stability of the orebed, enabling it to be stacked at greater depths
- The content and quality of the active leaching agents in the solutions and their subsequent regeneration
- The rate of solution spraying as well as the associated liquid and gaseous permeability
- The duration of the leaching cycles, and other less critical factors

Essentially, the effectiveness of the TL process is dependent upon the quality of the agglomeration achieved during the curing phase. To secure proper agglomeration, in addition to optimizing the doses of acid and the correct sequence in which they are added, a mechanical device needs to be used to rotate the particles and enable the smaller ones to adhere to the larger particles to form similar-sized spherical agglomerates. The preferred devices are a semihorizontal rotating drum, pelletizing disks, and steep belts that move upward in the opposite direction from the movement of the load. In general, the reactor of choice, thanks to both its effectiveness and simplicity of control and operation, is the rotating agglomeration drum.

The specific format in which the TL leaching process is applied varies from ore to ore, depending on the mineral composition and grade. Adjustments must also be made for the means by which the metal content will be recovered from the solutions. High concentrations of copper (20–25 g L^{-1}) are required if scrap iron is used, midrange concentrations (10–12 g L^{-1}) if a small solvent extraction is desired, such as when leaching high-grade copper feed or flotation concentrates, and low concentrations (3–6 g L^{-1}) if a conventional solvent extraction process is to be used. The last of these will be the required method if bacterial leaching of copper sulfide ores is considered.

Typically, most commercial TL process applications use finely crushed material, 80% smaller than 0.5 in., that following agglomeration is stacked into heaps of between 3 and 8 m high. The spraying flow rates vary from 6 to 40 L h^{-1} m^{-2}; with a total leach duration period, for processing copper oxide ores, of 15–90 days. For copper sulfide ores, the total leach duration is typically 10–18 months. In both cases, copper recovery from the copper-leachable fraction may be well above that of other leaching processes. Some 80–85% copper extraction can be commonly obtained and 90% recovery rates are not unheard of.

4.3   Current Status of Chilean Commercial Bioleaching Operations and Projects

At present, most if not all of the operations and the hydrometallurgical projects for copper recovery under way or planned in Chile use the basic concepts of agglomeration and heap leaching developed under the name of the TL Leaching process as described earlier. Not all of those projects are operated under license from Pudahuel, given that the original patent expired in January 1996 and has not been renewed. Nevertheless, this pioneering work, carried out 25 years ago, permitted the development and spread of this technology for the benefit of present day operations.

The primary exceptions to TL application in leaching were (1) mine water treatment at El Teniente, (2) the in-place *ripios* (coarse old vat leach tailings) leaching at Chuquicamata, (3) the ammonia leaching of flotation concentrates at La Escondida and (4) more recently, the thermophilic bioleaching of mixed copper sulfide/arsenide flotation concentrates, at Alliance Copper, the BHP Billiton/Codelco joint venture.

The following sections provide a review, update and a brief overview of the main current Chilean bioleaching operations and proposed projects, and an update on their operational capacity, future forecast and status of implementation, as most recently reported by Menacho et al. (2005), Riveros et al. (2003) and Domic (1998).

4.3.1   Lo Aguirre Mine

The Lo Aguirre operation was built and commissioned in 1980 by its owners, Pudahuel. The mine and the plant were located near Santiago, thus forcing the operation to strictly follow all the environmental regulations.

These included those related to dust production at the mine and the crushing plant, water effluents as well as acid mists and other air contamination. A nominal 15,000 t year^{-1} of higher-grade copper cathodes was produced from the time of startup until it was shut down in 2001 because of the total depletion of ore sources in the vicinity. The ore initially contained mainly copper oxide species, but over the years, a growing proportion of the feed consisted of chalcocite (Cu_2S) and in 1982 a significant secondary bioleach operation of the initial tailings was initiated. By 1985, increased sulfide leach production justified the construction of an expansion of the solvent extraction plant in order to process the increased amount of lower copper tenor pregnant leach solutions (PLS), more typical of a bioleaching process.

The implementation of continued improvements to the process over the 20 years of the Lo Aguirre operation served as an invaluable school for processes based on the concepts of TL leaching. This included the preparation of trained personnel, the choosing and proving of new materials, and many general aspects that resulted in the success of most of the future operations involving direct bioleaching of copper ores via controlled heap leaching operations, as reported by Beckel (2001) and recently recognized by *Minería Chilena* (Editor 2005).

The operation, its main design characteristics and metallurgical performance were duly recorded and published in a number of publications, which are summarized in Domic (2001).

4.3.2   Cerro Colorado Mine

Cerro Colorado, situated at an elevation of 3,200 m, was originally built by the Canadian mining company Río Algom. In 1993, the company began to operate the first stage of this project that brought together in a single operation the treatment of mixed copper oxide ores and fully copper sulfide ores using a leaching–solvent extraction–electrowinning application. The bacterial leaching process selected also uses agglomeration and TL processing under a licensing arrangement with Pudahuel.

For the first time in Chile's copper industry, the mobile heap stacking equipment was used. This consisted of a number of short portable conveyor belts forming an articulated chain, known as "grasshopper". Ore is crushed to 90% smaller than 12.5 mm agglomerated using water and sulfuric acid, stacked onto "on–off" reusable leaching pads and removed after treatment for future secondary leach. The operation commenced with a capacity of 45,000 t of copper cathode annually. By mid-1996, an additional capacity of 20,000 t was added, bringing total production to some 60,000–65,000 t year^{-1}. A second expansion, requiring an additional investment of about US $200 million, permitted the achievement of an increased production target of 100,000 t year^{-1} by late 1998. Billiton acquired Río Algom in 2000 and further expanded the capacity to record a production of 130,000 t year^{-1} of higher-grade cathode copper in 2004. By that time, total capital investment at Cerro Colorado had reached US $500 million.

Currently, Cerro Colorado belongs to BHP Billiton and continues to operate with a dominant sulfide copper ore feed at a grade of a little less than 1% total copper, and obtains an average 84–85% total copper recovery, equivalent to some 90–95% of the leachable copper, in a 450–500-day total leach period.

4.3.3   Quebrada Blanca Mine

The Quebrada Blanca operation is located at an altitude of 4,000 m and was the first TL process application on a solely sulfide copper ore of the chalcocite variety, containing about 1% total copper. The facility belongs to a consortium composed of the Canadian firm Aur Resources (76.5%) and the Chilean ENAMI (10%) and Pudahuel (13.5%). Pudahuel has provided the TL leaching technology. The plant came on-line in 1994 with a design capacity of 75,000 t year^{-1} of copper cathode, but it has recently reached 80,000 t year^{-1}. A detailed description of the plant and the studies performed to further improve its efficiency (a special challenge, in view of the adverse climatic conditions that the bioleaching process faces at the high-altitude location) have been summarized by Domic (2001).

In addition to the bacterial leach features mentioned before, modifications to the leaching operation include irrigation with preheated raffinate solutions, the use of semiburied drippers, aeration leaching from the bottom of the orebed and the use of "grasshopper"-type stacking equipment. Initially, the pads were designed for a multilift permanent stacking of the ore; however, after some years of operation, the effect of the overall height of the pad on pumping and further upward stacking made it more economical to consider other options. This resulted in a switch to an "on–off" reusable leach pad operation, which is the current practice.

Owing to the low ambient temperatures, a special concern has been to keep the leach operation as warm as possible, in practice, within the range 18–25°C. Operational experience has shown that the chalcocite sulfide ore leach kinetics may double with an increment in temperature of 6–7°C.

The design target is to reach 80% copper recovery, which requires an extended leach period that is dependent on temperature. For the lower range of 15–18°C the duration of leaching was up to 500 days. This period may shorten to 300–360 days if temperature is kept at 22–25°C.

4.3.4   Zaldívar Mine

The operation of Compañía Minera Zaldívar (CMZ) is fully owned by the Canadian mining company Placer Dome, and began operations in mid-1995. The facility is poised to produce some 125,000 t year^{-1} of copper cathode using nonflooded bacterial heap leaching followed by solvent extraction–electrowinning. In 2004 it reached a record production of 147,000 t year^{-1} of copper cathode. The ore is primarily copper sulfide (chalcocite), although considerable reserves of oxides have also been mined, which expedited the commencement of operations. For the first time in the copper industry, this facility is using a water-flush crushing system and a vertical-smooth-flow design in the mixer–settler equipment, designed for solvent extraction by Outokumpu. Prior to leaching, at the wet crushing, the fines are screened out, and it was planned to process these separately via flotation. The intention was that this concentrate would be returned to the main system for subsequent leaching. However, this part of the design never achieved success and the concentrates were, eventually, sold separately. The coarser particles (larger than 150–200 µm) which do not require agglomeration are processed in heaps. The separation of the fines has not proven to be beneficial, technically and economically, in comparison with the conventional agglomeration and curing as performed by the TL leaching approach. A design change to discontinue segregation of the fines and to incorporate drum agglomeration is in an advanced stage.

For stacking, a large capacity Rahco conveyor-tripper–stacker system is being used for the first time in a copper heap leaching application in Chile. The bacterial leach period extends to 365 days.

4.3.5 Ivan Mine

This facility is currently owned by the Peruvian mining group Compañía Minera Milpo, having been originally developed by the Canadian company Minera Rayrock, and has been in operation since 1994. The operation uses leaching–solvent extraction–electrowinning with acid curing, in a mixing drum agglomerator, and nonflooded heap leaching to produce a total of some 10,000 t year^{-1} of copper cathode.

The ore is primarily a mixture of oxides, containing atacamite [$Cu_2Cl(OH)_3$], and chalcocite-type sulfide copper. The latter ores are treated separately for bacterial leaching as they contain higher amounts of chloride ions than is usual. The facility is also of interest as the first application in South America of the Krebs compact-design reverse-flow solvent extraction mixer–settlers.

4.3.6 Chuquicamata Low-Grade Sulfide Dump Leach

This operation was the first investment in hydrometallurgy by the Chilean state-owned company Codelco for a number of years. The focus here is on bacterial leaching of very low grade material, primarily chalcopyrite, of typically 0.3% total copper or less, which has been mined at the Chuquicamata pit over a number of years. Dump leaching of run-of-mine size material is the only economically feasible procedure for treatment. Copper recovery using this technique was originally projected at 25%, but has been recently reported to be below 15% of the total copper content.

The solvent extraction–electrowinning units began operation in mid-1994 with a design production level of 12,500 t year^{-1} of copper cathode. Future expansions based on the giant low-grade resources accumulated in the Chuquicamata area are potentially under consideration.

4.3.7   Carmen de Andacollo Mine

This operation, well known since pre-Hispanic times, is owned by the Canadian mining company Aur Resources (63%), the Chilean iron producer Compañía Minera del Pacífico (27%) and the state-owned company ENAMI (10%). The operation is designed to produce some 22,500 t year^{-1} of copper cathodes using acid cure and agglomeration prior to the bacterial leaching–solvent extraction–electrowinning of an ore-feed with a grade of 0.58% total copper. The plant was successfully commissioned in late 1996 using the Davy (currently Kvaerner) "reverse flow" mixer–settlers design in the solvent extraction process with an open sides electrowinning tank-house.

4.3.8   Collahuasi Solvent Extraction–Electrowinning Operation

Compañía Minera Doña Inés de Collahuasi is 44% owned by the Canadian company Falconbridge (44%) and 44% by the Anglo-American Chile group of mining companies. The remaining 12% belongs to a Japanese consortium of smelting operators led by Mitsui. The deposit is located at an elevation of 4,300 meters in the environs of the current Quebrada Blanca leaching-solvent extraction-electrowinning operation. In addition to using flotation on both primary and secondary copper sulfide ore, a mixture of copper oxide and sulfide that warrants leaching has being mined since its commission late in 1997.

Agglomeration and non-flooded heap leaching followed by solvent extraction-electrowinning are used to generate some 50,000 t year^{-1} of copper cathode. This is the second solvent extraction-electrowinning operation in the world to be located at that extremely high altitude. In this case, no heat is added to the raffinate solutions used in leaching, and therefore bacterial activity is very slow.

4.3.9   Dos Amigos Mine

This is the smallest bacterial leaching-solvent extraction-electrowinning operation in Chile. It uses agglomeration and acid curing. Operations began early in 1997 producing some 5,000 t year^{-1} of copper cathode, which was subsequently expanded in 1999 to a current 10,000 t year^{-1} of cathodes. It belongs to the Chilean mining group Compañía Explotadora de Minas (CEMIN) and is located near Domeyko, in the Atacama Region. The plant receives the benefits of the warm climatic conditions dominant in that area.

The mineralogical composition of the ore is digenite (Cu_9S_5), chalcocite (Cu_2S) and chalcopyrite ($CuFeS_2$) as the only copper species, with digenite being dominant. As a result, leaching performance has been quite successful, reaching 80% total copper recovery in a period of 300–350 days.

4.3.10 Alliance Copper Concentrate Leaching Plant

With the prospect of the initiation of full-scale mining at the Mansa Mina operation, Codelco explored different approaches in an attempt to avoid the smelting of the high arsenide/sulfide flotation concentrates to be produced in that operation. During that search, it was decided to develop a joint venture with BHP Billiton to test the South African BioCop™ sulfide concentrate leaching technology, a subsequent development of the BIOX™ process for recovering refractory gold (Chap. 1) from arsenopyrite-rich ores. The copper leaching interests were acquired and developed as part of Billiton, which later merged with BHP.

This technology consists of an agitation leaching of the flotation concentrates in tanks using mesophilic, moderately thermophilic or themophilic microorganisms that operate at temperatures of 65–80°C. A benefit of the copper dissolution is the possibility of discarding the arsenic as inert scorodite ($FeAsO_4$), which is highly stable in aerated environments, and of producing as a by-product dilute sulfuric acid solutions. This acid may be used to close the circuit in a conventional oxide copper chemical heap leach operation treating any of the different types of ore available at the Chuquicamata mining complex.

After different bench-scale studies, a prototype plant with a production capacity of 20,000 t year^{-1} of cathode copper was operated successfully for about 18 months during 2004 and 2005. Following an evaluation of the results, a decision on whether to go ahead with the full-scale (100,000 t year^{-1} of cathode copper) is expected by mid-2006. According to Valenzuela (2003), the production-scale process will require an expenditure of some US $328 million for it to become operational early in 2009.

4.3.11 La Escondida Low-Grade Sulfide Ore Leaching

The La Escondida operation belongs to an international consortium composed of BHP Billiton (57.5%), Rio Tinto (30%), Mitsubishi (10%) and the IFC (2.5%). Since coming on-line in 1990, the facility has operated using traditional flotation techniques to produce concentrates for direct export. Since

1999 copper cathode has also been produced via a conventional copper oxide ore acid heap leaching–solvent extraction–electrowinning operation.

In addition, for the past 5 years, studies on the parameters required for heap leaching of the lower-grade (0.52%) copper sulfide ore have been conducted at a bacterial leach pilot plant. Construction for the new facility is in progress and early in 2006 the new operation is expected to start producing approximately 180,000 t year^{-1} of copper cathode, using acid agglomeration and bacterial leaching–solvent extraction–electrowinning.

4.3.12   Spence Mine Project

Spence is the largest recent "greenfield" discovery being developed in Chile. It was discovered in 1996 by the Canadian firm Río Algom as part of the exploration activities related to their Cerro Colorado operation. It was included in the purchase by Billiton of the Rio Algom assets in 2000, and in 2002 was absorbed into the global BHP Billiton mining consortium. An extensive pilot plant study led to the approval of a US $1 billion project to build a 200,000 t year^{-1} of copper cathode bacterial leach solvent extraction–electrowinning production facility, which is due to be commissioned by the second half of 2006 and to be at steady-state production by the end of 2007 (Costabal 2005).

The operation is located between Antofagasta and Calama. The deposit contains reserves of copper oxides (1.14% Cu, mainly as atacamite) and copper sulfide ores (1.12% Cu, primarily supergene chalcocite and some minor covellite, CuS). The reserves will be mined separately and processed in two parallel processing circuits, involving two separate "on-off" reusable dynamic leach pad areas, with a uniform height of 10 m. In this way, each type of ore will be treated using its optimum operational leaching parameters. This will permit the treatment of the copper oxide ores in what is effectively a 9-month leach period to reach 82.4% design copper recovery, and copper sulfide ores in the equivalent of a 22-month effective leach period to reach 80.8% design copper recovery. It is intended to use a leach solution application rate of 6 L h^{-1} m^{-2}, in two leach solution countercurrent passes to increase the final PLS copper concentration, as explained by M.M. Eamon (personal communication).

A key feature in the decision to separate the leach operations is that the atacamite leach will result in the oxide leach circuit containing up to 40 g Cl$^-$ L^{-1} in the PLS at chemical equilibrium. The oxide PLS is designed to contain 4.5–5.0 g Cu L^{-1}, with a pH of 1.65–1.8 and a redox potential (E_h) of 650–700 mV. The bacterial leach circuit will operate at a more conventional 0.8–1.5 g Cl$^-$ L^{-1} equilibrium level in the PLS. Sulfide PLS is designed to have 3.0–3.5 g Cu L^{-1}, with a pH of 1.7–1.9 and a redox potential of 700–750 mV.

With respect to the solvent extraction, this will operate with four equal interleaved series–parallel–series countercurrent solvent extraction trains, with both oxide and sulfide streams feeding each train at about 850–1,000 m^3

h^{-1} per train. The extracting reagent LIX-84 IC (concentrated version), supplied by Cognis, is to be used at a concentration of 17.1% v/v (equivalent to 24% standard concentration reagent) diluted in ORFORM SX-80 diluent, from Philips Petroleum. The raffinate is designed at 0.2–0.8 g Cu L^{-1}, with some 7–10 g L^{-1} sulfuric acid being returned to leaching.

At the electrowinning stage, three separate circuit sections, with a total of 378 cells, using stainless steel mother blanks and fully automated stripping machines, are being considered. The nominal current density is 362 A m^{-2}, with a maximum of 414 A m^{-2}. Air sparging in the electrowinning cells is used to reduce the boundary-layer effects at that high current density. A close-capture acid mist emission system is used over each electrowinning cell to avoid contamination of the environment.

In summary, the design parameters to be used at Spence will make this project state of the art in copper hydrometallurgy, more specifically, in the direct bacterial leach of copper sulfide ores. This project has benefited from the design, engineering and operation improvements introduced in Chile over the past 25 years since the Lo Aguirre facility first became operational.

4.4   Current Advances Applied Research and Development in Bioleaching in Chile

In 2002, the Chilean government, through Codelco, took the initiative to become a leader in the application of biotechnology in the processing of copper sulfide ores. A particular target was the establishment of technology for the direct leaching of chalcopyrite ores to enable the hydrometallurgical treatment of the vast reserve of otherwise subeconomical ores of this type that exist in this country.

The initial activities were channeled through BioSigma, a joint venture between Codelco (66.7%) and Nippon Mining & Metals (Nippon M&M) (33.3%), formalized in July 2002. The two companies agreed upon the need to incorporate the latest advances of biotechnology (genomics, proteomic and bioinformatics) to mining for bioleaching of low-grade sulfide ores and other secondary materials.

Nippon M&M provides access to Japan's latest advances in biotechnology in addition to the scientists and technologists working in Chile at BioSigma. Company plans include strategic alliances with leading university research groups, both in Chile and in Japan, in addition to technology development contracts with knowledge-based companies in the USA, Europe and Chile.

Amongst the achievements of BioSigma's bioleaching technology since 2002, according to Badilla (2005), are:

• Isolation and characterization of approximately 73 novel strains of bioleaching microorganisms

- Identification and quantification of leaching microorganisms: functional genomics, DNA sequencing and gene identification, including target genes in bioleaching
- Bioleaching amenability tests, which permit the establishment of process design criteria
- Identification of leaching microorganisms that are specific to minerals: covellite ore-specific bacteria identification and functional genomics
- On-site process monitoring and control

Future objectives for the development of scientific knowledge and the validation of technology for commercial applications include (1) the identification of chalcopyrite ore-specific bacteria and functional genomics, (2) microarray technology, design of genetic manipulation tools and the identification of interesting genes for cloning, (3) the production of genetically modified bioleaching microorganisms and (4) carrying out, in parallel, pilot plant validation tests using different Codelco operations low-grade copper sulfide ores at the scale of 5,000 t mini-piles and at 150,000 t pilot-size piles (Badilla 2005).

4.5   Concluding Remarks

In summary, the experience gained in Chile through the hydrometallurgical projects implemented in the early 1980s has provided a solid backup to the recent growth in Chile's copper industry. These include the introduction of the TL heap leaching techniques that made possible the successful bacterial heap leach directly on sulfide ores by Pudahuel, the dilute copper solutions solvent extraction operation at El Teniente and the highly concentrated solutions of the Ripios project at Chuquicamata, by Codelco. The rewards accruing from the development of the technologies of bacterial heap leaching, solvent extraction and electrowinning are likely to be even greater during the present decade.

Without the audacious and visionary step taken with the development and introduction of the TL heap leaching techniques, at the Lo Aguirre mine in 1980, the current expansion would be all but unthinkable, as described by Beckel (2001) and recently commented on by the journal *Minería Chilena* (Editor 2005). In all likelihood, the benefits of the Chilean experience will spill over into neighboring nations with similar levels of copper mining resources, such as Peru, Bolivia and Argentina.

Projections for the use of hydrometallurgy of sulfide ores in Chile are closely associated with the following areas of interest and priority applications:

- Intensification of the use of heap leaching, with the help of bacteria, including both oxide ores and chalcocite-type mixed sulfide ores.

- Extensive use of heap leaching techniques followed by solvent extraction and electrowinning in small-scale mining operations producing 5,000–15,000 t year^{-1} of copper cathode, and in large scale low-grade operations with grades around 0.4% copper and less.
- Stepped-up research and studies aimed at securing the dissolution in heaps of chalcopyrite sulfide ores at BioSigma, the collaboration between Codelco and Nippon M&M, may result in a major technological breakthrough.
- The use of leaching treatment for flotation concentrates (for either partial or total dissolution) will become more widespread, particularly when difficult impurities like arsenic are present in the concentrates; therefore, fewer new smelters will be built.

References

Badilla R (2005) Biotechnology for mining: BioSigma. Presentation at the technological innovation in Codelco symposium, Santiago, September 12–13

Beckel J (2001) Una innovación tecnológica en la minería cuprífera en Chile. In: Buitelaar RM (ed) Aglomeraciones mineras y desarrollo local en América latina. CEPAL–NU, Alfaomega, pp 107–137

Costabal F (2005) Construcción de Spence avanza. Min Chil 293:10–17

Domic EM (1998) Chilean projects in copper hydrometallurgy in the 1990s: a review and update. In: Bascur OA (ed) Latin American perspectives: exploration, mining and processing. SME, Littleton, pp 203–215

Domic EM (2001) Hidrometalurgia: fundamentos, procesos y aplicaciones. IIMCh, Santiago

Editor (2005) Lo aguirre el recuerdo de un hito. Min Chil 294:27–29

Menacho J et al (2005) Proceedings of the HydroCopper 2005 international conference. Universidad de Chile, Santiago, November 23–25

Riveros PA et al (eds) (2003) Proceedings of Copper 2003–Cobre 2003, vol 6. Hydrometallurgy of copper, MetSoc-IIMCh, Montreal-Santiago

Valenzuela I (2003) En operaciones planta de biolixiviación de concentrados. Min Chil 269: 111–115

5   The GeoBiotics GEOCOAT® Technology – Progress and Challenges

Todd J. Harvey, Murray Bath

5.1   Introduction

Minerals biooxidation is now accepted as a viable technology for the pre-treatment of refractory sulfidic gold ores and concentrates, and for the leaching of base metals from their ores and concentrates. Tank bioleaching or biooxidation is successful in achieving high metal recoveries, but both capital and operating costs are relatively high. Heap biooxidation has lower costs, but to date has suffered from low metal extraction rates and low ultimate metal recoveries. These disadvantages may outweigh the lower capital and operating costs of heap processes. GeoBiotics has developed and patented the GEOCOAT® biooxidation and bioleaching technology, which combines the high recoveries of tank processes with the low costs of heap-based processes. The process has been commercialized for the pretreatment of a refractory sulfidic gold concentrate. GeoBiotics is also developing the GEOLEACH™ technology for bioleaching and biooxidation of gold and base metal ores in heaps (Fig. 5.1).

5.2   The GEOCOAT® and GEOLEACH™ Technologies

The GEOCOAT® technology offers a unique approach to the application of bacterial minerals processing, combining the low capital and operating costs of heap leaching with the high recoveries obtained in agitated tank bioreactors (Harvey et al. 1998). Both of these technologies are well accepted in the minerals industry and both are in commercial operation worldwide (Brierley 1999). In the GEOCOAT® process, sulfide flotation or gravity concentrate is coated as a thickened slurry onto crushed and size-sorted support rock which may be barren or which also may contain sulfide or oxide mineral values. The coated material is stacked on a lined pad for biooxidation. The process is applicable to the biooxidation of refractory sulfide gold concentrates and to the bioleaching of copper, nickel, cobalt, zinc, and polymetallic base metal concentrates. Mesophilic or thermophilic microorganisms catalyze the sulfide oxidation reactions. In the processing of chalcopyrite concentrates, the higher temperatures associated with the use of thermophilic microorganisms

Biomining
(ed. by Douglas E. Rawlings and D. Barrie Johnson)
© Springer-Verlag Berlin Heidelberg 2007

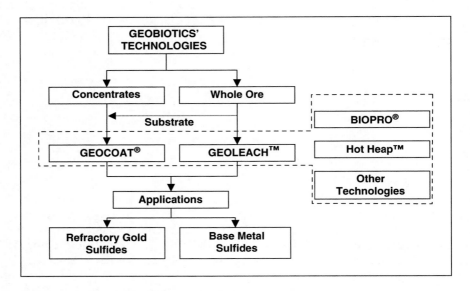

Fig. 5.1. GeoBiotics technology portfolio

have proven highly beneficial in increasing the rate and extent of copper leaching.

In the processing of refractory gold sulfide concentrates, the GEOCOAT® process offers significant cost advantages over established processes (roasting, pressure oxidation, and agitated tank biooxidation). In base metals operations, the process is particularly suited to the treatment of "dirty" concentrates, reduces transportation costs by allowing the on-site production of metal at remote operations, and can take advantage of the depletion of oxide reserves through the utilization of existing solvent extraction/electrowinning equipment. The process is simple, robust, and ideally suited to operation in remote locations.

The GEOLEACH™ technology is applicable to whole-ore systems where the metals occur as sulfides, or are occluded within sulfides, as with refractory gold. The incentive for the development of the process is the recognition that oxidation of the sulfides in most whole-ore leaching systems potentially can release enough energy to raise the heap temperature to very high levels; however, in practice, poor, or lack of any, heat management prevents a significant temperature rise. Unless heap temperatures can be raised above ambient, sulfide leaching kinetics is extremely slow; higher temperatures (above 70°C) appear to be particularly important for the successful bioleaching of chalcopyrite (Stott et al. 2000). The GEOLEACH™ technology is designed to maximize heat conservation through careful control of aeration and irrigation rates. GEOLEACH™ has built upon the best industry knowledge of bioleaching operations. The technology is very similar to that of conventional whole-ore acid heap leaching systems, but with the addition of an operational strategy to maximize heat conservation, raise temperature, and maintain bacterial activity.

5.2.1 Complementary GeoBiotics Technologies

The GEOCOAT® and GEOLEACH™ technologies, together with a wide variety of additional expertise and patents, constitutes the GeoBiotics technology suite, including high-temperature bioleaching, toxins removal, HotHeap™, BIOPRO™, and other complementary processes focused around pretreatment, aeration, stacking, and instrumentation. HotHeap™ is a bacterial heap leaching operating strategy coupled with an instrumentation and control system, that maximizes heat conservation, thereby enhancing bioleaching kinetics, while BIOPRO™ is an inoculation method for bacterial heap leaching systems licensed from Newmont Gold Company.

GeoBiotics has successfully commercialized the GEOCOAT® process for the treatment of refractory gold concentrates at African Pioneer Mining's (APM's) Agnes Mine near Barberton, Mpumalanga, South Africa. The plant was commissioned in June of 2003 to treat 12,000 t of concentrate per year, yielding approximately 25,000 oz of gold (Harvey and Bath 2003).

As with any new process, the development and implementation of the GEOCOAT® technology has taken considerable effort and time. This chapter provides an outline of the technology, its applications and advantages, a description of the Agnes GEOCOAT® plant, and a discussion of the challenges faced in the implementation of the new technology. It includes a brief description of GeoBiotics's plans for expansion into new biohydrometallurgical applications, particularly whole-ore chalcopyrite leaching.

5.2.2 The GEOCOAT® Process

The GEOCOAT® process uses iron- and sulfur-oxidizing microorganisms to facilitate the oxidation and leaching of sulfide minerals in an engineered heap environment. These microorganisms include the mesophilic (moderate-temperature) bacteria *Acidithiobacillus ferrooxidans*, *Acidithiobacillus thiooxidans*, and *Leptospirillum ferrooxidans*, and thermophilic (high-temperature) microorganisms such as the archaea *Sulfolobus* and *Acidianus*.

After concentration by conventional processes, typically flotation or gravity concentration, the sulfide minerals are coated as a thickened slurry onto a support rock, which may be barren (waste) rock or a low-grade sulfide or oxide material. The size of the support material is typically 6–25-mm diameter, allowing the concentrate to form a coating less than a 0.5-mm thick on the rock surfaces. The mass ratio of concentrate to support rock is typically in the range 1:7–1:10. The coating is applied by contacting the thickened concentrate slurry with the support as it discharges from the stacking conveyor onto the heap. This results in the formation of a thin, relatively uniform coating on the support rock surfaces. The adherent coating is not washed out of the heap by solution application or by heavy rain. The relatively uniform size of the support rock particles results in large interstitial spaces within the heap, offering very low resistance to air and solution flows. Low-pressure fans supply air through an engineered system of perforated pipes placed under the heap. The air-flow rate is varied to maximize bacterial activity and provide evaporative heat control.

The large interstitial spaces, combined with the thin concentrate layer, create ideal conditions for biooxidation. The sulfide mineral grains and the attached bacteria are constantly exposed to the downward-flowing solution and the countercurrent flow of air. The result is the efficient transfer of oxygen and thus rapid oxidation rates. Typically, oxidation is complete within 60–120 days, whereas in whole-ore heap biooxidation, oxidation may not be complete even after several hundred days. The larger void spaces and rigid support provided by the sized support rock particles in the GEOCOAT® heap prevent the compaction typical of whole-ore heaps and ensure uniform distribution of air and solution to all parts of the heap.

In the basic GEOCOAT® process for refractory gold concentrates, the concentrate is coated onto a barren, essentially inert, support rock. After biooxidation, the heap is reclaimed and the oxidized concentrate is separated from the support rock by wet screening. The washed support rock is recycled for recoating with fresh concentrate. A potentially attractive option is to coat the sulfide concentrate onto low-grade sulfide material which would otherwise be stockpiled or discarded as waste. The bacterial action in the concentrate coating is also effective in oxidizing the sulfide minerals in the support rock, making additional metal values available for recovery. This may allow sub-cutoff-grade material to be brought into the economic reserve. An alternative is to use a screened and sized portion of the ore as a support medium. The rest of the ore is ground and floated, producing the concentrate that is coated onto the support rock fraction for biooxidation.

Downstream processing operations depend on the purpose of the biooxidation or bioleaching process. In the treatment of refractory gold ores, the gold remains in the oxidized solid residue, which is removed from the pad for additional treatment, typically cyanidation. In the processing of copper and other base metal sulfides, the valuable metal is solubilized and is recovered from the leach solution, while the residue remains on the pad. An "on–off"-type pad is used for refractory gold ores, with the oxidized material being off-loaded for further processing, and the pad reused. However, for copper and other base metal ores, a permanent pad may be used, with the pad area being expanded as required. Alternatively, additional lifts of coated support rock may be stacked on top of the first. Figure 5.2 is a schematic representation of an "on–off" pad.

The GEOCOAT® process is also applicable to the treatment of concentrates containing both gold and copper values. The heap is bioleached to pro-

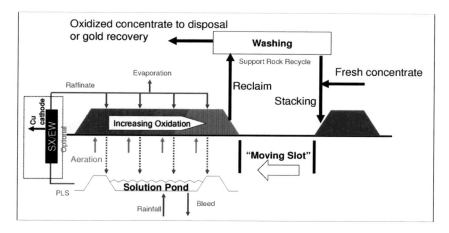

Fig. 5.2. Reusable pad configuration. *SX* solvent extraction, *EW* electrowinning, *PLS* pregnant liquor solution

duce a pregnant solution from which copper is recovered by solvent extraction and electrowinning. The residual copper-depleted material is unloaded from the pad and screened to remove the oxidized concentrate containing the gold values. The gold is typically recovered from the oxidized concentrate by cyanidation.

Figure 5.3 shows the flowsheet options for the application of the GEOCOAT® process for the treatment of refractory gold ores. Figure 5.4 is a schematic flowsheet for the processing of a base metal concentrate, in this case copper.

5.2.3   Advantages of the GEOCOAT® Process

The GEOCOAT® process has advantages over other refractory gold pretreatment options such as roasting, pressure oxidation, and stirred-tank biooxidation. Primary advantages are lower capital and operating costs, stemming from the simplicity of the process, the use of low-cost materials of construction, particularly plastics, the use of low-pressure air as the primary oxidant, and the relative ease with which most sulfides are biooxidized.

A recent independent assessment for GeoBiotics of the GEOCOAT® technology included an evaluation of the treatment options for refractory gold sulfides, showing the cost advantages of GEOCOAT®. Table 5.1 is an excerpt from the report showing selected capital and operating cost estimates for refractory gold GEOCOAT® plants.

As shown, the operating cost of the GEOCOAT® plant can vary substantially. The main reason for this, besides effects of scale, is the widely ranging cost of effluent neutralization. For the first project listed in Table 5.1, lime, at

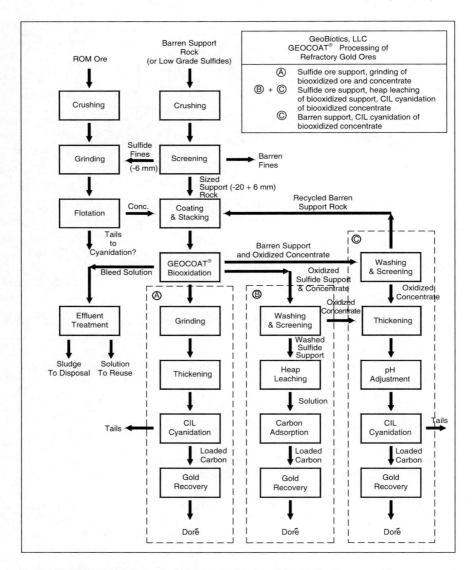

Fig. 5.3. GEOCOAT® flowsheet – refractory gold. *CIL* carbon in leach

a cost of US $120 per tonne, was employed for neutralization, whereas the other estimates were based on the use of much cheaper locally available lime-stone. It should be noted that the cost of neutralization is common to all sulfide oxidation processes. Unlike pressure oxidation and stirred-tank biooxidation, power costs for GEOCOAT® are extremely low, with consumption generally in the range 60–80 kWh t^{-1} of concentrate treated. Table 5.2 compares capital and operating costs of sulfide oxidation processes.

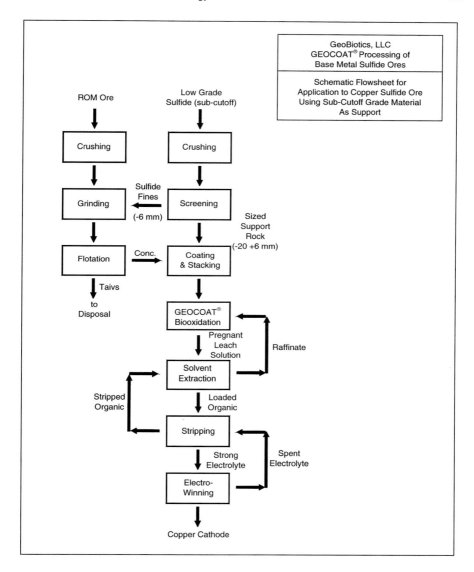

Fig. 5.4. GEOCOAT® flowsheet – base metals

5.3   The Agnes Mine GEOCOAT® Project

The Agnes orebodies, near Barberton in Mpumalanga Province, South Africa, were first exploited in 1893. The present owners and operators, APM, a subsidiary of Metallon Resources, acquired the property in 2002 from Cluff Mining. The refractory sulfide ore is mined by underground methods and a

Table 5.1. Selected GEOCOAT® capital and operating cost estimates

GEOCOAT® plant capacity (t day⁻¹)	Capital cost ratio (US $)	Operating cost ratio ($ US t⁻¹ concentrate)	Notes
80	1.00	1.00	Includes cost of neutralization, power $0.022 kWh⁻¹
1,000	2.62	0.12	Includes cost of neutralization, power $0.08 kWh⁻¹
4,400	9.32	0.07	Includes cost of neutralization, power $0.025 kWh⁻¹

Table 5.2. Relative costs of sulfide oxidation processes (tonnes per day)

Process	100 t day⁻¹ concentrate feed		1,000 t day⁻¹ concentrate feed	
	Capital cost ratio	Operating cost ratio	Capital cost ratio	Operating cost ratio
Roasting	1.67	1.63	2.14	5.56
Pressure oxidation	2.50	2.50	3.21	8.33
Agitated-tank biooxidation	1.33	1.63	2.86	6.11
GEOCOAT®	1.00	1.00	1.00	1.00

conventional milling and flotation plant produces a gold-bearing sulfide concentrate. In 2000, Cluff Mining signed a license agreement with GeoBiotics for the use of the GEOCOAT® heap biooxidation process for pretreatment of the concentrate. Column biooxidation testwork at SGS Lakefield confirmed the amenability of the concentrate to biooxidation, and APM prepared a feasibility study based on the use of the GEOCOAT® process. Design and construction followed, with commissioning starting in the first quarter of 2003. Table 5.3 lists the general design criteria for the GEOCOAT® facility at the Agnes Mine.

A double pad liner system of synthetic geomembranes ensures that the process solution is fully contained. A leak detection system is installed between the liners to give an early indication of potential solution loss and allow remedial action before any discharge to the environment occurs.

Three low-pressure fans installed along the edge of the heap provide process air via a system of high-density polyethylene headers, subheaders, and perforated stringers. The stringers are buried in a 1-m-deep layer of crushed rock to assist in the uniform distribution of the air. One of the fans and the air distribution headers and piping are shown in Fig. 5.5. The rock layer also provides a path for the leach solution draining from the heap to reach the pad liner and the solution collection trench.

Table 5.3. Key design criteria of the Agnes Mine GEOCOAT® process

	As-built design specifications
Stacking rate	34.5 t h^{-1}
Concentrate rate	4.6 t h^{-1}
Biooxidation time	60 days
Irrigation rate	10–30 L m^{-2} h^{-1} (80 m3 h^{-1})
Solution application	Wobbler sprinklers
Aeration equipment	Centrifugal fan 3×360 m^3 min^{-1} at 2.5 kPa
Air distribution	Perforated pipes in drain rock base
Pad dimensions	50 m×120 m
Heap dimensions	6 m×45 m×60 m
Solution pond capacity	7,000 m^3
Stacking method	Slewing radial stacker with automated materials handling
Concentrate recovery	Front-end loader, trommel, thickener
Gold recovery	Carbon in leach 6×20-m^3 tanks
Effluent disposal	Heap bleed solution is neutralized by mixing with carbonate flotation tails. Cyanide in carbon-in-leach residue is destroyed using excess acid bleed
Performance monitoring	Solution analysis, solids sampling, and temperature monitoring

Fig. 5.5. Air supply fan and distribution system. Note the pad liner

Support rock was initially prepared by crushing and screening waste rock from an old dump. Recycled support rock is fed to a horizontal conveyor which runs alongside the heap. This conveyor transfers the support via a tripper to a "grasshopper" conveyor, which in turn feeds the heap-stacking conveyor. The grasshopper and stacker operate on the surface of the drain rock layer and the heap is stacked to a height of 6 m above this surface. The "moving slot" method is used to stack and reclaim the heap in an "on–off" configuration. Freshly coated rock is stacked at the advancing face of the slot and oxidized material is reclaimed from the opposite, retreating face. Once sufficient new heap area has been stacked, solution distribution piping is installed and irrigation is started. Solution is applied via sprinklers at a rate of 10–30 L $m^{-2} h^{-1}$. Figure 5.6 is a general view of the Agnes GEOCOAT® heap and materials handling system. Figure 5.7 shows the surface of the heap with solution application in progress; note the stream leaving the heap.

Solution is recirculated to the heap via a lined pond. A stainless steel pump delivers solution from the pond to sprinklers on the heap. A portion of the circulating solution is bled off to maintain the iron concentration within design limits. The bleed stream is pumped to the neutralization circuit, a series of agitated tanks, where flotation tailings are added to neutralize acid and precipitate iron. Flotation tailings at the Agnes operation contain carbonate minerals and provide an inexpensive and convenient source of neutralizing agent. The neutralized solution, containing the precipitated iron,

Fig. 5.6. General view of the Agnes GEOCOAT® plant

Fig. 5.7. Irrigation of the heap surface

is pumped to a tailings impoundment. A separate impoundment is provided for the cyanide residue to ensure that no cyanide is returned to the GEO-COAT® circuit in recycled process water. Cyanide and its decomposition products are highly toxic to bioleaching microorganisms.

Support rock, with its oxidized concentrate coating, is reclaimed from the heap by a front-end loader and conveyed to a trommel screen where the concentrate is separated (Fig. 5.8). The concentrate slurry underflow from the trommel screen is pumped to a stainless steel high-rate thickener. Thickener underflow is transferred to the pH-adjustment tank, after secondary screening for rock chip removal, and lime is added to raise the pH in preparation for cyanidation. The concentrate slurry is then pumped to the carbon-in-leach (CIL) plant located adjacent to the GEOCOAT® heap. The washed support rock is returned to the stacker for recoating with fresh concentrate. Support losses are made up by the addition of fresh rock.

The original as-built GEOCOAT® flowsheet for the Agnes operation is shown in Fig. 5.9. This reflects a conventional approach to refractory gold pre-treatment, in that the concentrate is oxidized before cyanidation. The flotation concentrate is coated onto the support rock and stacked on the GEOCOAT® heap. After biooxidation for 60–75 days, the coated rock is reclaimed from the heap and the concentrate separated by screening. The pH of the oxidized concentrate slurry is adjusted with lime, and the slurry is subjected to CIL for gold recovery. However, various circumstances have resulted in the

Fig. 5.8. Trommel for concentrate separation

evolution of the flowsheet to that shown in Fig. 5.10. This flowsheet is non-conventional in that the flotation concentrate is biooxidized only after initial cyanide leaching. Since the baseline cyanide gold recovery is relatively high at 60–70%, economic benefits were seen in removing the cyanide-leachable gold before biooxidation.

Although the original Agnes flowsheet was operated successfully, it was modified as a result of difficulties in development of the underground mine, leading to cash-flow shortfalls. Additionally, teething problems associated with the GEOCOAT® plant exacerbated the cash-flow situation.

The reason for the decision to change the flowsheet was APM's failure to ramp up mine production to the design level in the expected timeframe. The resulting lack of ore led to the decision by the owner to reclaim the GEOCOAT® heap prematurely, "robbing" the inventory to maintain cash flow. It was always expected that the heap inventory would quickly be replenished but the underground production ramp-up took much longer than expected. Furthermore, APM started cyanidation of the unoxidized concentrate, producing enough cash flow to cover costs. The tailings from cyanidation of the flotation concentrate, containing 15–25 g gold/t^{-1}, was discarded. The intermittent supply of concentrate to the GEOCOAT® plant created commissioning difficulties and complicated the identification and resolution of commissioning issues.

When, after a period of several months, it became evident that a return to the original flowsheet was unlikely, GeoBiotics and APM embarked on a

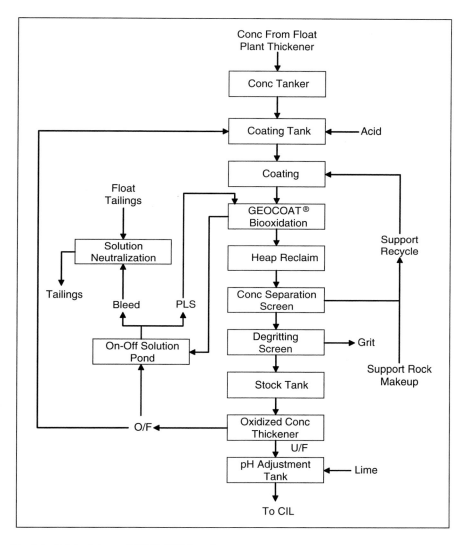

Fig. 5.9. Original Agnes GEOCOAT® flowsheet

program to recover the residual gold in the CIL tailings by treatment in the GEOCOAT® plant. A comprehensive test program demonstrated that the CIL tailings would require thorough washing and acid pretreatment to remove cyanide and reduce toxicity to the bacteria sufficiently to allow biooxidation.

Several issues associated with the operation of the GEOCOAT® plant revealed during commissioning required attention to optimize biooxidation. The main problems were the unexpectedly high level of carbonates in the concentrate and support rock, and control of the coating system. The high

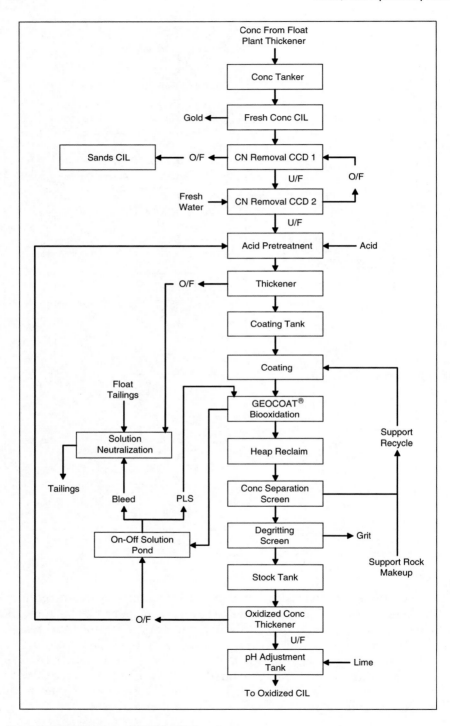

Fig. 5.10. Modified Agnes GEOCOAT® flowsheet

carbonate content of the support rock resulted from the decision to use a different rock from that tested in the initial design work. Carbonate levels in the concentrate were high because the flotation circuit was operated with a high concentrate mass yield in an attempt to maximize gold recovery; much of the additional concentrate mass was made up of carbonate gangue minerals.

The high carbonate content of the concentrate was addressed by the installation of an acid pretreatment stage, but the support rock issue was more difficult to rectify. Wholesale replacement of the support rock was not feasible, so operational controls were developed to minimize the impact of its carbonate content. It was necessary to maintain the irrigation solution pH sufficiently low to allow for acid consumption by the support rock. A bulk-acid storage tank was installed to facilitate control of the pH in the solution pond and to allow the use of lower-cost bulk acid. Poor control of the concentrate slurry coating density and uneven concentrate distribution resulted in the migration of some of the concentrate from the support rock into the drainage layer beneath the heap. The coating system was modified through the addition of a "coating contactor" to intimately mix the concentrate and support, ensuring a more uniform distribution of concentrate on the support rock surfaces. Additionally, a new high-density concentrate thickener was installed as part of the acid pretreatment circuit, replacing the existing undersized flotation concentrate thickener. Subsequent testwork has demonstrated that the use of finer support rock is highly beneficial in preventing migration of the concentrate. This change is expected to be implemented in the future.

GeoBiotics continues to have a significant presence at the Agnes plant, working closely with APM to further development of the GEOCOAT® process. A planned expansion of the Agnes Mine is under way to bring a second orebody on-stream and to treat the highly refractory concentrate by GEOCOAT®.

5.4   Developing Technologies

GeoBiotics is continuing to develop novel biotechnologies for the minerals industry. The GEOLEACH™ technology is expected to revolutionize the copper industry by allowing whole-ore chalcopyrite heap leaching to yield what is predicted to be unprecedented copper extractions in relatively short leach cycles. Traditionally, whole-ore heap leaching of primary copper sulfides has been plagued by low copper extractions and very long leach times. The ability to leach chalcopyrite ore in a heap promises to change the way in which these ores are processed. Additionally, GeoBiotics continues to work on refractory gold, with promising results coming from research on double refractory gold ores such as those commonly found in the Carlin Trend in Nevada, and the Ashanti Trend in Ghana. GeoBiotics expects to test the GEOLEACH™ technology at scale at a copper mine in Chile in late 2006.

References

Brierley CL (1999) Commercial concentrate and ore biooxidation plants. In: Randol copper hydrometallurgy roundtable, pp 55–58

Harvey TJ, Adutwum KO, Potter GM (1998) Heap biooxidation for the treatment of low grade refractory ores at Ashanti Goldfields' Obuasi operations. In: Randol gold and silver forum, Denver, pp 267–274

Harvey TJ, Bath MD (2003) Development of the first commercial GEOCOAT® heap leach for refractory gold at the Agnes Mine, Barberton South Africa. In: Tsezos M, Hatzikioseyian A, Remoundaki E (eds) Biohydrometallurgy: a sustainable technology in evolution (part I). National Technical University of Athens, pp 387–398

Stott MB, Wattling HR, Franzmann PD, Sutton D (2000) Role of iron-hydroxy precipitates in the passivation of chalcopyrite during bioleaching. Miner Eng 13:1117–1127

6 Whole-Ore Heap Biooxidation of Sulfidic Gold-Bearing Ores

THOMAS C. LOGAN, THOM SEAL, JAMES A. BRIERLEY

6.1 Introduction

Nearly three decades have passed since the first two conferences of an ongoing series focused on bioleaching fundamentals and practices (Schwartz 1977; Murr et al. 1978). Subsequently, there have been significant advancements in understanding microbiological phenomena of bioleaching (Rohwerder et al. 2003) and the application of technologies for metals recovery (Brierley and Brierley 2001; Olson et al. 2003). Other chapters in this book describe advances in commercial, stirred-tank biooxidation/bioleaching applications. This chapter describes commercial application of whole-ore heap biooxidation of refractory sulfidic gold ores as exemplified by Newmont Mining Corporation's BIOPRO™ technology currently employed at the Gold Quarry mine near Carlin, Nevada, USA.

Since inception in late 1999, Newmont has successfully heap-biooxidized more than 12 commercial batches containing 8.8 Mt of sulfidic ore and recovered approximately 12.2 t of gold by carbon-in-leach (CIL) processing. Newmont heap-biooxidizes low-grade (1–3 g Au t^{-1}) sulfidic refractory ores containing pyrite and arsenian pyrite with a preferred range of 1–2.5% sulfide-S content. Whole ore with higher sulfide-S content has also been demonstrated to be amenable to heap biooxidation. The BIOPRO™ process incorporates patented inoculation of crushed ore with iron-grown microbial consortium prior to placement on a heap. By distributing microbes in the ore before placement, inoculation fosters early initiation and a more uniform rate of biooxidation (Brierley and Hill 1993) when compared with surface application of microbes after a heap is formed.

Progress and results of Newmont's efforts, including development history, process design, operating results, and lessons learned from a decade of experience in successful utilization of whole-ore heap biooxidation on refractory gold-bearing ores, are presented.

6.2 History of BIOPRO™ Development

Anticipating depletion of oxide reserves and the increasing presence of sulfides at Gold Quarry, Newmont began laboratory investigation of biooxidation in 1988. Two years of fundamental research work built a foundation of

Biomining
(ed. by Douglas E. Rawlings and D. Barrie Johnson)
© Springer-Verlag Berlin Heidelberg 2007

an innovative, microbial-heap-oxidation pretreatment technique that liberates sulfide-locked gold from pyrite and arsenian pyrite, exposing gold for chemical extraction. Field pilot testing began in 1990. From 1990 through 1994, six field whole-ore heap biooxidation tests ranging from 360 to 23,000 t took place (Brierley et al. 1995). By mid-1994, an independent prefeasibility study validated the economic viability of heap biooxidation, and Newmont reclassified 22.5 Mt of sulfidic waste containing 30.7 t of gold to reserves.

In late 1994, Newmont completed a US $13,500,000 two-phase demonstration project utilizing all necessary components of whole-ore heap biooxidation. The first phase, put into service in mid-December 1994 (Shutey-McCann et al. 1997), featured a 708,000 t year^{-1} heap-biooxidation facility with five pads of varying designs. Two methods of ore placement were incorporated—mine haul-truck end-dump or stacking conveyor ore placement. Biooxidized ore removal used mine front-end loaders. The second phase, a separate thiosulfate leach pad and a unique gold recovery plant to recover gold from carbonaceous preg-robbing ore, was completed in September 1995. Front-end loaders removed biooxidized ore and conveyors then transferred the ore to pugmill lime neutralization, followed by conveyor stacking on the thiosulfate leach pad. A 163 m^3 h^{-1} Merrill Crowe gold recovery plant, modified to use copper precipitation (Wan and Brierley 1997), was constructed to test the use of ammonium thiosulfate for gold extraction.

Multiple biooxidation cycles from 1995 through 1999 proved the feasibility of inexpensive whole-ore heap biooxidation on a near-commercial scale. On the basis of the success of the demonstration facility, an additional 43 Mt of unmined sulfidic ore containing 62 t of gold was declared reserves in 1997.

Commercial-scale, whole-ore heap biooxidation facilities were authorized in early 1997, aiming for heap biooxidation starting in late 1998 and reaching full gold production levels in late 1999. This project was conceived to heap-biooxidize 32,300 t day^{-1} (11.7 Mt year^{-1}) of sulfidic and carbonaceous preg-robbing ore on reusable pads, and to produce 8.4 t of gold annually, thus sustaining overall production levels at Gold Quarry as oxide ore production diminished. Work was begun on upgrading existing crushing facilities for the finer crush size required to effectively biooxidize refractory ores. Preparation of the heap-biooxidation site by engineered placement of selected mine waste using company mining equipment had previously begun in 1996 and was completed in 1997.

The original Gold Quarry plan envisioned constructing a 12-pad biooxidation facility. This multipad approach gave flexibility to segregate ore on the basis of royalty status, ore grade, preg-robbing nature, or biooxidation characteristics. In the midst of declining gold prices in late 1997, the biooxidation project was put on hold. At the same time, a planned pit expansion was deferred, thereby delaying production of oxide ore needed to feed the existing CIL mill.

Rather than idling the CIL mill stand idle, an option of milling biooxidized ore from existing refractory sulfide stockpiles was proposed. Laboratory investigation indicated that minor modifications to the mill would permit biooxidized ore treatment. A full-scale plant trial in 1998 confirmed the mill capabilities and identified necessary changes to effectively treat low-pH biooxidized ore. To match the lower daily CIL processing rate of 9,620 t day^{-1} (3.5 Mt year^{-1}) instead of the original rate of 32,000 t day^{-1} (11.7 Mt year^{-1}), the heap-biooxidation facilities were scaled back to three heap-biooxidation pads, plus continued use of the demonstration facility as a fourth pad. Newmont's downsized commercial heap-biooxidation facility was constructed in 9 months and commissioned with ore placement on December 27, 1999. The whole-ore heap biooxidation facility is still in operation nearly 6 years later (Brierley et al. 1995; Brierley 1997; Tempel 2003) although the demonstration pad has been decommissioned as part of ongoing concurrent reclamation.

6.3   Commercial BIOPRO™ Process

6.3.1   Biooxidation Facilities Overview

The commercial biooxidation facility with pads in various stages of a biooxidation cycle is seen in Fig. 6.1. On the far right, a biooxidized heap is being offloaded. The middle pad has been offloaded and is being regraded prior to the next batch placement. On the far left, the third pad is undergoing active biooxidation. The pond in the foreground is a surface-water pond associated with the CIL mill tailings impoundment. Also seen in the foreground are haul truck access unloading ramps. The biosolution pond with the pond aeration system, a small acid addition system with sulfuric acid storage tanks, a power distribution center, and biosolution pumps are in the background. Perimeter solution piping and aeration headers surround each pad. Located at the front left of each pad is an aeration fan.

6.3.2   Biooxidation Process Description

Refractory sulfidic ore is segregated from oxide and carbonaceous preg-robbing ore through a complex ore control matrix. Ore with a cyanide-soluble gold to fire-assay gold ratio less than 0.3, low preg-robbing potential, sulfide-S content above 0.2%, and maximum calcium carbonate less than 4.2% is a potential feed to heap biooxidation.

Figure 6.2 illustrates the generalized biooxidation process flow diagram. Run-of-mine sulfidic ore with a crushing work index of 10–12 kWh t^{-1} is either directly fed from the pit or stockpiled, reclaimed, and truck-fed through an Allis Chalmers 5474 primary gyratory crusher. Crushed ore (smaller than 150 mm) is transported to the secondary crushing plant feed stockpile. Ore is fed to secondary crushing at 685 t h^{-1} (16,460 t day^{-1}). Before

Fig. 6.1. Newmont whole-ore heap-biooxidation facility

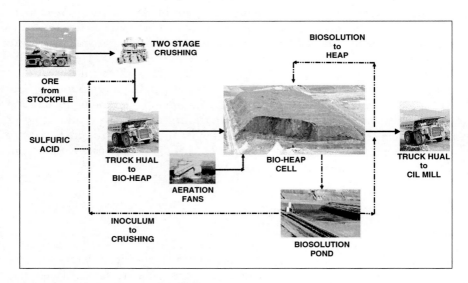

Fig. 6.2. Whole-ore biooxidation heap flow diagram. *CIL* carbon in leach

crushing, ore is prescreened over three parallel, single-deck, multislope, vibrating screens. One is a Seibtechnic 3-m wide by 10-m long screen; the other two are Metso 3-m wide by 8-m long screen. All are fitted with 20-mm×80-mm slotted-polyurethane decking. Screen oversize is fed to two Metso MP1000 crushers. Secondary crushing reduces ore to 80% smaller than19 mm. Screen undersize is recombined with crusher discharge and then conveyed to a haul-truck loading station.

Biosolution is pumped through a buried, double-contained pipeline to the load-out conveyor head pulley where biosolution is applied at a nominal 40 kg t^{-1} ore to inoculate the ore and start biooxidation. The biosolution is applied as crushed ore falls into 140-t-capacity haul trucks. Inoculated ore is hauled 2.4 km and immediately placed on a pad.

Each pad holds approximately 810,000 t of ore when built to the original 10-m design height. On occasion, this height has been increased to 16 m. Biooxidation cycles typically last about 435 days and include 90 days to crush and load ore, 165 days for biooxidation, 60 days for draining down, 90 daysfor offloading, and 30 days offline for cleanup or pad base/piping repairs.

Figure 6.3 shows the pad design. Pads are underlain with a 150-mm-thick secondary clay liner, compacted to achieve less than 10^{-9} m s^{-1} hydraulic conductivity. Pads are lined with 1.0-mm-thick high-density polyethylene (HDPE) liner. The HDPE primary liner is covered with 900 mm of crushed drain rock over the biosolution drain collection system.

Figure 6.4 shows installation of drain rock over solution collection piping normally found in pad bases. After the drain rock has been placed, an air distribution system, visible in Fig. 6.5, comprising a 100-mm-diameter, N12 CPT (smooth internal wall) HDPE pipe on 3-m centers is laid laterally. Three air holes, varying in diameter from 3 to 5 mm along the length of the pipe, are drilled circumferentially on 1-m spacing to ensure uniform air distribution. Air is distributed peripherally along both sides of each pad through 610-mm-diameter HDPE header pipes connecting each 100-mm-diameter lateral pipe to the opposite side header. Pads are individually ventilated with 150-kW fans, producing 43,500 m^3 h^{-1} at 1–2 kPa. Aeration is constant and continuous during biooxidation. Heap air permeability is typically about 10^{-8} m^2 (Pantelis 1994).

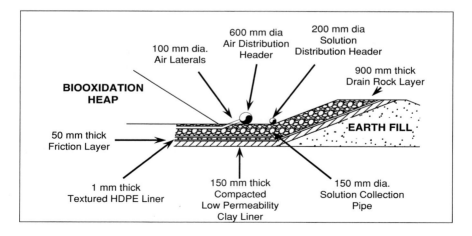

Fig. 6.3. Biooxidation heap pad base cross section. *HDPE* high-density polyethylene

Fig. 6.4. Drain rock layer

Fig. 6.5. Aeration laterals and header

Inoculated ore is end-dumped from the top to form the 13.7-m-high heap. To offset truck compaction, each heap is dozed down to 12.8 m and cross-ripped to a depth of 1.5 m. In the past, pads were segmented into three sections (cells) and drip emitters were placed as soon as sufficient distance from haul-truck traffic provided safe access. After the first three heaps had been built, this practice changed. The loading pattern was modified so that each heap could be dozed down and cross-ripped. Heaps are now piped only after the heap has been completed and dozed to the final height. Pads are irrigated with 2.3 L h^{-1} drip emitters spaced at 760 mm along each emitter line and lines are placed on 760-mm centers to provide a biosolution application rate up to 4 L m^{-2} h^{-1}. Heap hydraulic conductivity ranges between 10^{-2} and 10^{-4} m s^{-1}. As pads are filled, oxygen- and temperature-monitoring devices are placed. Bundles of four 50-mm-diameter HDPE pipes of various lengths, sealed at each end, are banded together and vertically spaced in three locations along the central longitudinal axis. A 9.5-mm-diameter 316 stainless steel tube is clamped to each pipe and commercially available thermocouples are inserted. Operators monitor temperatures weekly, or every other week, visiting each bundle and collecting data with a portable meter. Each tube is also monitored weekly for oxygen content, which can change rapidly. Oxygen measurements are made by evacuating each tube and measuring the oxygen concentration in air pumped from within the heap. Each cell's irrigation schedule is monitored and adjusted as necessary to maintain optimal conditions. Weekly plots of data are made to display trends.

Biosolution primarily acts to keep microbes in an acidic environment and remove biooxidation products. Ferric iron is reduced to ferrous iron as pyrite is oxidized. Microbes then reoxidize ferrous iron to ferric iron within the heap. For the first portion of a cycle, biosolution is applied continuously from a 117,200-m^3-capacity pond double-lined with HDPE (1-mm top layer and 0.67-mm bottom layer) sandwiching a continuous leak detection layer and underlain by a 150-mm layer of clay. Two down-hole, inclined 316 stainless steel well pumps each with 94-kW motors provide a total pumping capacity of 710 m^3 h^{-1}. Heap solution piping is laid out and valved to subdivide a heap into three cells that can be individually controlled. Target biosolution characteristics are pH between 1.3 and 2.2, E(SCE) above +550 mV (where SCE is the standard calomel electrode), dissolved oxygen level about 2 mg L^{-1} (minimum of 0.5 mg L^{-1}), and ferric iron content between 5 and 25 g L^{-1}. After the initial period, biosolution application is discontinued during rest periods and then it is intermittently applied to different areas of a heap. Over the biooxidation period, biosolution application rates decrease from 431 m^3 h^{-1} at the start of biooxidation to 272 m^3 h^{-1} at the end. Biosolution percolates through the heap and drains into a collection pipe system routed back to the biosolution pond. Biosolution chemistry is monitored from each heap discharge to gauge biooxidation activity. Heap drains are monitored for solution color, temperature, pH, E(SCE), and iron content. The biosolution pond is covered with "bird balls" (hollow 100-mm-diameter HDPE thin-walled dimpled spheres). These prevent avian access to low-pH biosolution and also provide limited insulation to retain heat. Low-pressure air is blown into the pond to continue microbial oxidation of residual ferrous iron to ferric iron before the solution is recycled or used for inoculation of fresh ore.

For mesophiles and moderate thermophiles, optimal temperatures range between 35 and 50°C. Higher temperatures of 60°C or more are preferred for extreme thermophilic archaea. Measured temperatures have ranged between 25°C and almost 80°C. These temperature variations and differing stages of biooxidation in different heaps insure microbial consortium diversity in biosolution.

When biooxidation duration is less than designed and temperatures are rising, heaps are continued on biooxidation. If heap temperatures decrease and biooxidation durations are shorter than planned, heaps are placed on drain-down and rest until temperatures begin climbing again. If tempera-tures decline and biooxidation has reached the design duration, heaps enter final drain-down, are offloaded, and biooxidized ore is fed to the CIL mill.

Compromises to preferred design criteria have affected biooxidation cycles. With changes in heap-building methods, the first ore placed is the last ore offloaded rather than first ore on being the first offloaded. Pads are unloaded by hydraulic shovel and haul trucks, eliminating the former prac-tice of using front-end loaders. Currently, the hydraulic shovel mines the heap down to 2–4 m above the base to protect the aeration piping from mobile equipment loads. Then a dozer pushes off an additional 1–2 m of biooxidized ore. Unless aeration pipes are scheduled for replacement, the remaining 2–3-m layer is left as a protective layer. After unloading and doz-ing, pads are inspected, and surfaces are cross-ripped to restore permeability before reloading with the next batch.

Upon reclaiming, biooxidized ore is hauled to the CIL mill. Ore is neutralized on the SAG mill feed belt with 7.5–20 kg t^{-1} pebble lime. The 8.5-m-diameter SAG mill is coupled with a 4.4-m-diameter ball mill. Standard steel mill liners and grinding media are used with no noticeable corrosion. Ore is ground to 80% smaller than 44 μm and stored in two surge tanks sparged with oxygen. The original two-train CIL circuit of four tanks each has been revamped to a single eight-tank train, which has reduced soluble gold losses. Oxygen is added into the first tank to overcome oxygen depletion caused by high ferrous iron levels. The minimum oxygen concentration is now set at 4 mg L^{-1}. The sodium cyanide concentration is raised when processing biooxidized ore to about 0.15–0.3 g L^{-1} in the first CIL tank. The cyanide concentration is monitored to maintain residual concentrations around 0.01 g L^{-1} NaCN in the CIL discharge.

6.4 Commercial BIOPRO™ Operating Performance

Since commissioning the commercial biooxidation facilities, heap-biooxidation performance has been variable. Dedicated operations crews have made contin-uous improvements over the course of 6 years, yet heaps have responded less predictably than expected. Over the course of the first 3 years, heaps were well instrumented and routine data collection was more rigorous. As experience was gained, some parameters were monitored less frequently or their monitoring was discontinued. The following discussion focuses on early performance.

6.4.1 Collecting Data and Monitoring Performance

Recurrent data collection and routine analysis of biooxidation is required to gauge efficacy. Daily measurements of biosolution chemistry are crucial to spot early signs of biooxidation impairment. Measurements of other parameters are also needed to monitor biooxidation progress. Weekly data presentation is important, but as important are biooxidation trends that occur over the life cycle of a pad. One such chart, Fig. 6.6, shows pertinent data over an entire cycle. These responses are typical for whole-ore heap biooxidation of Gold Quarry sulfidic refractory ore. Of note are the two top curves recording internal heap temperatures in the upper-central and lower-central portions of the heap. These temperature curves track each other, with the lower heap portion at slightly lower temperatures. Another curve shows daily ambient temperatures from May through early December. The small diamond symbols near the top of the chart show biosolution on–off status and application rate (scale for application rate on right-hand side of chart).

6.4.2   Original Facility Design/As-Built Comparison

A decade of laboratory and field experimentation defined and validated the necessary process design criteria. As noted, serious design compromises were made when the commercial facilities were constructed. The downsized facility cut the biooxidation time, reduced scale-up factors, sacrificed controlled placement of ore by stacker, coarsened the crush size, and changed loading/unloading sequencing. The final facilities are compared with the original design criteria in Table 6.1.

When the conveyor stacking concept was eliminated, operating parameters were altered appreciably. The overall biooxidation cycle was revised to 428 days as shown in Table 6.2.

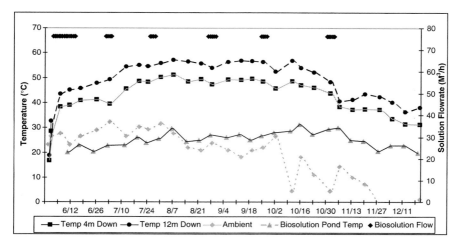

Fig. 6.6. Typical whole-ore heap-biooxidation temperature response

Table 6.1. Original design vs. as-built biooxidation facilities

Design component	Original design	As-built facility
Number of pads	12	3 plus demonstration pad
Pad size	305 m × 147 m	305 m × 147 m
Crushing	Tertiary P80 = 10 mm	Secondary P80 = 19 mm
Agglomeration method	Drum	Belt transfer
Bacteria	*Acidithiobacillus ferrooxidans, Leptospirillum ferrooxidans, Sulfobacillus thermosulfido-oxidans*	*At. ferrooxidans, L. ferrooxidans, Sb. thermosulfidooxidans,* extreme-thermophilic archaea
Material transport	Conveyor	Haul truck
Ore stacking method	Radial stacker	Haul-truck end dumping
Ore reclaim	Bucket wheel excavator	Shovel and haul trucks
Pad base air piping	Replace – every cycle	Replace – 3 cycles
Biosolution treatment	Lime precipitation	Bleed to mill tailings
Cyanide recovery process	Heap leach	Carbon in leach

Table 6.2. Original design vs. commercial design cycle time

Element	Original design	Commercial design
Crushing rate (t day^{-1})	30,136	8,606
Crushing/loading (days)	15	94
Biooxidation (days)	270	150
Drain-down (days)	30	60
Rinse (days)	15	None
Unloading (days)	15	94
Pad repair/schedule buffer (days)	9	30
Total cycle time (days)	354	428
Biooxidation time (%)	76	35

6.4.3 Performance History

Eleven batches of ore have completed an entire biooxidation cycle, been offloaded, and processed through the mill. The remaining pad is about to be loaded with another batch. Table 6.3 summarizes 6 years of performance and compares the design criteria for the original project with those for the smaller commercial facility.

Global sulfide-S oxidation was first set at 40% based on average sulfide-S feed head grade of 1.67%. For the downsized commercial facility, this was

Table 6.3. Original operating criteria vs. project-to-date operating performance

Operating criteria	Original design	Commercial design	Project to date through May 2005
Daily throughput (t day^{-1})	30,136	9,616	8,606
Annualized capacity (t year^{-1})	11,000,000	3,509,900	3,141,200
Biooxidation time (days)	270	150	164
Placed heap height (m)	10.0	11.6	12.8
Placed ore density (kg m^{-3})	1.36	1.60	1.67
Heap settlement (%)	15	5	4
Gold grade (g t^{-1})	1.71	2.49	2.64
Sulfide-S content (%)	1.67	1.80	1.58
Sulfide oxidation (%)	38	30	21.9
Average carbonate content (% $CaCO_3$)	0.58	3.33	2.40
Carbonate destruction (%)	40	40	66.2
Inoculum addition (wt%)	4.0	4.0	2.4
Sulfuric acid consumption (kg H_2SO_4 t^{-1})	0.38	2.18	0.92
Heap temperature range (°C)	35–50	35–60	25–81
Biosolution application rate (L m^{-2}h^{-1})	10	4	4
Biosolution pond E(SCE) (mV Ag/AgCl)	>550	>550	519
Dissolved oxygen (mg L^{-1})	2	4	>2
Biosolution iron concentration (g L^{-1})	13	13	26.9
Gold recovery (%)	55.0	71.3[a]	53.6

SCE standard calomel electrode

[a] Projected recovery included ore from another property that exhibited better biooxidation response.

revised down to 30% oxidation of 1.8% sulfide-S after review of prior demonstration plant experience. This change reflected oxidation rates increasing from 0.24% per day to 0.36% per day.

Recognizing the neutralization effects of carbonates, the maximum calcium carbonate content – as calcite and minor amounts of dolomite – was originally set at 0.6% and averaged 0.3% for both stockpiles and ore to be mined. When plans were deferred for mining the pit expansion, the calcium carbonate criterion was relaxed to accommodate the stockpiled ore. Table 6.4 tabulates biooxidation performance by year for the 11 completed batches.

CIL mill performance is listed in Table 6.5. A detailed analysis of the first year's performance has been previously published (Bhakta and Arthur 2002). Through May 2005, 8.8 Mt of sulfidic ore was processed with an average gold head grade of 2.64 g Au t^{-1}. Overall gold recovery was 53.6%. Total gold production was 12,200 kg (379,500 troy oz).

Figures 6.7–6.9 illustrate CIL mill recovery variability. Figure 6.7 shows CIL mill recovery as a function of biooxidation time. Biooxidation improves

Table 6.4. Annual heap-biooxidation operating results

Description	2000	2001	2002	2003	2004	Through May 2005
Placed (10^6 t)	2.703	2.381	1.974	0.471	1.208	0.845
Sulfide (%)	1.82	1.45	1.46	1.82	1.45	1.82
Calcium carbonate (%)	4.21	2.26	1.59	1.15	1.70	0.68
Gold grade (g t^{-1})	2.57	2.99	2.53	2.15	2.84	2.17
Pad cycle time (days)	364	372	475	628	474	352
Biooxidation period (days)	160	157	142	299	170	129
Sulfide oxidation (%)	27.4	28.4	15.3	8.0	19.3	13.4
Carbonate destruction (%)	76.3	61.3	51.4	64.3	76.2	69.6

Table 6.5. Newmont annual carbon-in-leach mill operating results

Description	2000	2001	2002	2003	2004	Through May 2005
Milled (10^6 t)	1.635	2.290	2.888	0.690	0.337	0.388
Processing rate (t day^{-1})	6,960	10,138	10,165	4,970	9,162	11,190
Mill feed (%)	71.0	68.2	85.8	42.0	11.1	25.3
Mill operating time (days)	235	226	284	139	37	35
Carbon-in-leach recovery (%)	53.8	50.2	53.1	59.7	55.0	63.4[a]
Au produced (kg)	2,355	3,552	4,012	914	543	796
Lime (kg t^{-1})	8.3	12.1	12.8	18.7	14.6	3.6
Cyanide (g t^{-1})	374	373	389	379	372	357

[a] Flotation circuit and commingling may bias recovery attributed to biooxidation.

CIL mill baseline recovery from approximately 30% to approximately 55%. Gold-bearing frambiodal pyrite oxidized to a large extent within 300 days, thus biooxidation times as long as 470 days showed little improvement in recovery.

Figure 6.8 shows monthly CIL mill recovery through 2003. Data for 2004 and 2005 to date are not plotted owing to the minor quantities of biooxidized ore processed in conjunction with oxide ore, which hampered recovery determinations.

Recovery increased slightly during the first 2 years as operations personnel made improvements and modifications to the mill. Recovery significantly improved when the two-train CIL mill configuration was changed to a single train, which dropped high solution losses. Other changes included oxygen aeration to overcome oxygen-depleted conditions in CIL leaching and modifications in lime addition capabilities to control pH.

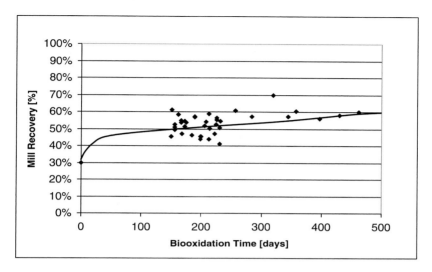

Fig. 6.7. Mill recovery vs. biooxidation time

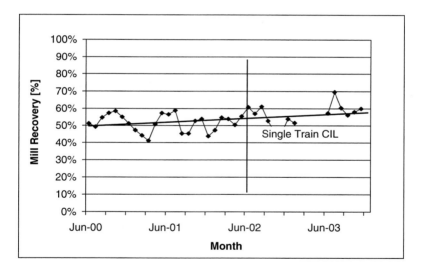

Fig. 6.8. Mill recovery by month

Figure 6.9 illustrates the chronological history of biooxidation and mill recovery. Particularly noteworthy is the apparent sudden decrease in sulfide-S oxidation without a consequent loss of gold recovery. This effect is under investigation. Known causes in the past have been use of an unsuitable sulfide-S assay determination method, sampling errors, and improper drying of samples before assaying. Both pads 6-4 and 8-4 have been biooxidized and sampled, but not processed; thus, only oxidation levels are known.

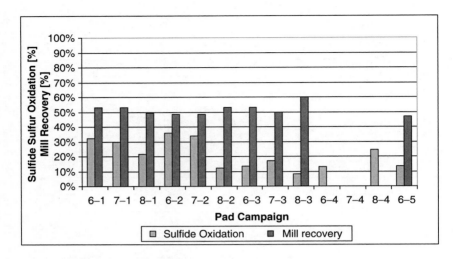

Fig. 6.9. Sulfide-S oxidation and mill recovery vs. pad campaign

Evaluation of biooxidation performance has recently been complicated by addition of flotation to the mill. All ore is now floated, including oxide ore. Sulfide recovery is impacted by oxide gold floating into the sulfide concentrate, giving the appearance of improved biooxidation recoveries. Metallurgical accounting now depends on conducting confirmatory bench flotation tests followed by bottle-roll cyanidation of the flotation tailings to allocate gold response accurately.

6.4.4   Microbial Populations

Initially, test and demonstration heaps were inoculated with a consortium of *Acidithiobacillus ferrooxidans*, iron-oxidizing *Leptospirillum*, and moderately-thermophilic *Sulfobacillus* species. Archaea were found in neither fresh ore nor pilot heaps. When field trials were conducted, larger-sized heaps had sporadic areas of heating to around 80°C, indicating that addition of thermophilic iron-oxidizing archaea of *Acidianus*, *Metallosphaera*, and *Sulfolobus* would facilitate the oxidation process. Laboratory experimentation demonstrated good response of a mix of mesophilic, moderately-thermophilic, and thermophilic iron-oxidizing microbes to changes in heap temperatures (Brierley 2003). As noted, thermophilic microbes proliferated as temperature increased or decreased to the preferred range for growth. Mesophilic and moderately thermophilic bacteria also either increased in a favorable range of growth or decreased with temperatures exceeding those of respective growth ranges. In the basis of these confirming results, addition of thermophilic archaea was considered.

Biosolution from the demonstration pond was pumped to the new commercial biosolution pond in late 1999. This adapted biosolution was used for inoculation. Microbial cell count data for commercial plant operation are presented in Table 6.6.

The largest population consisted of mesophilic iron-oxidizing *At. ferrooxidans* and *Leptospirillum*-like bacteria. Moderately thermophilic *Sulfobacillus* species were present in the original inoculum. As with prior experience, no thermophilic archaea were detected. Six months after start-up, thermophilic archaea, including *Acidianus* and *Metallosphaera*, were added to the inoculum. Thereafter, these microbes were detected in all samples. The apparent increases in the *At. ferrooxidans* and *Leptospirillum*-like bacteria populations occurring in 2001 may reflect addition of an aeration system to the commercial pond. Studies were not conducted to determine a correlation among changes in population densities and heap operation. Microbial populations were monitored for viable densities using end-point serial dilution techniques in a ferrous iron nutrient medium. Incubation was conducted at temperatures appropriate for each iron-oxidizing microbe population.

6.4.5 Process Advances

Detailed mineralogical investigation of mill losses indicated residual recoverable gold still locked in sulfides. Testwork demonstrated flotation mitigated part of these losses. One process available to accept flotation concentrates is a 9,000 t day^{-1} roaster for high-grade sulfidic ore. This roaster could use additional sulfide flotation concentrates for blending purposes. The addition of flotation in 2005 boosted gold recovery by about 10%.

Current bench and pilot plant testwork revealed the advantages of blending biooxidized ore with a large resource of high-carbonate ore in the Gold Quarry pit through the new flotation/CIL circuit. Direct flotation/CIL of biooxidized ore requires lime for neutralization. Direct flotation/CIL of high-carbonate ores requires reducing the pH with sulfuric acid for flotation, then adding lime for leaching. Blending of biooxidized ore with high-carbonate ore, at

Table 6.6. Range of viable cell counts for biosolution pond samples (cells per milliliter)

	2000	2001	2002	2003	2004	2005
Number of samples	6	13	4	3	12	4
Microbe cell counts						
30°C	10^3–10^6	10^4–>10^8	10^7	10^5–10^6	10^4–10^6	10^6
50°C	10^1–10^5	10^1–10^4	10^2–10^5	10^1–10^2	10–10^4	10^2–10^3
65°C	0–10^6	10^4–10^6	10^5	10^3	10^2–10^4	10^2–10^4

laboratory scale, show improved reagent utilization when compared with milling the ores separately, with limited deleterious effect on gold recovery.

6.5   Lessons Learned

Successfully commercializing whole-ore heap biooxidation technology has been an ambitious undertaking for Newmont. More than a decade of experience has rendered crucial lessons about whole-ore heap biooxidation. In the main, reliable and consistent control of global biooxidation is desirable and not easy. Numerous aspects of biooxidation require attention.

6.5.1   Ore Control

Ore control is critical for proper biooxidation response. Managing ore control for gold grade, a usual practice for milling, is unimportant for biooxidation, although not for profitability. Ore characteristics affecting microbial activity must be measured and managed.

Favorable mineralogical deportment is necessary. In Gold Quarry refractory sulfidic ore, gold is predominately found in framboidal pyrite rimming massive gold-barren cubic pyrite, which is favorable for biooxidation. Only the easily oxidized framboidal pyrite needs to be decomposed; therefore, partial oxidation is acceptable.

Sulfide-S content is a vital ore characteristic to be controlled on each pad. While biooxidation is quite capable of eventually degrading most sulfides, the wide variability in feed sulfide content means it is difficult to assess the achievement of sufficient global biooxidation. Variability in the extent of biooxidation is manifested as an unpredictable CIL mill processing response.

Calcium carbonate content affects biooxidation responsiveness, hampering formation of optimal biooxidation conditions. Gold Quarry host rock can contain significant quantities of acid-consuming calcareous minerals. Carbonate reacts with sulfate and produces insoluble gypsum, thus increasing global solution pH. Once the pH exceeds about 2.5, ferric iron is precipitated and is no longer available for sulfide oxidation reaction.

Clay can hinder biooxidation by lowering heap hydraulic conductivity when too much clay is present. Excess clay, which adversely affects the rate and extent of biooxidation, can be compensated for by coarser crushing and improved agglomeration.

Blending practices may be beneficial; but, as with most deleterious characteristics, blending may not improve the biooxidation response. A decision to blend in less than 10% of high-carbonate ore together with low carbonate-ore, for example, was made to permit addition of high-carbonate material to production schedules. The resultant additional carbonate hindered biooxidation of low-carbonate ore because high-carbonate ore consumed acid and iron necessary for biooxidation. Effective biooxidation was delayed while the carbonate was neutralized.

6.5.2 Crush Size

The biooxidation level achieved in a cycle is a function of the proportion of refractory sulfide-S available to the microbial consortium. When ore is crushed finer, improved access results in increased sulfide oxidation. Only oxidation of framboidal pyrite is needed at Gold Quarry to effect a substantial improvement in extraction. Decreasing the crush size from P80 smaller than 19 mm to P80 smaller than 9 mm by installing tertiary crushing would improve gold recovery by about 10%.

Further crushing effort does not incrementally improve gold recovery. Heap-biooxidation feed is now secondary-crushed, not tertiary-crushed, in deference to truck-dumping compaction issues. Although it is economically attractive to crush more finely, the simplified facility eliminated drum agglomeration and conveyor stacking. Both are necessary to maintain adequate heap permeability when biooxidizing fine particles. When the decision was made to truck-dump ore on a heap, the drum agglomeration concept was dismissed, which forced a reversion to coarser crushing.

6.5.3 Compaction and Hydraulic Conductivity

Heap biooxidation depends on unrestricted percolation of biosolution and adequate air ventilation. Heap biooxidation does not respond optimally if crushing is too fine, clay content is high, or placement is by any method compacting a heap. As noted, during demonstration facility operations, two methods of heap construction were tested – conveyor stacking and truck dumping. Conveyor stacking from the pad base places ore optimally for permeability and has the added benefit of low ground-bearing pressure on the pad base. Heap construction by end-dumping ore requires building a ramp to the top of a heap, then advancing the face by dumping ore along the top edge of the heap. Wheel compaction from truck traffic affects the top 2–3 m of the heap. Comparable compaction occurs at the base when mobile equipment removes biooxidized ore.

Not surprisingly, conveyor stacking proved preferable in the demonstration. Conveyor-stacked ore maintained a high permeability of 10^{-3} –10^{-4} m s^{-1}. Measured hydraulic conductivities in truck-dumped ore at the demonstration facility were lower by 2–3 orders of magnitude or 10^{-5}–10^{-6} m s^{-1} when compared those for with conveyor-stacked ore. Low hydraulic conductivity of end-dumped ore proved to slow sulfide-S oxidation rates by as much as 50%, once extensive evaluation of the demonstration data was completed.

Even though ore is now crushed more coarsely to 80% smaller than 19 mm for end-dump placement on commercial cells, the biooxidation rate is still affected by lower heap permeability.

Early attempts to restore commercial heap hydraulic conductivity by dozing the surface with a Caterpillar D10 dozer and ripping with a 1.0-m-long shank ripper appeared to work, but trenching and in situ measurements showed compaction extending past the 1.5-m ripping depth to as deep as 2.8 m. Overbuilding heaps to 13.7 m, dozer-cutting the heap to 12.8 m, and cross-ripping has proved to be effective in restoring hydraulic conductivity.

6.5.4   Inoculum/Acid Addition and Carbonate Destruction

Newmont patented the technique of inoculating ore with microbes prior to placement on a heap after early field experience showed this practice accelerated global biooxidation. Ore can tolerate addition of 4% moisture by weight when secondary-crushed. Biosolution is pumped to the crushing plant and added as ore is loaded into trucks from a conveyor. With low calcium carbonate content (less than 0.6%) anticipated, biosolution could be used to neutralize carbonate, inoculate, and moisture-condition the ore for agglomeration.

Complete destruction of carbonate is not required. Portions of carbonate are not initially exposed, thus requiring no immediate neutralization. The design criterion for carbonate destruction was originally set between 35 and 50%. The ore criterion changed from low calcium carbonate levels (maximum of 0.6%) to high carbonate levels (above 4.2%) in the actual stockpiled ore. Biosolution inoculum addition rates higher than 4% by weight were needed to neutralize the increased carbonate content, which disrupted formation of agglomerates. Where no additional sulfuric acid addition was planned initially, a significant quantity of sulfuric acid is now added to overcome the high carbonate carbon content when encountered. Operating practices now target addition of about 2 kg concentrated sulfuric acid per tonne of ore per percent of calcium carbonate. This amount of sulfuric acid is sufficient to destruct enough calcium carbonate to maintain the pH below 2 and keep biosolution application near the optimal 4% target.

Review of laboratory testwork illuminated interesting biooxidation performance in columns of varying carbonate content. When the calcium carbonate content was kept less than 2.5% by weight, column biooxidation proceeded immediately. Calcium carbonate levels above 2.5% showed increasingly long delays and sluggishness in biooxidation as the carbonate content increased. In many cases, the global solution pH did not decrease for weeks, extending the biooxidation cycle by 20–50% and lowering ultimate gold extraction by one third.

Changes to the design criteria to coarsen the crush size reduced the amount of biosolution needed to form agglomerates. Overapplication of inoculum caused by spillage from truck beds resulted in cutting back on inoculum addition. Occasional biosolution pump malfunctions when crushing caused some ore to miss inoculation. Prior field experiences with the small trial heaps and the demonstration facility did not encounter this condition.

When the first commercial heap was offloaded, operators noted areas where ore had not been wetted. Nearly one tenth of the ore was estimated to be dry. Biooxidation had not occurred in these pockets, nor was gold extraction improved.

6.5.5   Biosolution Chemistry

Microbes are robust and can tolerate wide variations in solution chemistry related to pH, E(SCE), and iron concentrations. Newmont practice maintains the biosolution pH between 1.3 and 2.2. Excursions above pH 2.2 cause operators to add sulfuric acid to the biosolution pond. When the pH decreases, no action is typically taken until the pH drops below 1.3. Then, excess biosolution is pumped from the system and neutralized by commingling biosolution with mill tailings. The volume removed is replaced with fresh water, increasing the pH.

Similarly, the iron content is monitored and maintained. The ideal total iron concentration is between 8 and 15 g L^{-1}, balancing biooxidation performance with CIL mill costs of neutralization and cyanide consumption. Total iron concentrations have ranged between 8 and 36 g L^{-1}, with individual pads reaching up to 60 g L^{-1} total iron when initial solution breakthrough occurred. Ferrous iron concentration less than 1 g L^{-1} are sought. In practice, most heaps have operated with 2–5 g L^{-1} ferrous iron. The best biooxidation rates have been observed when the solution E(SCE) is above 550 m, but acceptable activity is achieved when E(SCE) is above 475 mV. E(SCE) below 400 mV is a clear indication that biooxidation is hindered and corrective action is required. In the early stage of operation, the pond biosolution E(SCE) hovered near 450–475 mV with undetectable dissolved oxygen concentration. A pond aeration system was added and within 2 weeks pond chemistry improved. E(SCE) jumped from near 450 to over 600 mV, ferrous iron concentrations dropped to less than 1 g L^{-1}, and dissolved oxygen concentration increased to 4–6 mg L^{-1}.

6.5.6   Impacts of Precipitates

Biooxidation destructs solid pyrite, creating soluble ferrous iron and sulfate ions. Microbial oxidation also solubilizes other mineral constituents present in the host rock. Eventually, solubility limits are reached and precipitates form within the biooxidation heap. Additionally, unreacted calcite from the acidic inoculation step reacts with fresh biosolution to form gypsum precipitates.

Of particular note is the formation of scorodite and ferricopiapites (Chen and Dutrizac 2000) during column simulation of the BIOPRO™ technology (Brierley 2003) instead of jarosites. Excellent studies (Dutrizac and Harris 1996) have been compiled on iron control that relate to mineral species precipitation in acidic systems. These studies may help to understand complex biooxidation precipitation mechanisms.

During field trials, evaporates formed on heap surfaces. These were determined to be iron sulfate crystals that formed when heap surfaces were alternately wetted and dried. Surface evaporation residues are not of concern. Occasional precipitation in or near the pad base in the drain rock layer was observed in the field trials. Because these trials were single-use, no detrimental effects were seen. Experience at the demonstration facility indicated widespread precipitation at the base surface and surrounding the aeration pipes. On the basis of this, the commercial facility was designed to replace aeration piping every two or three cycles. In commercial practice, precipitates in the heap and near the pad base are common and a hindrance. After 11 complete biooxidation cycles, current practice is to replace aeration piping every cycle. Precipitation formation around aeration piping may be exacerbated by low-humidity air injection. Future incremental improvements may address humidification of air prior to introduction.

Another consideration of carbonate neutralization is carbonate armoring. Formation of gypsum tends to coat (armor) carbonate, slowing further reaction with acid. Evidence to date indicates gypsum precipitates do not coat sulfides, but there may be some effects of gypsum precipitates affecting biosolution movement by blocking interstices.

6.5.7   Pad Aeration

Sulfide-S oxidation demands oxygen in larger amounts than that available interstitially in a heap (Bartlett 1998); thus, induced-oxygen addition by aeration is required to maintain oxidation. In small field heaps, temperatures did not rise enough to establish effective convection to self-aspirate a heap, but the large surface area to volume and the shallow height allowed sufficient aeration. As field and demonstration heap sizes increased, artificial aeration was induced by low-pressure blowers to insure microbial activity. Oxidation of readily available ore-particle surface sulfides and liberated sulfide particles demands the highest quantities of oxygen. Aeration requirements diminish as addressable, residual sulfide quantities decrease. Empirical calculations of oxygen requirements for an entire biooxidation period, assuming 150 days and 30% sulfide-S oxidation, were made to determine the stoichiometric amount of oxygen required. Industrial practice indicated oxygen utilization was much lower than stoichiometric calculations suggested. An overall oxygen utilization efficiency of 20% was selected for fan and piping sizing. This factor may be conservative as minimal oxygen depletion in interstitial air has been measured when the fans are operating. Oxygen depletion occurs within hours when aeration is discontinued or interrupted by power failure. Unlike stirred-tank bioreactors where oxygen demand is extremely high requiring emergency backup generators to prevent cessation of aeration and immediate microbial damage, whole-ore heap biooxidation tolerates oxygen limitation for long periods. The only consequence is reduced microbial ability to effectively reoxidize ferrous iron within the heap.

Once the demonstration heap had been built, heap air conductivity was measured in full-height field columns and was determined to be about 10^{-8} m^2. The aeration pressure to ventilate the heap was calculated to be very low and low-pressure fans were selected. The design criterion for the aeration flow rate was set at 1.5×10^{-4} m^3 m^{-2} s^{-1} (Ritchie 1996) using the oxygen utilization efficiency proposed. Constant aeration is maintained over the life of the heap. The aeration piping grid is designed to connect laterally under the pad from one distribution header to the opposite header. This design feature permits any crushed pipe to be fed from the either side to the point of blockage.

In Nevada, ambient air humidity is usually low (less than 30%) and humidification of dry air as it enters the heap base occurs in the lower portion of the heap. Observed dry spots in the heap base suggest air has dried out portions of the ore, effectively stopping biooxidation locally. As much as 3–5% of the total ore may be affected by drying.

6.5.8 Cell Irrigation and Temperature Response

Over the past decade, heap biooxidation has been conducted under a wide range of biosolution application conditions. No single method has proven substantially better or worse than others, although indications are that higher application rates may improve biooxidation (Ritchie 1996). Early field trials used continuous solution application at the common oxide-heap leach rate of 10 L m^{-2} h^{-1}. As experience was gained, the average application rates on some heaps were as low as 1 L m^{-2} h^{-1}. Various application regimes from continuous irrigation, infrequent solution application, and cycling of solution application have been tested. Continuous application for up to 2 months was initially practiced. Solution-to-ore ratios exceeding unity and as low as 0.1 have been attempted. Lately, biosolution application has been applied for 1–2 weeks with a rest period. Biooxidation response in commercial heaps has shown the deleterious effects of decreasing biosolution application to the point where localized drying within the heap has occurred.

Temperatures vary from ambient to in excess of 80°C in a full-height heap. High temperature in the presence of mesophiles and moderate thermophiles was worrisome when the first demonstration pads were biooxidized. Attempts to moderate high temperatures by adding biosolution decreased local temperatures, but the results were variable and inconsistent. On at least two occasions, excess solution built up in a heap or on top of a heap, flooding the ore and stifling biooxidation. Remedial action of ripping the heap did not reestablish acceptable permeability. Since the establishment of archaea in the biosolution, current practice allows each heap to biooxidize without regard to maximum temperature.

After the initial six commercial heaps had been biooxidized, temperature monitoring was discontinued; however, gauging heap performance was hampered by the lack of data. New heaps are being constructed with monitoring capabilities again as beneficial practices are reinstituted. Heap-building practices formerly allowed careful placement of monitoring pipe bundles along the face. Cutting down each heap to restore permeability precludes this method, so now monitoring pipe bundles are installed in specially dug trenches.

6.5.9   Pad Base Conditions

Of all the design issues faced, pad base durability has been the most troublesome. Experience gained with on–off pad bases in the uranium industry and copper vat leaching was helpful in identifying problems with mobile equipment working on the base. Copper heap leaching experience in South America proved the benefits of mechanical unloading with bucket wheel excavators to offset compaction and plugging problems. When first designed in 1997, the preferred concept was pad offloading with low ground bearing pressure bucket wheel reclaiming. As noted, in 2000, this method was eliminated in favor of mobile equipment offloading. The base has not been able to withstand movement of loaders and trucks, which cause high ground pressures, resulting in serious damage to the pad base drain layer.

A small mining shovel was brought from another Newmont operation in 2001, which reduced damage. While planned for 10 days, drain-down has been lengthened to 30–60 days in an attempt to dry out the base and reduce damage. Finally, a 2-m layer of biooxidized ore has been left over the base for later removal by dozing. Special care was taken to accurately construct the base and liner with laser-guided equipment, and GPS was installed on Newmont equipment to maintain a safe buffer of drain rock between the equipment and the HDPE liner. In spite of GPS, two HDPE liners have been damaged during the unloading cycle by rippers and have required substantial repairs.

Pad base repair is difficult and disrupts production. Aeration pipe collapse from mobile equipment movement is problematic. Such damage must be repaired. Residual biooxidized ore, the 100-mm-diameter aeration pipe, 900 mm of drain rock, and the biosolution collection pipe have to be accurately removed down to the original HDPE liner to expose the liner for repair. When HDPE liner repairs have been completed, a new solution collection system is placed, fresh drain rock is dozed onto the cell, and new aeration piping is positioned. The 100-mm-diameter aeration pipe is replaced with each loading cycle to insure restoration of optimal aeration.

The original drain rock was prepared by crushing and screening low-carbonate rock through the rebuilt commercial crushing plant. The permanent plant with high capacity and low operating cost made drain rock available at favorable unit rates. Since then, replacement of drain rock has required contracted crushing and screening at substantially lower crushing rates and consequently higher cost, influencing operating costs.

6.5.10   Carbon-in-Leach Mill Experience

The CIL mill was converted to process biooxidized ore in 2000. The SAG mill was lined with polyurethane epoxy backing to prevent shell and head corrosion. This corrosion barrier failed, and the SAG mill was rubber-lined with 40 durometer 6-mm-thick rubber in 2002. This liner also failed. Normal steel liners are used today without a noticeable increase in wear due to corrosion. The lime-addition system was upgraded to increase capacity to add pebble lime onto the SAG mill feed conveyor. To accommodate processing biooxidized ore, the SAG mill grates were changed to smaller openings with less open area to retain smaller-diameter balls. Gold recovery does not appear to be particularly grind-sensitive, as long as the final grind is 50% 44 μm or greater. Over the past 4 years operating practices have migrated back to 125-mm-diameter grinding balls and larger grates.

When biooxidized ore was first offloaded and fed to the CIL mill, gold recoveries were unexpectedly low. After processing biooxidized ore with lower-than-expected recoveries, plant performance was intensively studied and many changes were made. Biooxidized ore appears to require additional retention time. This may be due to competing reactions that consume cyanide and oxygen. To overcome the biooxidized ore oxygen demand in CIL as competing reactions occurred, the circuit was modified to use grinding mill surge capacity to condition slurry with aeration. This move also added needed retention time. Oxygen was added into the first two CIL tanks, raising dissolved oxygen levels from less than 0.5 mg L^{-1}. Air injection to the other CIL tanks improved dissolved oxygen levels to nearly 5 mg L^{-1}. Slurry densities were also decreased to optimize oxygen solubility. The two four-tank CIL trains were converted to a single eight-tank train to decrease soluble losses.

Lower carbon loading of gold in the latter part of the CIL circuit resulted. Calcium sulfate scale from reaction of residual biooxidation sulfates with lime added for pH adjustment hampered carbon kinetics. Improvements were made to the acid wash cycle in the stripping circuit that reduced the carry-over of the gypsum salts blinding the carbon. The thickener in front of CIL was operated with a deeper bed to increase retention time. To aid kinetics, the cyanide concentration was increased to 0.15 g L^{-1} NaCN.

The biooxidation process dissolves nickel and other metals from ore that complex with cyanide to make relatively stable compounds. Caro's acid concentrations for tailings cyanide destruction were adjusted for these complexes.

6.5.11   Expectations

Heap biooxidation is a biologic process by living organisms that are extremely robust, but sensitive to operating conditions. Anticipation grew as development moved closer to production. Success in scaling the process from the laboratory bench to field trials of increasing size and then on to a demonstration of the technology led to expectations that commercial facilities would operate according to design criteria and past experiences. Then, gold prices weakened. Circumstances demanded major compromises in design and capital allocation. Innovative changes were made once operations had begun, but biooxidation performance suffered owing to capital limitations.

Another aspect of heap biooxidation known to cause difficulty is the slow response of biooxidation to changes. Unlike a mill, where effects of operating changes are known immediately, changing any biooxidation variable can take days, weeks, or months for the effect to be revealed. Depending upon circumstances, further biooxidation projects deserve attention to be paid to predictions of the extent of biooxidation and subsequent metal recovery to partially account for the difficulty in controlling biooxidation conditions as closely as in milling.

In retrospect, despite diligent attention and dedication of operations personnel and management, biooxidation performance has not consistently achieved predicted levels because of cumulative effects of changes in the constructed facility compared with design requirements.

6.6   Final Thoughts

Development and extension of whole-ore heap-biooxidation technology has been rewarding. Where metallurgists expect to control inorganic milling processes reliably, attempting fine control of a large biooxidation heap is challenging. Newmont operations personnel achieved the implementation of new technology, produced significant gold production profitably, and now strive for consistency and continued incremental improvements. Consistency in ore characteristics, uniformity in crush size, evenness in moisture conditioning to a preset level, regularity in placed-ore permeability, and steadiness in operating are keys to achieving optimal performance. Lastly, continuity of personnel and transfer of knowledge is crucial. Learning the nuances of heap biooxidation by constant observation over several cycles is essential for decision making.

Biooxidation heap applications are becoming commonplace for copper where specialized equipment (typically mobile track-mounted stackers and bucket wheel excavators) is used to load and unload on–off pads to maintain heap permeability and control potential damage to the pad base. Economic circumstances pushed Newmont away from known operating techniques toward compromised techniques of uncertain viability. The diligence of operations personnel overcame serious compromises of the design criteria to a large extent.

In summary, moving to commercial operation has been successful. Significant quantities of sulfidic ore, untreatable economically any other way, have yielded recovered gold. Full-scale operations have expanded our understanding of whole-ore heap biooxidation and this will be applied in future biooxidation projects. Techniques to grow substantial quantities of inoculum for start-up had to be developed and the knowledge is transferable. Whole-ore heap biooxidation is now considered a viable option for future opportunities.

No presentation of Newmont's experience with BIOPRO™ whole-ore heap biooxidation would be complete without an expression of acknowledgement and gratitude for the cooperation and support of the legion of people who have worked on development and implementation of new, unfamiliar technology. Thanks to you, Newmont has achieved a milestone in the application of biooxidation technology and continues to learn important lessons about biooxidation.

References

Bartlett RW (1998) Solution mining – leaching and fluid recovery of materials. Gordon and Breach, Amsterdam

Bhakta P, Arthur B (2002) Heap bio-oxidation and gold recovery at Newmont mining: first-year results. J Met 54:31–34

Brierley JA (1997) Heap leaching of gold-bearing deposits: theory and operational description. In: Rawlings DE (ed) Biomining: theory, microbes, and industrial processes. Springer, Berlin Heidelberg New York, pp 103–115

Brierley JA (2003) Response of microbial systems to thermal stress in heap-biooxidation pretreatment of refractory gold ores. Hydrometallurgy 71:13–19

Brierley JA, Brierley CL (2001) Present and future commercial applications of biohydrometallurgy. Hydrometallurgy 59:233–239

Brierley JA, Hill DL (1993) Biooxidation process for recovery of gold from heaps of low grade sulfidic and carbonaceous sulfidic ore materials. US Patent 5,246,486

Brierley JA, Wan RY, Hill DL, Logan TC (1995) Heap-biooxidation pretreatment technology for processing lower grade refractory gold ores. In: Vargas T, Jerez CA, Wiertz, JV, Toledo, H (eds) Biohydrometallurgical processing, vol 1. University of Chile, Santiago, pp 253–262

Chen TT, Dutrizac JE (2000) Identification of iron precipitates in biooxidized gold ore. CANMET report MMSL 2000-048, Ottawa

Dutrizac JE, Harris GB (1996) Iron control and disposal. In: Proceedings of the 2nd international symposium on iron control in hydrometallurgy. Ottawa, 20–23 October 1996

Murr LE, Torma AE, Brierley JA (1978) Metallurgical applications of bacterial leaching and related microbiological phenomena. Academic, New York

Olson GJ, Brierley JA, Brierley CL (2003) Bioleaching review part B: progress in bioleaching: applications of microbial processes by the minerals industries. Appl Microbiol Biotechnol 63:249–257

Pantelis G (1994) FIDHOX: Description of mathematical model and users guide. Australian Nuclear Science and Technology Organization report C394.

Ritchie AIM (1996) Assessment of the performance of the batch 1 demonstration biooxidation heap. Australian Nuclear Science and Technology Organisation report C468

Rohwerder T, Gehrke T, Kinzler K, Sand W (2003) Bioleaching review part A: progress in bioleaching: fundamentals and mechanisms of bacterial metal sulfide oxidation. Appl Microbiol Biotechnol 63:239–248

Schwartz W (1977) Conference on bacterial leaching 1977. Verlag Chemie, Weinheim

Shutey-McCann ML, Sawyer F-P, Logan T, Schindler AJ, Perry RM (1997) Operation of Newmont's biooxidation demonstration facility. In: Hausen, DM (ed) Global exploitation of heap leachable gold deposits. The Minerals, Metals and Materials Society, Warrendale, pp 75–82

Tempel K (2003) Commercial biooxidation challenges at Newmont's Nevada operations. In: 2003 SME Annual Meeting. Society of Mining, Metallurgy and Exploration, Littleton, preprint 03–067

Wan RY, Brierley JA (1997) Thiosulfate leaching following biooxidation pretreatment for gold recovery from refractory carbonaceous-sulfidic ore. Min Eng 49:76–80

7   Heap Leaching of Black Schist

Jaakko A. Puhakka, Anna H. Kaksonen, Marja Riekkola-Vanhanen

7.1   Introduction

Bioleaching is proven technology for recovering metals. In its simplest form it has been used for copper recovery from low-grade materials since the 1500s in Spain. At present, approximately 20% of the world's annual copper production is recovered by bioleaching (Anonymous 2002). The heap leaching of low-grade sulfidic ores is the widest application of mining biotechnology (for a review, see Brierley and Brierley 2001). A typical solution from heap leaching has a low pH and not more than a few grams per liter of valuable base metals. In addition, the solution can have small amounts of rare and precious metals. Simplicity, low-cost and applicability to low-value ores are the main benefits of biohydrometallurgy. Bioleaching has the potential to be used for obtaining metals from mineral resources that have not been accessible by conventional mining (for reviews, see Bosecker 1997; Brandl 2001; Brierley and Brierley 2001; Hsu and Harrison 1995). The understanding of the number and kind of biocatalysts in bioleaching environments has advanced from the days when *Acidithiobacillus ferrooxidans* and *At. thiooxidans* were the only microorganisms considered (for reviews, see Johnson 1998; Hallberg and Johnson 2001; Rawlings 2002).

On the other hand, many significant ore deposits are not accessible by conventional methods or even by bioleaching. Many of the world's largest nickel deposits are very low grade, often occur in lateritic deposits, are located in warm climates (Anonymous 2004) and are typically very acid consuming (Brierley 2001). The acid consumption makes biological applications very challenging (Salo-Zieman et al. 2006). On the other hand, very significant and yet-unexploited low-grade mineral deposits exist in boreal and subarctic locations. For example, the largest European nickel deposit is located in Sotkamo, Finland.

7.2   Significance and Potential of Talvivaara Deposit

The geochemistry and the genesis of the Talvivaara deposit have been carefully described (Loukola-Ruskeeniemi 1996). The deposit is one of the largest in the world by volume with low metal grades: 340 Mt of ore with an average

Biomining
(ed. by Douglas E. Rawlings and D. Barrie Johnson)
© Springer-Verlag Berlin Heidelberg 2007

of 0.27% Ni, 0.14% Cu, 0.02% Co and 0.56% Zn. In addition to vein sulfides, finely grained and disseminated spheroidal sulfidic compounds contain Ni, Cu, Zn and Mn. There are also horizons with elevated Au, Ag and Pd concentrations.

The black schist ore has a bulk concentration of quartz (25%), aluminium silicates (potassium feldspar and plagioclase 38%), iron sulfides (16%, of which approximately two thirds of Fe is in pyrrhotite and one third is in pyrite), graphite (10%), magnesium–iron silicates (8%) and sphalerite–pentlandite-violarite (3.2%). The pentlandite and violarite contain 80–90% of the nickel with the rest in pyrite and pyrrhotite. Zn is in sphalerite and copper in chalcopyrite. Cobalt is distributed in pyrite and Ni-containing minerals. The long-term environmental impacts of this deposit have been demonstrated (Loukola-Ruskeeniemi et al. 1998), indicating possible long-term intrinsic biogenic oxidation of the sulfidic minerals.

The black schist ore and the possible utilization of the deposit have been extensively studied for over 20 years. Conventional methods for recovering the valuable metals, including Ni, Zn, Cu and Co, were not feasible. Chemical and biological leaching methods have been considered as potential options. The studies in the 1980s focused on evaluation of tank leaching of finely ground ore, whereas the recent and on-going studies aim at showing the viability of heap leaching. Heap bioleaching would make Talvivaara the most attractive source of Ni in the whole of Europe. This chapter reviews the research conducted over the years for bioleaching of this black schist.

7.3 Biooxidation Potential and Factors Affecting Bioleaching

The first leaching trials with Talvivaara ore material using enrichment cultures from the acidic water samples from the site resulted in 80–90% yields of Ni, Zn and Co and 27–34% of Cu, these being significantly higher than by solely chemical solubilization (Fig. 7.1; Puhakka et al. 1985). This promising result encouraged the continuation of studies to explore the bioleaching potential further.

The factors affecting bioleaching of the black schist were studied in the 1980s and early 1990s, mainly with finely ground ore material and slurry-phase applications in mind. Of the inorganic nutrients, phosphate was sufficiently solubilized from the ore, whereas ammonium enhanced and nitrate inhibited bioleaching (Niemelä et al. 1994). A low concentration of organic carbon amendment was a prerequisite for the microbiological leaching of the black schist as demonstrated both by a pure culture of *At. ferrooxidans* and by three mixed cultures of acidophiles (Puhakka and Tuovinen 1987).

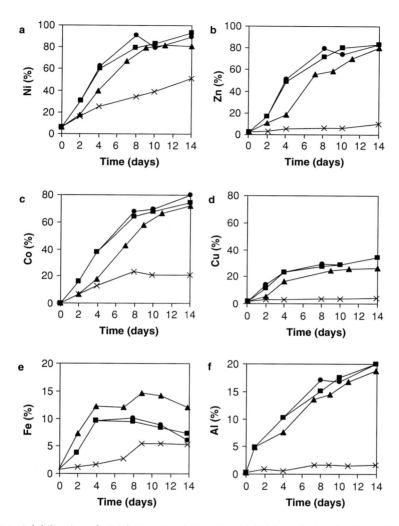

Fig. 7.1. Solubilization of **a** Ni, **b** Zn, **c** Co, **d** Cu, **e** Fe and **f** Al from the black schist with three enrichment cultures derived from the mine site waters (*triangles, squares, circles*) and the uninoculated control (*crosses*) (from Puhakka et al. 1985)

7.4 Leaching of Finely Ground Ore with Different Suspension Regimes

Several different experimental approaches have been used to evaluate the microbiological leaching of the black schist. These include shake-flask, air-lift percolators, aerated columns, air-lift reactor, stirred-tank reactors and ore

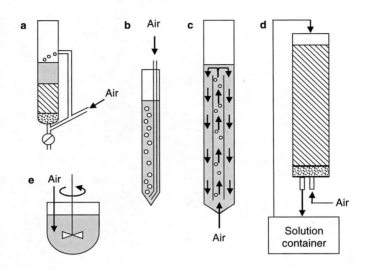

Fig. 7.2. Schematic models of different reactor types used in black schist bioleaching experiments. **a** Air-lift percolator, **b** aerated column, **c** air-lift column, **d** ore-column stirred-tank reactor. (Modified from Puhakka and Tuovinen 1986b; Kinnunen et al. 2005)

columns (Fig. 7.2) inoculated with enrichment cultures of indigenous mineral-associated microorganisms. A typical leachable metal profile of black schist from a shake-flask experiment is shown in Fig. 7.3 (Puhakka and Tuovinen 1986b). The concentrations of iron increased for 2 weeks, followed by precipitation for the remaining time. The recoveries for Ni and Zn were 100%, 73% for Co, 29% for Cu and 22% for Al.

Black schist bioleaching in aerated-column and air-lift reactors at 10–30% (wt/vol) suspensions resulted in longer leaching times and complete solubilization of valuable metals was not achieved (Puhakka and Tuovinen 1986a, b). The leaching in both aerated-column and air-lift reactor types had extended lag periods that were associated with the initial evolution of hydrogen sulfide. These results showed that bioleaching of finely ground black schist with indigenous enrichment cultures results in high percentage recoveries of valuable metals. Typical examples of bioleaching yields with different techniques are shown in Table 7.1.

7.5   Heap Leaching Simulations

Heap leaching potential of the black schist was studied in air-lift percolators and columns of various sizes. In air-lift percolators, the leach solution was recirculated with different percolation regimes with the aim of simulating flood and trickle leaching (Puhakka and Tuovinen 1986c). The highest rates

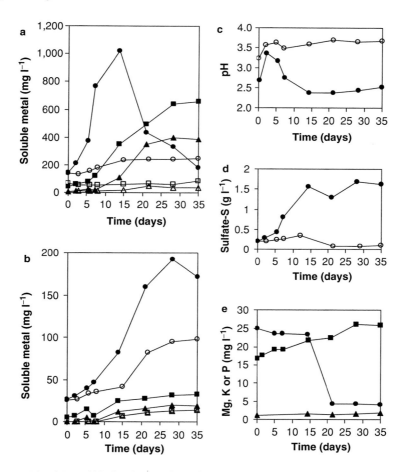

Fig. 7.3. Bioleaching of black schist in shake-flasks. **a** Solubilization of Fe (*closed circles*), Al (*closed squares*) and Zn (*closed triangles*) in inoculated flasks and in uninoculated flasks (*open circles, open squares* and *open triangles,* for Fe, Al and Zn, respectively). **b** Solubilization of Ni (*closed circles*), Cu (*closed squares*) and Co (*closed triangles*) in inoculated flasks and in uninoculated flasks (*open circles, open squares* and *open triangles* for Ni, Cu and Co, respectively). **c** pH changes in inoculated (*closed circles*) and uninoculated (*open circles*) flasks. **d** The formation of sulfate in inoculated (*closed circles*) and uninoculated (*open circles* ○) flasks. **e** Solubilization of K (*circles*), Mg (*squares*) and PO_4-P (*triangles*) in inoculated flasks. (From Puhakka and Tuovinen 1986b)

of leaching from black schist of smaller than 1-mm size were obtained with flood leaching (Table 7.1). These early experimental designs in light of current understanding of air/liquid flow requirements do not simulate actual heap leaching conditions.

Further studies were conducted to better simulate heap leaching. For example, Riekkola-Vanhanen and Heimala (1999) used 450-kg black schist/

Table 7.1. Bioleaching metal yields from black schist with different leaching regimes

Leaching regime and time (days)	Ni (%)	Zn (%)	Cu (%)	Co (%)	Reference
Shake flask (35)	100	100	30	73	Puhakka and Tuovinen (1986b)
Aerated column (69–97)	53–81	45–55	27–32	30–52	Puhakka and Tuovinen (1986a)
Air-lift reactor (100)	>84	>91	31–39	65–79	Puhakka and Tuovinen (1986b)
Air-lift percolators (170)	22–44	14–25	1–8	9–18	Puhakka and Tuovinen (1986c)
Stirred-tank reactor (17)	64–99	26–100	6–24	12–65	Riekkola-Vanhanen et al. (2001)
Ore column (331–338)	57	63	100	34	Kinnunen et al. (2005)
Ore column (150–210)	59	60	12	14	Rahunen (2005)
Pilot ore column (460)	92	80	66	65	Riekkola-Vanhanen and Heimala (1999)

columns with a grain size of 70% 0.5–2 mm and an acidophilic enrichment culture SB/P-II in long-term leaching tests. The pH of the leach solution was maintained at around 3 in order to prevent leaching of silica and subsequent leach liquor gelling, and the columns were efficiently aerated (300 L h^{-1}) with water-moisturized air. High yields of all valuable metals were obtained in the 460-day experiment (Table 7.1).

The most important design criteria of black schist heap leaching include the pH and the temperature. The selection of the operation pH is a compromise of high leaching yield of valuable metals, slow leaching of other ore constituents and the cost of pH control.

The experimental system provided a liquid flow rate of 10 L m^{-2} h^{-1} together with a sufficient air-flow rate for the columns. The black schist was agglomerated (8-mm grain size), an enrichment culture was used as described in Sect. 7.6 and the columns were equipped with on-line pH control (Rahunen 2005; A.-K. Hakala, N. Rahunen A.H. Kaksonen and J.A. Puhakka, unpublished results). The leaching rates and yields of Ni, Zn, Cu, Co and Fe were directly proportional to the leach liquor pH used, with highest yields at pH 1.5 (Fig. 7.4). Most of the iron was precipitated during leaching at pH

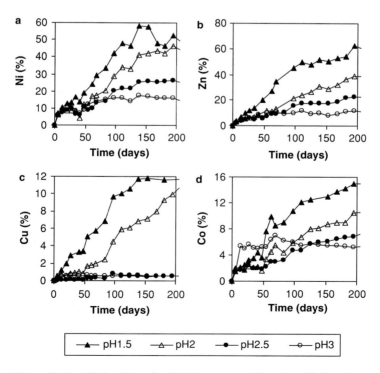

Fig. 7.4. Effects of pH on bioleaching of **a** Ni, **b** Zn, **c** Cu and **d** Co from black schist in column experiments (data obtained from Rahunen 2005; A.-K. Hakala, N. Rahunen A.H. Kaksonen and J.A. Puhakka, unpublished results)

2.5–3.0. The highest Al, Ca, Mg, Mn and Si concentrations of about 10, 0.7, 1.6, 22 and 3 g L^{-1}, respectively, were also obtained at pH 1.5 with decreasing concentrations at higher solution pH (Fig. 7.5). These results demonstrate that pH 1.5, although providing the highest yields of valuable metals, cannot

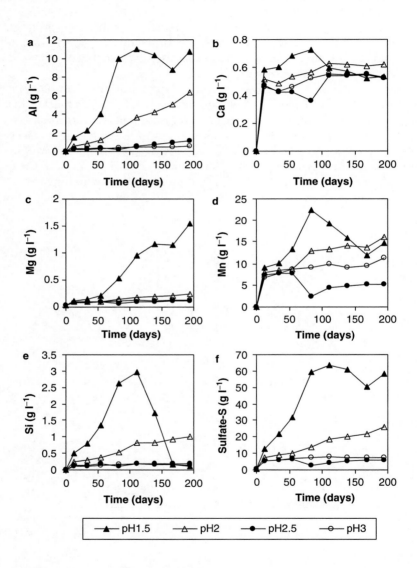

Fig. 7.5. Effects of pH on **a** Al, **b** Ca, **c** Mg, **d** Mn and **e** Si from black schist and **f** SO$_4^{2-}$ in column experiments. At pH 1.5, 2.0, 2.5 and 3.0 the corresponding sulfuric acid consumptions in 200 days were 195, 55, 12 and 4 kg kg^{-1} ore, respectively. (Data obtained from Rahunen 2005; Hakala et al., unpublished results)

be applied because of elevated Fe, Si and Al concentrations that interfere with valuable metal recoveries.

The temperature range of heap leaching is controlled by the boreal environmental conditions, the heat generation during leaching and the dimensions of the heap. In order to characterize bioleaching at various temperatures, new column experiments were designed. Temperature effects on the black schist bioleaching was studied in columns at temperatures of 5, 21, 35 and 50°C and the results were as shown in Fig. 7.6. The highest yields of Ni, Zn and Co were obtained at 21°C. The leaching rates and yields were relatively high also at 5 and 35°C. The 50°C column was inoculated after 50 days of operation with a thermophilic VS2 culture dominated by a *Sulfolobus* species (Salo-Zieman et al. 2006) with subsequent increases in metal leaching rates. These results demonstrate that bioleaching occurred at all temperatures studied from 5 to 50°C and that temperature changes in this range would allow heap leaching at varying temperature conditions.

In the summer 2005, a 50 000-t pilot heap of agglomerated ore was constructed at the mine site (Fig. 7.7). The leaching demonstration was started in August 2005.

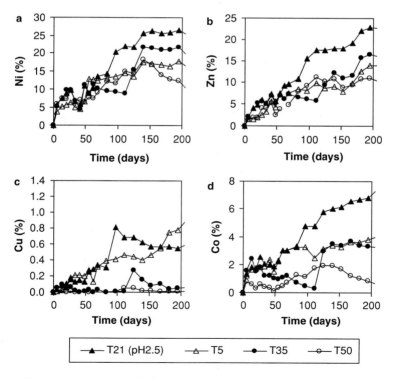

Fig. 7.6. Effects of temperature on bioleaching of **a** Ni, **b** Zn, **c** Cu and **d** Co from black schist in column experiments (data obtained from Rahunen 2005; Hakala et al., unpublished results)

Fig. 7.7. The pilot heap in Talvivaara, Sotkamo, Finland

7.6   Dynamics of Biocatalyst Populations

The ore deposit site was shown to harbor iron-, sulfur- and glucose-oxidizing acidophiles, including the presence of *At. ferrooxidans* and *At. thiooxidans*, by using conventional microbiological methods (Puhakka et al. 1985;

Puhakka and Tuovinen 1986c). During the time of these studies, limited qualitative or quantitative methods to evaluate the heterogeneity of mixed microbial cultures existed.

In recent studies, modern molecular methods were available to describe the black schist bioleaching communities and factors affecting the community dynamics in column tests (Rahunen 2005; Hakala et al., unpublished results). The mixed cultures for column tests were enriched from mine site water samples with a mixture of black schist ore, elemental sulfur and ferrous iron. The enrichment culture as revealed by PCR denaturing gradient gel electrophoresis (DGGE) followed by sequencing contained *At. ferrooxidans*, *At. thiooxidans* or *At. albertensis*, *At. caldus* and *Leptospirillum ferrooxidans* (Fig. 7.8). Depending on the leaching conditions, various chemolithotrophs became enriched during the progress of leaching in the column leaching liquors. At 5°C, *At. ferroxidans* and *At. thiooxidans* or *At. albertensis* became dominant. At 21°C, *At. ferroxidans* and *L. ferrooxidans* had the strongest DGGE bands. At 35 C, *L. ferrooxidans* and *At. caldus* dominated, whereas at 50°C, *At. caldus* was dominant. These successions in the enrichment culture composition with operation temperature are in line with the effects of temperature on the rates of iron and sulfur oxidation by bioleaching organisms (Franzmann et al. 2005). In the columns at 21°C with pH varying from 1.5 to 3.0, no significant changes were apparent in the DGGE profiles. These results demonstrate that the mine site waters harbor a microbial community consisting of several members and that depending on the leaching conditions and

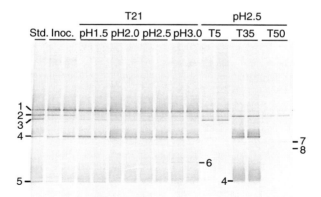

Fig. 7.8. Denaturing gradient gel electrophoresis (DGGE) profiles of partial 16S rRNA gene fragments retrieved from enrichment culture (Inoc.) and bioleaching column effluents after 27 days of bioleaching (data obtained from Rahunen 2005; Hakala et al., unpublished results). Std Standard; labeled bands: 1 = Acidithiobacillus ferrooxidans (99%), 2 = At. caldus, 3 = At. thiooxidans / At. albertensis (100%), 4 = Leptospirillum ferrooxidans (99–100%), 5 = Sulfobacillus thermosulfidooxidans (96–97%), 6 = mixotrophic iron oxidizing bacterium (93%), 7 = Thiobacillus delicates (99%), 8 = Alicyclobacillus acidocaldarius (98%)

especially the leaching temperature, the dominant biocatalysts become different. As the actual leaching heaps will have temperature gradients owing to the heat released in the exothermic reactions and the varying boreal conditions, succession of different organisms at different depths of the heap and at different times is obvious.

References

Anonymous (2002) Billiton: creating value through innovation biotechnology in mining. http://www.imm.org.uk/gilbertsonpaper.htm

Anonymous (2004) Minerals exploration and mining technology. http://www.iom3.org/materialsworld/Mar04/news2.htm

Bosecker K (1997) Bioleaching: metal solubilization by microorganisms. FEMS Microbiol Rev 20:591–604

Brandl H (2001) Microbial leaching of metals. In: Rehm H-J (ed) Biotechnology, vol 10. Wiley-VCH, Weinheim, pp 2181–2186

Brierley CL (2001) Bacterial succession in bioheap leaching. Hydrometallurgy 59:249–255

Brierley JA, Brierley CL (2001) Present and future commercial applications of biohydrometallurgy. Hydrometallurgy 59:319–326

Franzmann PD, Haddad CM, Hawkes RB, Robertson WJ, Plumb JJ (2005) Effects of temperature on the rates of iron and sulfur oxidation by selected bioleaching Bacteria and Archaea: application of the Ratkowsky equation. Miner Eng 18:1304–1314

Hallberg KB, Johnson DB (2001) Biodiversity of acidophilic prokaryotes. Adv Appl Microbiol 49:37–84

Hsu C-L, Harrison RG (1995) Bacterial leaching of zinc and copper from mining wastes. Hydrometallurgy 37:169–179

Johnson DB (1998) Importance of microbial ecology of acidophilic microorganisms. FEMS Microbiol Ecol 27:307–317

Kinnunen PH-M, van der Meer T, Nevatalo LM, Riekkola-Vanhanen M, Kaksonen AH, Puhakka JA (2005) Bioleaching of a complex sulfide ore in boreal conditions. Proceedings of the international biohydrometallurgy symposium IBS 2005, Cape Town

Loukola-Ruskeeniemi K (1996) Geochemistry and genesis of the black shale-hosted Ni-Cu-Zn deposit at Talvivaara, Finland. Econ Geol 91:80–110

Loukola-Ruskeeniemi K, Uutela A, Tenhola M, Paukola T (1998) Environmental impact of metalliferous black shales at Talvivaara in Finland, with indication of lake acidification 9000 years ago. J Geochem Explor 64:395–407

Niemelä SI, Riekkola-Vanhanen M, Sivelä C, Viguera F, Tuovinen OH (1994) Nutrient effect on the biological leaching of a black-schist ore. Appl Environ Microbiol 60:1287–1291

Puhakka J, Tuovinen OH (1986a) Microbiological solubilization of metals from complex sulfide ore material in aerated column reactors. Acta Biotechnol 6:233–238

Puhakka J, Tuovinen OH (1986b) Biological leaching of sulfide minerals with the use of shake flasks, aerated column, air-lift reactor and percolation techniques. Acta Biotechnol 6:345–354

Puhakka J, Tuovinen OH (1986c) Microbiological leaching of sulfide minerals with different percolation regimes. Appl Microbiol Biotechnol 24:144–148

Puhakka J, Tuovinen OH (1987) Effect of organic compounds on the microbiological leaching of a complex sulfide ore material. MIRCEN J 3:429–436

Puhakka J, Hiltunen P, Tuovinen OH (1985) Patterns of substrate utilization and metal solubilization from sulfide ore material by mixed cultures of microorganisms enriched from mine waters. Syst Appl Microbiol 6:302–307

Rahunen N (2005) Heap bioleaching of complex sulfide ore in boreal conditions. MSc thesis, Tampere University of Technology, Tampere

Rawlings DE (2002) Heavy metal mining using microbes. Annu Rev Microbiol 56:65–91

Riekkola-Vanhanen M, Heimala, S (1999) Study of the bioleaching of a nickel containing black-schist ore. In Amils R, Ballester A (eds) Biohydrometallurgy and the environment toward the mining of the 21st century. Proceedings of the international biohydrometallurgy symposium IBS'99 held in San Lorenzo de el Escorial, Madrid, 20–23 June 1999, part A, pp 533–554

Riekkola-Vanhanen M, Sivelä C, Viguera F, Tuovinen OH (2001) Effect of pH on the biological leaching of a black schist ore containing multiple sulfide minerals. In: Ciminelli VST, Garcia O (eds.) Biohydrometallurgy: fundamentals, technology and sustainable development. Proceedings of the international biohydrometallurgy symposium IBS 2001 in Ouro Preto, Minas Gerais, 16–19 September 2001, part A. Elsevier, Amsterdam, pp 167–174

Salo-Zieman VLA, Kinnunen PH-M, Puhakka JA (2006) Bioleaching of acid-consuming low-grade ore with elemental sulfur addition and subsequent acid generation. J Chem Technol Biotechnol 81:34–40

8 Modeling and Optimization of Heap Bioleach Processes

Jochen Petersen, David G. Dixon

8.1 Introduction

Heap bioleaching is emerging as the predominant technology route for the recovery of copper from low-grade ores. In terms of revenue generated, it is therefore by far the most significant industrial application of biohydrometallurgy.

In a way, this development is returning the discipline to its original roots-a number of iron- and sulfur-oxidizing microorganisms were first discovered in dump leach operations, and in a sense, microbial oxidation of pyrite and other sulfides in deposits of mineral tailings and coal spoils, producing acid rock drainage, can be perceived as a form of involuntary heap bioleaching, making the technology essentially a natural phenomenon.

Heap bioleaching aims to create conditions in a pile of ore material that are conducive for these natural phenomena to proceed and thus achieve the desired mineral dissolution. The attractiveness of using heap bioleaching as an industrial process is that it is technologically extremely simple and involves relatively minor capital and operating expenses. A less obvious drawback is that relying on a natural process may also require allowing for long process times and thus low rates of recovery. In an economic context this can significantly reduce the attractiveness of the technology owing to the large investments of capital required, even within an average-sized operation.

Heap bioleach technology has evolved from conventional heap and dump leaching of oxide ores, with the fundamental departure being the introduction of aeration from underneath the heap to ensure an aerobic environment in which microbes can thrive, and increasing the portion of copper in the feed ore that is associated with sulfide minerals. Operating practice, however, has remained strongly influenced by that of abiotic oxide heaps, which is concerned primarily with the consumption of acid per tonne of ore or per tonne of copper produced. Systematic monitoring of microbial populations within an operating heap is rarely practiced and there are few tools available to assess and promote microbial activity in heaps. Nevertheless, as significant portions of the sulfide-associated copper can be leached over time, the technology is, by and large, considered successful.

Historically, the development of heap leach technology has been accompanied by the development of mathematical models, primarily as a tool to monitor stockpile inventories and forecast production figures. However,

Biomining
(ed. by Douglas E. Rawlings and D. Barrie Johnson)
© Springer-Verlag Berlin Heidelberg 2007

even at an early stage, formulation of these models revealed that the processes underlying even basic oxide heap leaching are physically complex, that their mathematical description is not trivial and that simulation requires elaborate computer codes. The move to bioleaching has further increased the level of complexity of the process and hence that of associated models. It is perhaps this complexity of describing an apparently simple process that has hindered the acceptance of modeling as a tool to guide the development of the technology in this particular industry.

As a result, the questions of whether heap bioleaching practice is operating anywhere near optimal conditions, and of what scope exists for improving the technology to make it more efficient and thus more economically viable, remain unresolved. It is postulated that these questions can only be answered through systematic study of heap bioleaching, centrally involving modeling in combination with well-designed experimental studies. It is with this premise in mind that the work described in this chapter has been conducted.

The first part of this chapter outlines current understanding of the physical, chemical and biological subprocesses that take place within bioleach heaps. This is followed by an overview of the various modeling approaches that have been used to describe heap and dump leaching as well as the formation of acid rock drainage. The remainder of the chapter describes in some detail how the HeapSim model, developed by the authors, has been used in a number of experimental studies to understand the key factors that drive and limit heap bioleaching in chalcocite, sphalerite and chalcopyrite heap bioleaching, and the recommendations for improved heap operation that have resulted from these studies.

8.2   Physical, Chemical and Biological Processes Underlying Heap Bioleaching

To unravel the processes underlying heap bioleaching it is useful to distinguish between phenomena taking place at different scales within the heap. Beginning at the heap scale, we can distinguish a number of transport effects, as illustrated in Fig. 8.1.

8.2.1   Solution Flow

The solution applied to the heap surface migrates through the porous matrix of the stacked ore in some path, primarily in the downward direction. Not all liquid is in motion, however - a significant portion will remain trapped in pores and crevices between the ore particles. Flowing solution will pass the stagnant moisture and thus can exchange dissolved constituents with it, but

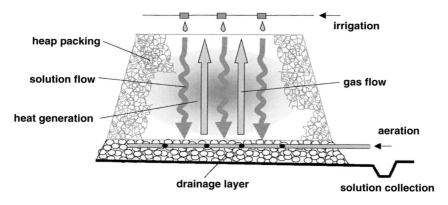

Fig. 8.1. A heap indicating the major heap-scale transport effects

this depends strongly on the distribution of solution flow channels and the relative size of stagnant zones.

8.2.2 Gas Flow

Moisture in the heap will not fully saturate the porous space between particles, except potentially near the bottom, where poor drainage may cause a hold-up of fluid and consequent full saturation. The residual gas space provides an interconnected pore network through which air injected from underneath the ore bed can be distributed throughout the heap in a reasonably homogeneous manner. In operations which rely on natural convection for air ingress through the heap or dump side walls, air distribution follows more complex patterns, strongly related to heat generation.

8.2.3 Heat Flow

The chemical reactions involved in the exothermic oxidation of sulfide minerals result in the generation of substantial amounts of heat. This is transported through and from the heap in a number of ways:

- Transport within the heap through conductive heating of the rock and stagnant liquid
- Transport downward with the warm solution
- Transport upward with gas flow, primarily in the form of humidity (water vapor released from warmer zones of the heap which condenses in cooler zones)
- Heat exchange at the heap surface by the mechanisms of transpiration and radiation

The temperatures achieved at any particular point within the heap are directly related to the interplay between these effects.

A closer examination of a cluster of individual ore particles within a heap bed (as illustrated in Fig. 8.2) identifies a further set of subprocesses contributing to the overall process.

8.2.4 Diffusion Transport

A significant portion of solution is trapped as stagnant moisture in pore spaces between the solid particles, creating a network of liquid pathways. Here, transport of dissolved constituents, especially metal ions, will proceed by molecular diffusion, while the solution itself remains stagnant. By this mode, dissolved constituents can move along concentration gradients between channels of flowing solution and particle surfaces. In similar fashion, the particles themselves can be perceived as matrices of micropores through which dissolved constituents can migrate by diffusion.

The rate of diffusion is strongly related to distance. In this sense the geometry of stagnant solution zones relative to flow channels is of importance, as is the distribution of particle sizes.

8.2.5 Microbial Population Dynamics

In heap bioleach operations microorganisms populate the solution within the stagnant pore network. Exposure to the gas phase provides a good supply of oxygen and carbon dioxide needed for metabolic processes and the key reactions of ferrous iron and sulfur oxidation. However, it is not as yet clearly understood exactly how microbial populations are structured in these spaces, and what governs their growth and propagation. As described elsewhere (e.g., Chaps. 3, 5–7), most work in this regard has focused on direct microbial attachment to mineral surfaces within exopolysaccharide layers. However, in

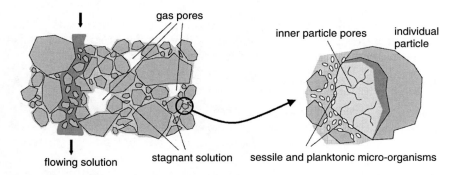

Fig. 8.2. A cluster of particles within a heap bed (*left*) and the microbial colonies inhabiting the moist pore space between particles (*right*)

heaps, the quantity of mineral grains exposed to the stagnant solution phase is comparatively small and prevails only in the initial stages of leaching. Furthermore, the micropore network within particles is likely to be too small to allow deep microbial penetration, forcing a fully indirect interaction between microbes and minerals.

8.2.6   Solution Chemistry

In addition to biologically catalyzed reactions, other abiotic chemical reactions can take place in the solution phase of a mineral heap, especially the hydrolysis of various dissolved ionic species. As the medium is sulfate-based and will contain ferric ions, the formation and possible precipitation of basic ferric sulfate minerals, such as jarosites, is an important reaction to be considered, not only as it affects the balance of ferric ions and acid required for sulfide oxidation, but also as the precipitates formed can alter the structural characteristics of the ore bed.

Finally, the smallest scale at which subprocesses of heap bioleaching need to be analyzed is that of the individual mineral grain. Here leaching is governed by the electrochemical interactions between the mineral grains and reagents in solution.

8.2.7   Mechanism of Mineral Leaching

Sulfide minerals generally leach by electrochemical interaction with ferric ions in sulfate solution, which are reduced to ferrous ions as the mineral is oxidized, either partially or completely, to metal ions and sulfur species. For example, chalcocite is oxidized to copper ions and elemental sulfur by a two-stage reaction, the net result of which is as follows:

$$Cu_2S + 4Fe^{3+} \rightarrow 2Cu^{2+} + 4Fe^{2+} + S^0. \qquad (8.1)$$

The reaction kinetics are primarily a function of temperature (as characterized by the activation energy) and the concentration of key reagents (such as ferric iron, ferrous iron and acid) in solution. The principal mechanisms of such reactions are reasonably well understood, except, perhaps, for chalcopyrite. Nonetheless, the exact values of the kinetic parameters remain subject to measurement in each case. Further complexities may arise when different minerals are galvanically coupled, particularly with pyrite, which is present in many base metal sulfide ores.

8.2.8   Grain Topology

As the rate of leaching is usually proportional the total mineral surface available to leaching, a further degree of complexity is added by the distribution of grains of different size and accessibility within the ore. This aspect is referred to as "mineral topology" and must be accounted for in a suitable description of mineral leach kinetics, for example, by using the shrinking core concept.

The different scales of the foregoing description are interlinked through the dynamics of the overall leach process. Figure 8.3 illustrates, in simplified form, the reaction network and associated mass exchanges found in heap bioleaching. This is described in the context of ferric leaching of chalcocite as described in Eq. 8.1 and the biocatalyzed oxidation of ferrous iron to ferric iron:

$$4Fe^{2+} + O_2 + 4H^+ \rightarrow 4Fe^{3+} + 2H_2O. \qquad (8.2)$$

For this reaction to proceed at any particular location in the heap, oxygen must be transferred from the gas phase into the solution phase. Associated with this is the transport of air at the heap scale as well as the gas–liquid mass transfer kinetics at the solution–gas interface. Acid, on the other hand, must be supplied with the solution fed to the heap and then transported to reaction sites by solution flow and interparticle diffusion. Given the reagent concentrations, microbial population characteristics and prevailing temperature, the oxidation reactions will proceed at a certain rate. The ferric iron generated will have to migrate by diffusion to mineral grains, which may be located deep inside a particle far from the location of the microbial oxidation. The mineral is now oxidized at a rate again depending on prevailing concentrations and temperatures. The released heat of reaction determines, in interaction with all global gas and liquid flow phenomena, the local temperature. Finally, the dissolved copper migrates through the particle and stagnant solution into the flowing solution to be transported out of the heap.

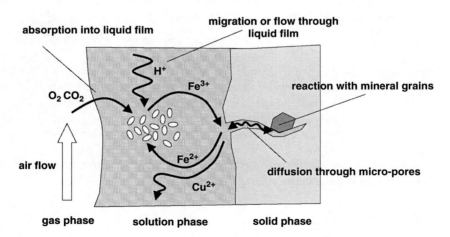

Fig. 8.3. The mineral biooxidation reaction-transport network between gas, liquid and solid phases in heaps

Thus, the overall rate of copper leaching from a given heap depends on a long and complex chain of subprocesses. In general, the slowest of these determines the overall rate. This is referred to as the rate-controlling step or subprocess. Thus, for example, if diffusion through large particles is the slowest step, then intraparticle diffusion is the rate-controlling step, which will determine the overall rate, regardless of the intrinsic rates of gas–liquid mass-transfer, biological oxidation or mineral dissolution.

The real complexity of this lies in the fact that different subprocesses can be rate-controlling at different locations in the heap at different times. For example, temperatures in a heap tend to be low near the surface, but higher in the center. Consequently, the kinetics of mineral dissolution, which are often a strong function of temperature, may limit the rate of leaching near the surface, but not in the center. Conversely, the rate of gas absorption is determined by the solubility of oxygen in the aqueous phase, which decreases with increasing temperature, and hence it may be this process that is rate-limiting in the center of the heap. Furthermore, this dynamic may shift as the heap heats up at the beginning or cools down near the end of its cycle.

It is therefore fair to say that the assessment of the optimal conditions for an efficient heap bioleach operation is not a trivial exercise and is impossible to accomplish on the basis of operator experience alone. It is the contention of the authors that this task requires the help of a comprehensive mathematical model which appreciates the complex dynamics and interactions that have been laid out in the preceding paragraphs.

8.3   Mathematical Modeling

Mathematical modeling of heap leaching dates back some 30 years to the early work of Roman et al. (1974) on modeling copper oxide dump leaching. Since then, new model formulations for heap and dump leach models, both biotic and abiotic, of increasing complexity have been presented. Models for acid rock drainage have also been developed, although in general without reference to the parallel efforts in heap leach modeling. A detailed review and comparison of some 15 heap and dump leach and acid mine drainage models was presented by Dixon (2003).

In essence one can distinguish between two classes of model approaches. One class focuses on particle-level phenomena with varying degrees of sophistication, while processes at the heap scale are taken as 1D plug flow, and mostly isothermal. Validation of these models has mostly occurred with data from column studies at various scales. The other class of models has focused on heap-scale processes, especially 2D gas and (to a lesser extent) solution flow in nonisothermal dumps, while using only simple reaction terms at the level of particles. Most of these models have been used in qualitative studies only, and have not been validated against full-scale heap or dump data.

These two classes approach the overall process of heap leaching from two directions, in each case making certain model assumptions with respect to the overall rate-limiting phenomena. For the particle-level models these are either intraparticle diffusion or reaction kinetics (biotic or abiotic), and for the heap-scale models mostly the availability of oxygen for sulfide mineral oxidation at some depth within the dump or heap.

The HeapSim model developed by the authors is by far the most comprehensive of the models reviewed by Dixon (2003), explicitly accounting for almost all of the subprocesses and effects discussed in the previous section. At the particle level, it accounts for mineral reaction kinetics, biooxidation kinetics, gas–liquid mass transfer and diffusion effects separately. At the heap scale it includes advective solution transport with diffusion into stagnant zones, flow of humid air and a comprehensive energy balance to determine temperature profiles. A limitation is that it remains a 1D model and thus cannot account for 2D or 3D gas advection and diffusion effects as they are likely to occur in unaerated heaps or dumps. Also, the 1D advective solution flow with stagnant zone diffusion is a somewhat simplistic approach to what is essentially a complex 2D or 3D phenomenon in real heaps. This is discussed in more detail later.

The mathematics of the HeapSim code have been described in some detail by Ogbonna et al. (2005). The following is a summary of the key mathematical aspects and their integration into the HeapSim modeling package.

8.3.1 Mineral Kinetics

The intrinsic kinetics of leaching are represented by an equation for the rate of mineral conversion X of the general form

$$\frac{dX}{dt} = k(T, d_0)\, f(C)\, (1-X)^\varphi, \tag{8.3}$$

where $k(T,d_0)$ is a rate constant which is a function of temperature (i.e., Arrhenius's law) and initial mineral grain size (or surface area), $f(C)$ is a function of solution composition, and $(1-X)^\varphi$ is a semi-empirical function of the fraction of unreacted mineral, which represents the changing topology of the mineral surface over the course of leaching. The "topological exponent" φ is equal to or greater than 2/3 (the shrinking sphere model) and may be as high as 3 when the distribution of the effective grain size is particularly wide (Dixon and Hendrix 1993). Work by Bouffard and Dixon (2002) has demonstrated the applicability of this simple empirical approach to describe the leaching kinetics of pyrite in various ore assemblages, and has shown further that the effect of the particle size distribution can essentially be represented through the parameter φ alone.

8.3.2 Microbial Kinetics

Microbial growth is represented by the following equation:

$$\frac{dc_x}{dt} = c_x k_g \{ f_g(T) [\Pi(k_e + 1) - k_e] - k_d f_d(T) \}, \tag{8.4}$$

where c_x represents the microbial biomass concentration, k_g the maximum growth constant, k_e the endogenous decay constant and k_d the death rate. $f_g(T)$ and $f_d(T)$ are functions that represent the responses of growth and death rates to temperature, respectively. The variable Π represents the product of growth/substrate limiting and inhibitory terms as follows:

$$\Pi = \prod_i \left(\frac{c_i}{K_i + c_i} \right) \prod_j \left(\frac{K_j}{K_j + c_j} \right), \tag{8.5}$$

where each counter i represents any limiting substrate, such as Fe^{2+}, H^+, S, O_2, and or CO_2, and each counter j represents any inhibitory factor, such as toxic ions, Fe^{3+} and total microbial biomass concentration.

Microbial oxidation kinetics is linked to microbial growth via the Pirt equation:

$$r_{ox} = c_x \left(\frac{k_g f_g(T) \Pi}{y_g} + k_m \right), \tag{8.6}$$

where y_g is the yield constant and k_m is the maintenance (nongrowth) coefficient.

The deportment of sessile (attached) to planktonic (free-swimming) microorganisms is represented through a Langmuir isotherm. These correlations can be formulated for any number of microbial species separately, thus enabling the separate modeling of Fe- and S-oxidizing microbes active under different solution conditions and temperatures.

8.3.3 Gas–Liquid Mass Transfer

The exchange of oxygen (and likewise CO_2) between the gas and the solution phase is accounted for by a simple mass transfer model analogous to Newton's law of cooling:

$$r_{o_2} = k_L a (c_{O_2}^* - c_{O_2}), \tag{8.7}$$

where $k_L a$ represents the gas–liquid mass transfer rate constant determined for given heap hydrodynamic conditions, and $c_{O_2}^*$ represents the solubility of oxygen in aqueous solution at a given temperature and partial pressure.

8.3.4   Diffusion Transport

Transport of dissolved species through a pore containing stagnant solution by Fickian diffusion is represented by the following differential equation in terms of the concentration of dissolved species i:

$$\frac{dc_i}{dt} = D_e\left(\frac{d^2c_i}{dr^2} + \frac{m_s}{r}\frac{dc_i}{dr}\right) + \frac{s_i}{\varepsilon_s}, \tag{8.8}$$

where D_e represents the effective diffusivity accounting for the tortuosity of the diffusion path, s_i is a source term accounting for the net generation of i by chemical reactions and ε_s is the mass ratio of solution to ore solids. The factor m_s represents a shape factor that describes whether the diffusion proceeds linearly ($m_s=0$) or radially in a cylindrical ($m_s=1$) or spherical ($m_s=2$) fashion. Equation 8.8 is also valid for the transport of microbial species, if an additional term is added on the LHS to account for microbial adsorption onto the solid phase.

8.3.5   The Combined Diffusion–Advection Model

In a departure from most other heap models, HeapSim approaches solution transport at the heap scale in a pseudo-2D fashion. The underlying assumption is that solution flow proceeds through the heap in discrete flow channels, transporting dissolved species primarily by advection (Fig. 8.4):

$$\frac{Dc_{f,i}}{Dz} = \frac{\partial c_{f,i}}{\partial t} + \frac{u_f}{\varepsilon_f}\frac{\partial c_{f,i}}{\partial z} = S_z, \tag{8.9}$$

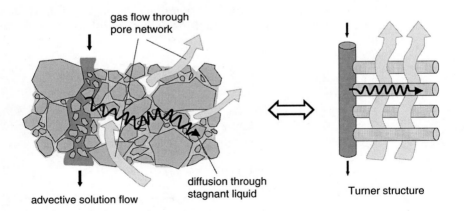

Fig. 8.4. Transport effects through the particle cluster (*left*) and their representation by a Turner structure (*right*)

where $c_{f,i}$ represents the concentration of species i in the flowing solution, u_f is the superficial velocity of the flowing solution, ε_f is the portion of the bed volume taken up by the flowing solution, S_z is the local source term and D/Dz denotes a material derivative. The remainder of the bed consists of a network of moist pores between particles (Fig. 8.4), but the pore solution is stagnant. Transport of dissolved species within these pores therefore proceeds exclusively by diffusion as described by Eq. 8.8.

Transport at the heap scale is thus a combination of the two domains, in a construct known as a Turner structure, shown in Fig. 8.4. In the present model this is represented by Eq. 8.8 providing the source term for Eq. 8.9 at every height level in the heap:

$$S_z = (m_s+1)\frac{D_e \varepsilon_s \rho_s}{RG_l}\frac{\partial c_i}{\partial r}\bigg|_{r=R}, \qquad (8.10)$$

where G_l represents the solution mass flux (irrigation rate) in the heap, R is the effective stagnant solution pore length, ε_s is the solution to dry ore mass ratio and ρ_s is the solution density. m_s is the shape factor as before, with the value $m_s=1$, representing a cylindrical column, most commonly used.

The present formulation of the heap-scale solution transport model may be conceptually simplistic, but has thus far proved well capable of describing full-scale heap bioleaching processes, as discussed further later. Nonetheless, it is acknowledged that it remains only an approximation.

A more rigorous model of water and solute transport would use Darcy's law to describe the flow of water through a variably unsaturated medium, along with constitutive relationships for capillary head and relative permeability as functions of effective saturation, such as the well-known van Genuchten and Mualem formulations. This would then provide the velocity field for a generalized advection–dispersion equation. Efforts are currently under way to incorporate such an approach into HeapSim.

8.3.6 Gas Transport

Transport of gaseous species would be described by a simple 1D advection equation. However, the partial pressure of oxygen, which drives the rate of oxygen gas absorption into the liquid phase, varies depending on the heap height and the local heap temperature. It may decrease significantly from local atmospheric levels as the air moves upward through the heap, owing to depletion, and (often more importantly) the sharp increase in absolute humidity (water vapor) with increasing heap temperature. These phenomena are captured by the following equation (Ogbonna et al. 2005):

$$\frac{\partial p_{O_2}}{\partial z} = \frac{\frac{P_g \rho_h \varepsilon_s r_{O_2}}{\rho_s G_a} - \frac{\psi'(T)}{M_v}\frac{\partial T}{\partial z}p_{O_2}}{\frac{1}{M_a}+\frac{\psi(T)}{M_v}}, \qquad (8.11)$$

where p_{O_2} is the partial pressure of oxygen, P_g is the total gas pressure, ε_s is the heap solution to ore mass ratio as before, r_{O_2} is the rate of oxygen absorption as represented by Eq. 8.7, ρ_h represents the heap packed density, ρ_s is the solution density, G_a is the dry air mass flux (aeration rate), $\psi(T)$ is the absolute local humidity, and the prime denotes differentiation with respect to temperature. M_a and M_v represent the molecular masses of dry air and water vapor, respectively.

8.3.7   Heat Balance

An independent 1D heat conservation model for heap leaching was developed by Dixon (2000), which is summarized in the following equation:

$$\overline{\rho C_p}\frac{\partial T}{\partial t} = k\frac{\partial^2 T}{\partial z^2} - [G_1 C_{p,1} f_1(T) - G_a C_{p,a} f_v(T)]\frac{\partial T}{\partial z} + \widehat{q}, \qquad (8.12)$$

where z represents the heap height variable, $\widehat{q}$ the volumetric rate of heat generation, $\overline{\rho C_p}$ is the average volumetric heat capacity; k is the thermal conductivity of the solid matrix, G_1 and G_a are the mass fluxes of water and dry air, and $C_{p,1}$ and $C_{p,a}$ are their respective heat capacities. The functional $f_1(T)$ incorporates boundary effects such as liquid-phase evaporation at the heap surface and the functional $f_v(T)$ combines effects of latent and sensible heat carried in water vapor in the gas phase. The influences of solar radiation and diurnal temperature cycles are incorporated through the boundary conditions to Eq. 8.12.

The heat-generation term $\widehat{q}$ is determined by the rates of reactions as determined by Eq. 8.3, acknowledging that the only substantial heat of reaction is generated from those reactions which reduce oxygen, such as the oxidation of ferrous iron to ferric iron (regardless of whether facilitated by bacteria or not) and of sulfur to higher oxidation states (usually facilitated by bacteria). This rate of heat generation corresponds to roughly 100 kJ mol^{-1} electrons transferred (i.e., 400 kJ mol^{-1} of O_2 reduced), irrespective of what this electron eventually oxidizes (Petersen and Dixon 2002).

8.3.8   The HeapSim Package

The detailed numerical algorithm of solving the mathematical models presented here is beyond the scope of this chapter. In essence, the heap is divided into vertical segments and a number of horizontal" nodes at each height level. Stepping down the heap at each time interval, the algorithm solves the chemical and biological kinetics equations at each node to provide the source term for the diffusion equation (Eq. 8.8) at each level. The equation is solved, providing the boundary condition for the next level through Eq. 8.10. This process is repeated for each height level, delivering the effluent concentrations of all dissolved species. Heat generation at each node is calculated from the progress of all chemical reactions and Eq. 8.12 is then integrated over the heap to give a new temperature profile. Finally, the gas-transport equation is solved before the next time step to provide the new oxygen profile through Eq. 8.11.

The HeapSim modeling package has been set up such that it interacts with an Excel™ spreadsheet. All parameters are defined in a user interface and all modeling results are reported in a multitude of graphs representing the changes in space and time of all relevant variables. Additional features include the definition of time-dependent parameter changes (such as changes in feed concentrations and flow rates) and the direct comparison of model outputs with experimental or operational data. This is especially useful during model calibration with laboratory data.

8.4 Application of Mathematical Modeling – from Laboratory to Heap

Any model is only as good as the data that it is fed, and is only as good as the assumptions that are made in its formulation. In general, it could be claimed that by increasing the level of detail in the modeled subprocesses, the number of model assumptions decreases, but the number of required model input parameters increases. Regardless of the balance between the two, any specific model application needs to be preceded by a systematic assessment of the scenario to be modeled, the model to be chosen and the parameters to be determined.

8.4.1 Model Parameters

The HeapSim model is the most comprehensive of all existing models, explicitly accounting for all potential subprocesses with a minimum of predetermined assumptions regarding which subprocess is rate limiting and which is not. However, this level of model detail requires a larger set of parameters than models with built-in simplifications.

Model parameters fall into essentially three categories:

1. *System parameters*, which describe the operational features of the particular system to be modeled, such as heap height, irrigation rate, aeration rate, particle size distribution and geographical information. At the operational level many of these parameters can be changed to effect modifications or control for improved heap performance.
2. *Physical and chemical parameters*, which are all constants and quantities that remain valid regardless of the specific conditions in a heap, such as diffusivities, chemical equilibrium constants, thermal conductivities and

thermodynamic properties. These are effectively built into the model and do not need to be reassessed for any specific application.

3. *Specific parameters*, which constitute the parameters that need to be assessed for specific intrinsic conditions, relating to
   - Ore characteristics, such as particle size distribution, mineral composition and grades
   - Mineral characteristics, such as kinetic constant for mineral dissolution
   - Biological characteristics, such as growth and yield constants for the relevant microbial strains

The set of specific parameters needs to be determined through systematic laboratory study involving a specific ore, concentrates of the relevant minerals and cultures of the relevant microorganisms. For ore characterization, standard assaying techniques can be applied to determine the particle size distribution and the overall grades in terms of Cu, Fe and S. Increased level of detail, such as mineralogical analysis, or relating grade distribution to particle size distribution, is useful for increased model accuracy.

The kinetic characteristics of the dissolution of individual minerals need not be determined in each case, as these are reported widely in the literature; however, a wide range of parameter values (e.g., activation energies), are commonly reported, and it is recommended to reassess these for each specific ore type. Techniques to do so revolve around constant-potential leach tests for sulfide minerals and acid-consumption tests for oxide minerals. Details of these are beyond the scope of this chapter. A more in-depth discussion of such tests can be found, for example, in Bolorunduro and Dixon (2006) or Bouffard et al. (2006).

Likewise, the growth and oxidation kinetics characteristics of specific microbial strains have been studied in pure cultures, and are available in the literature, although large gaps still remain. It is nevertheless useful to confirm the validity of these parameters in the context of the ore to be leached, and in fact to ensure the microbial strains considered do indeed thrive on the ore under given conditions.

8.4.2   Model Calibration and Laboratory-Scale Validation

A common technique for assessing heap leach characteristics of an ore at the laboratory scale is leaching in narrow-bore columns tests, with column diameters characteristically ranging from 10 to 30 cm and column lengths from 0.5 to 6 m. Typically these are operated at conditions similar to those expected in the heap, often including a solvent extraction/electrowinning step to remove metals from solution, with raffinate recycle.

If it can be assumed that solution flow through a column essentially resembles that through an unconfined heap bed, then the results from such tests should be fairly representative of heap performance and can hence serve as an excellent test case for model validation. As will be shown later, this assumption is not always valid and has been one of the main reasons for the limited success of heap modeling at the industrial scale. However, even if the assumption is valid, a column experiment represents the comprehensive interplay of all the heap subprocesses described earlier. Hence, modeling of a column test is not meaningful in the absence of the set of specific parameters pertaining to the ore, minerals and microorganisms used in the specific case. It is surprising, therefore, that often only very limited data are available in this regard.

In principle, any model can be set up to simulate a column experiment if the system parameters are set to the specific column conditions. If all specific parameters are predetermined to sufficient detail, the simulation should produce a reasonably good correlation with experimental data. If not, the discrepancies can be used to investigate model shortfalls (assuming that the experimental data are always correct). These may be incorrect model assumptions or incorrect model parameter values, especially where these have been estimated from generic data rather than measured for the specific case. Correcting estimates or remeasuring parameters, and thus improving on model prediction of column data, is an elaborate iterative process, which has been presented elsewhere (Dixon and Petersen 2004; Petersen and Dixon 2006). Done in this rigorous manner, it is the correct approach to model calibration.

Once calibration has been achieved, the model can be tested against data from other column experiments that are operated under slightly different conditions (e.g. irrigation rate, temperature or feed composition). Changing the relevant system parameters in the model should instantly result in predictions of similar quality to those of the initial calibration. If this is successful, the model has been fully validated at this scale, but if not, further loops of the iterative calibration process may be required.

8.4.3   Extending to Full Scale – Model Applications

As indicated previously, there is a remarkable absence of comprehensive full-scale data from heap leach operations other than production figures. This makes validation of models at the full scale rather difficult. Furthermore, full-scale heaps, which are often constructed in lifts and exhibit varying ore grades and crush sizes, and which pool the pregnant solution collected over a wide area into a few large pipes and basins, do not lend themselves to simple simulation as 1D beds in the way one would model laboratory columns. So, even if effluent solution data exist, it would represent the mixture of a large number of such columns of different ages with different materials, operated under different conditions.

For these reasons, up to now full-scale modeling has been restricted to a few special cases, some of which are detailed in subsequent sections. As will become clear, extending models from the level of laboratory column or even pilot plant to the full scale requires a change not only of the relevant system parameters, but also of two key aspects relating to model assumptions: solution flow in unconfined beds, which in general is not uniform and 1D, and temperature profiles in large heaps, which are not isothermal as are columns.

If these departures are accounted for, a state-of-the-art model, such as HeapSim, can provide a powerful tool for analysis of heap operations, even within the limited scope within which it has been validated. The model can be used to diagnose weaknesses of any given system and to find optimized operating conditions, under which production is substantially enhanced. Examples of this are illustrated in the following case studies.

8.5   Case Study I – Chalcocite

Chalcocite is oxidized in ferric sulfate medium in a two-step mechanism with very different kinetics, as mentioned before. The first step involves the formation of a pseudo-covellite intermediate sulfide product, also known as *blaubleibender*,

$$Cu_2S + 1.6Fe^{3+} \rightarrow 0.8Cu^{2+} + 1.6Fe^{2+} + Cu_{1.2}S, \qquad (8.13)$$

which is subsequently oxidized to elemental sulfur according to

$$Cu_{1.2}S + 2.4Fe^{3+} \rightarrow 1.2Cu^{2+} + 2.4Fe^{2+} + S^0. \qquad (8.14)$$

The rate of the first stage is very rapid, being controlled by the solid-state diffusion of copper in the sulfide lattice and the mass transport of ferric ions to the mineral surface, and therefore exhibits a low activation energy (Bolorunduro and Dixon 2006). The second stage is much slower, controlled by the rate of charge transfer in the anodic decomposition process, and has an extremely high activation energy ($80–100$ kJ mol^{-1}). In the ore under investigation, chalcocite was the only copper sulfide mineral, with brochantite [$Cu_4(OH)_6SO_4$] as the only other cupriferous oxide mineral. Pyrite occurred in minor quantities.

Extensive testwork was undertaken in 0.5-m and 6-m laboratory-scale columns. The following observations are of note:

- The acid-soluble fraction of copper was leached from the column rapidly.
- The sulfide fraction leached slowly initially, but then with increasing speed, leveling only near the end of the leach period.
- The solution redox potential was low initially, but increased rapidly after a few days of leaching. This effect propagated along the length of the column like a wave.
- The breakthrough of the sudden increase in the redox potential corresponded to a peak in pH and the breakthrough of a similar increase in bacterial numbers.
- Experiments with a column operated with a bacterial culture obtained from the production site in the production raffinate (as opposed to laboratory

cultures in solution containing only acid and iron) yielded significantly retarded leaching of the chalcocite phase. Shake-flask experiments of these cultures revealed serious microbial retardation owing to very high levels of Mg and Al in the production raffinate.

- A 6-m laboratory column operated under conditions similar to those for the actual heap with the same material (but with a laboratory culture in artificial feed solution) yielded 90% copper extraction within 120 days of leaching, while the actual heap yielded an estimated 75–80% copper extraction within 14 months.

The mechanism of chalcocite bioleaching in narrow-bore columns has been described by the authors (Petersen and Dixon 2003). Chalcocite stage I leaching proceeds so rapidly that any ferric iron produced by bacterial oxidation is consumed as fast as it can be produced. According to Eq. 8.2, acid is required to achieve ferrous iron oxidation. At the rapid consumption rate, the system becomes depleted in acid, the only source of which is the feed solution; hence chalcocite leaching proceeds in a narrow front which migrates down the column. Within this front chalcocite is converted by a thriving bacterial population, which is consuming all available acid. Ahead of this zone there is a shortage of acid (i.e., high pH) and oxidation cannot proceed. In its wake, stage II leaching proceeds much more slowly, resulting in much higher residual levels of ferric iron, and hence, higher solution potentials. The product elemental sulfur can potentially be oxidized by sulfur-oxidizing microorganisms, resulting in the in situ production of acid. Thus the system is limited by acid delivery during stage I and by the slow mineral kinetics of the *blaubleibender* in stage II.

The column run with native bacteria in production raffinate, on the other hand, was limited not by acid delivery but by the severe retardation of microbial oxidation kinetics caused by the extremely high concentrations of Mg and Al ions in the production solution, which in turn originated from the acid leaching of gangue minerals (Petersen and Dixon 2004). It was shown that reduction by half of the concentration of these ions dramatically enhanced microbial kinetics; however, during chalcocite stage II leaching the slow mineral kinetics remained rate-limiting and no significant difference in rate as compared with those for the column run with the laboratory culture was observed.

A modeling exercise involving HeapSim was conducted. In the initial stage, the model was calibrated using ore assay information and literature values for mineral and microbial kinetics against a set of data from one laboratory column. The fits achieved were reasonable from the outset, but could be substantially improved by fine-tuning some of the estimated (literature) data, especially with respect to the effect on microbial kinetics of pH and ferric ion concentration (i.e., the relevant parameters K_i and K_j in Eq. 8.5). This calibration exercise is described in more detail elsewhere (Dixon and Petersen 2003). The calibrated model was then used to predict the results from the 6-m column, adjusting only for the relevant system parameters. The immediate fit of data, as shown in Fig. 8.5, is remarkable and was seen as validation of the approach.

The same model was also used to simulate the column operated with the native culture, again adjusting only for system parameters as well as the kinetics constant k_g of the microbial growth and oxidation model (Eq. 8.4) to mimic retardation of the high ionic strength solution. Again, the result was a remarkably good fit, confirming that retarded growth was only limiting the initial chalcocite stage I phase, but not the stage II phase (Petersen and Dixon 2004).

An attempt was then made to use the model to simulate conditions in the heap. For this the temperature model was changed from isothermal to autothermal (i.e., moving from constant temperature to the heat model of Eq. 8.12) and the system parameters were adjusted. This simulation suggested the heap would heat up fairly rapidly and leach to completion even faster than the laboratory-column experiment. It was then discovered (during a field trip to the mine site) that the common practice of irrigating via drip emitters, at the time on a fairly large grid, resulted in solution flow along preferential channels, leaving large portions of the bed moistened by only capillary action, and not exposed to flowing solution (Figs. 8.2, 8.6). This was subsequently accounted for in the model by changing the pore length in the diffusion model (parameter R in Eq. 8.10) from 4 cm representing effective radius of the laboratory column to 35 cm, which was the dripper half spacing, but retaining all other parameters. The result was remarkable, as it suggested 80% copper extraction after 420 days (14 months; Fig. 8.5) and only marginal heating of the heap, closely approximating what was experienced in the industrial operation.

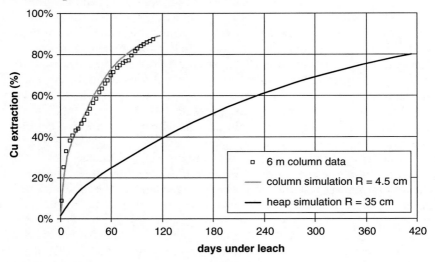

Fig. 8.5. Experimental data and simulation of a chalcocite column leach experiment and simulation of the corresponding heap under widely spaced drip emitters

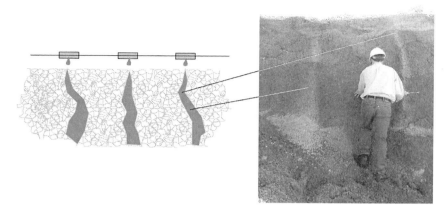

Fig. 8.6. Likely solution flow patterns in beds under drip irrigation. Periodic vertical patterns found on the cut face of a reclaimed heap support this interpretation. (Photograph courtesy of Amado Guzman, first published in Petersen and Dixon 2003)

This result needs to be discussed in more detail. Acid supply remains the rate-limiting step, even in the full-scale heap. Different from columns, however, acid is delivered from the flowing solution channels sideways into the stagnant solution, primarily by diffusion. In addition, the acid diffusing into the stagnant pores tends to interact with gangue minerals faster than with the biooxidation reactions, where it is required, and hence is wasted. Consequently, concentrations of soluble Al and Mg in this system are constantly increasing, resulting in increasing retardation of microbial oxidation.

8.6   Case Study II – Sphalerite and Pyrite

The HydroZinc^{TM} process developed by Teck Cominco Metals Ltd. uses heap bioleaching of zinc from sphalerite ores and subsequent recovery of zinc by solvent extraction/electrowinning. Prior to the solvent extraction step, removal of iron is achieved by coprecipitation with gypsum upon lime addition (Lizama et al. 2003). A HeapSim study was conducted in order to correlate laboratory columns with a pilot heap study and subsequently to gauge optimal operating conditions for full-scale operation. This work is summarized with some detail in Dixon and Petersen (2004) and Petersen and Dixon (2006).

The model calibration again follows the steps described before, with fitting of selected specific parameters to match simulation with data from a laboratory column. The calibrated model is then taken to predict the pilot scale data, adjusting only for system parameters such as heap height and irrigation rate. From the understanding gained from the chalcocite work, it was also realized that diffusion pore length is closely linked to dripper spacing, and hence should effectively be considered as a system, rather than specific, parameter.

In this instance the ore characteristic parameters, especially mineral grades, were significantly different from those of the ore used in the laboratory column, against which the model was calibrated, and needed to be adjusted. Nevertheless, the model fit of data from the pilot heap was reasonably good without any further parameter adjustments being needed (Fig. 8.7).

A closer analysis of the results indicated the following key aspects:

- Sphalerite and pyrite compete for ferric iron, with the apportionment directly linked to their relative rates of leaching. Thus, sphalerite leaches more rapidly at the beginning, but leaching of sphalerite steadily decelerates upon depletion in favor of pyrite. This is important for the downstream process in which dissolved iron requires removal through lime addition, such that there may be a point beyond which the leaching of iron renders further leaching uneconomic.
- The generation of ferric iron is initially limited by the supply of acid into the heap, at least until the rate of pyrite leaching (which is acid-generating) becomes significant relative to that of sphalerite (which is acid-consuming).
- Heat generation is substantial and temperatures as high as 60°C have been observed in the pilot heap. The model captures this well, but predicts that the maximum temperature is directly linked to the maximum temperature tolerated by the microorganisms. Where this temperature is reached in the heap, microbial activity declines as do the heat-generating oxidation reactions. As a result, the model predicts that a significant portion of the

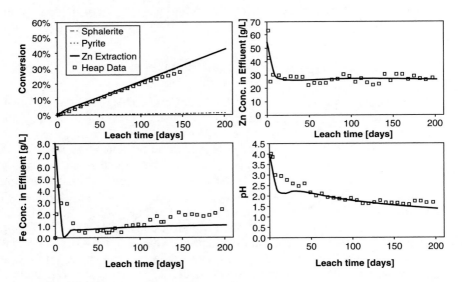

Fig. 8.7. Comparison of model simulation (*lines*) with sphalerite pilot heap data (*squares*) in terms of Zn extraction, Zn and Fe concentrations and pH in the heap effluent

heap may experience reduced leaching because of overheating (Fig. 8.8) and thus the microbial temperature tolerance becomes the rate-limiting process in this phase.

The optimization study was therefore aimed at ensuring rapid delivery of acid into all zones of the heap, while maintaining operating conditions so as to maintain a favorable temperature profile throughout the heap. Acid distribution is improved by low heap height, relatively high feed acid concentration and tight dripper spacing, whereas heat management is best in low heaps at moderate irrigation and high aeration rates. It should be noted that internal acid generation improves with time as pyrite leaching becomes dominant and hence the initial acid feed concentration may be reduced at a later stage. The recommendations for temperature management only apply for heaps with high internal heat-generation rates and may in fact be quite different for those with low internal heat generation, such as chalcocite heaps. Where heat management is critical, a simple temperature-control system – using aeration and irrigation rates as the manipulated variables – is feasible in principle (Dixon 2000; Petersen and Dixon 2002).

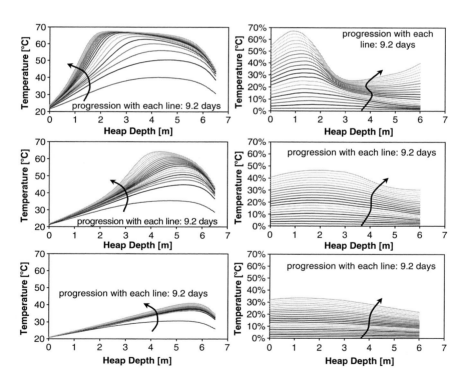

Fig. 8.8. Temperature profiles and ZnS conversion profiles with depth for operating conditions causing local overheating (*top*), ideal heat distribution (*middle*) and excessive cooling (*bottom*)

8.7   The Route Forward – Chalcopyrite

Low-grade chalcopyrite represents the largest resource of copper throughout the world. Conventional concentration by comminution and flotation for further processing (through either pyrometallurgical or hydrometallurgical routes) has an economic cutoff grade (usually around 0.6% Cu), which disqualifies many ore bodies. Chalcopyrite has shown considerable resistance to hydrometallurgical treatment under atmospheric conditions, but appears amenable to thermophile bioleaching, as applied in the BioCop™ process (Batty and Rorke 2005). It would thus appear to be an extremely attractive option to treat low-grade chalcopyrite ores in a thermophile heap bioleach process.

Petersen and Dixon (2002) demonstrated that bioleaching of a chalcopyrite concentrate in the heaplike Geocoat™ process (Chap. 5) is well possible at the laboratory scale, and modeling suggests thermophilic temperatures (above 60°C) are easily achievable in full-scale heaps. However, the Geocoat™ process effectively creates a high-grade ore with all mineral grains situated on the particle surface, and thus cannot immediately be used to predict the feasibility of leaching from a low-grade material with grains embedded within an inert rock matrix.

The hydrometallurgy and biohydrometallurgy of chalcopyrite leaching are not fully understood and continue to be the subjects of numerous investigations (a selection of key papers includes those by Kametami and Aoki 1985; Dutrizac 1989; Hackl et al. 1995; Stott et al. 2000; Third et al. 2000; Hiroyoshi et al. 2004). The location of microbes relative to the mineral grains and the presence of competing, less refractory minerals (especially pyrite) may in fact be key to the success of chalcopyrite leaching in a heap context (Petersen and Dixon 2005). Only with a clearer understanding of these interactions at the micro (grain) scale and the meso (particle/agglomerate) scale, and a suitable model description thereof, can the development of a low-grade chalcopyrite heap bioleach process be developed further.

8.8   Conclusions

The complex physical, chemical and biological interactions in the context of heap bioleaching have been gradually unraveled over the years, and development of systematic modeling tools, such as the HeapSim model presented here, have been ongoing in parallel with this process. In the experience of the authors, limited benefit can be derived from mathematical modeling in the absence of experimental study, both at laboratory and at operational scales. Model development must go hand in hand with model calibration and validation against measured data. It is this interaction which has historically been the weak link in the acceptance of modeling work into practice in this particular industry.

Many insights that have been gained through a rigorous pursuit of this systematic approach. Some of these appear to have had an impact on industrial practice; for example, the spacing of drip emitters in heap irrigation systems is now typically 0.5 m or less, whereas it used to be as much as 1 m. Other aspects, especially biological health within heap systems, are beginning to be taken more seriously as factors that can potentially limit the efficiency of an operation.

However, there remains a wide scope for further implementation of insights gained from systematic modeling and assessment studies. Owing to the large number of parameters that influence the behavior of a given heap scenario, it is unlikely that there will ever be a "generic" heap leach process. Rather, each existing or new operation should be assessed individually, at least in part, so that for each case an optimized operating window can be established. Modeling would be an integral part of this approach, and should therefore be embraced as an important tool for the development of industrial heap bioleaching practice.

References

Batty JD, Rorke GV (2005) Development and commercial demonstration of the BioCop™ thermophile process. In: Harrison STL, Rawlings DE, Petersen J (eds) Proceedings of the 16th international biohydrometallurgy symposium (IBS 2005), Cape Town, pp 153–162

Bolorunduro SA, Dixon DG (2006) An electrochemical model for the leaching of chalcocite by ferric sulfate. Hydrometallurgy (in press)

Bouffard SC, Dixon DG (2002) On the rate-limiting steps of pyritic refractory gold ores heap leaching. Miner Eng 15:859–870

Bouffard, SC, Rivera-Vasquez BF, Dixon DG (2006) Leaching kinetics and stoichiometry of pyrite oxidation from a pyrite-marcasite concentrate in acid ferric sulfate media. Hydrometallurgy (in press)

Dixon DG (2000) Analysis of heat conservation in copper sulfide heap leaching. Hydrometallurgy 58:27–41

Dixon DG (2003) Heap leach modeling – the current state of the art. In: Young C, Alfantazi A, Anderson C, James A, Dreisinger DB, Harris B (eds) Hydrometallurgy 2003, proceedings of the 5th international symposium honoring Professor Ian M. Ritchie, vol 1: leaching and solution purification. The Minerals, Metals and Materials Society, Warrendale, pp 289–314

Dixon DG, Hendrix JL (1993) Theoretical basis for variable order assumption in the kinetics of leaching of discrete grains. AIChE J 39:904–907

Dixon DG, Petersen J (2003) Comprehensive modelling study of chalcocite column and heap bioleaching. In: Riveros PA, Dixon DG, Dreisinger DB, Menacho J (eds) Hydrometallurgy of copper, proceedings of Copper 2003, vol 6. CIM-MetSoc, Montreal, pp 493–516

Dixon DG, Petersen J (2004) Modeling the dynamics of heap bioleaching for process improvement and innovation. In: Hydro-Sulfides 2004, proceedings of the international colloquium on hydrometallurgical processing of copper sulfides, pp 13–45

Dutrizac JE (1989) Elemental sulfur formation during the ferric sulfate leaching of chalcopyrite. Can Metall Q 28:337–344

Hackl RP, Dreisinger DB, Peters E, King JA (1995) Passivation of chalcopyrite during oxidative leaching in sulfate media. Hydrometallurgy 39:25–48

Hiroyoshi N, Kuroiwa S, Miki H, Tsunekawa M, Hirajima T (2004) Synergistic effect of cupric and ferrous ions on active-passive behavior in anodic dissolution of chalcopyrite in sulfuric acid solutions. Hydrometallurgy 74:103–116

Kametami H, Aoki A (1985) Effect of suspension potential on the oxidation rate of copper concentrate in a sulfuric acid solution. Metall Trans B 16:695–705

Lizama HM, Harlamovs JR, Belanger S, Brienne SH (2003) The Teck Cominco Hydrozinc™ process. In: Young C, Alfantazi A, Anderson C, James A, Dreisinger DB, Harris B (eds) Hydrometallurgy 2003, proceedings of the 5th international symposium honoring Professor Ian M. Ritchie, vol 2: electrometallurgy and environmental hydrometallurgy. The Minerals, Metals and Materials Society, Warrendale, pp 1503–1516

Ogbonna N, Petersen J, Dixon DG (2005) HeapSim – unravelling the mathematics of heap bioleaching. In: Dry M, Dixon DG (eds) Computational analysis in hydrometallurgy, 35th annual hydrometallurgy meeting. CIM-MetSoc, Montreal, pp 225–240

Petersen J, Dixon DG (2002) Thermophilic heap leaching of a chalcopyrite concentrate. Miner Eng 15:777–785

Petersen J, Dixon DG (2003) The dynamics of chalcocite heap bioleaching. In: Young C, Alfantazi A, Anderson C, James A, Dreisinger DB, Harris B (eds) Hydrometallurgy 2003, proceedings of the 5th international symposium honoring Professor Ian M. Ritchie, vol 1: leaching and solution purification. The Minerals, Metals and Materials Society, Warrendale, pp 351–364

Petersen J, Dixon DG (2004) Bacterial growth and propagation in chalcocite heap bioleach scenarios. In: Tzesos M, Hatzikioseyan A, Remoundaki E (eds) Biohydrometallurgy – a sustainable technology in evolution, IBS 2003. National Technical University of Athens, pp 65–74

Petersen J, Dixon DG (2005) Competitive bioleaching of pyrite and chalcopyrite. In: Harrison STL, Rawlings DE, Petersen J (eds) Proceedings of the 16th international biohydrometallurgy symposium (IBS 2005), Cape Town, pp 55–64

Petersen J, Dixon DG (2006) Modelling and optimisation of zinc heap bioleaching. Hydrometallurgy (in press)

Roman RJ, Benner BR, Becker GW (1974) Diffusion model for heap leaching and its application to scale-up. Trans SME-AIME 256:247–252

Stott MB, Watling HR, Franzmann PD, Sutton DC (2000) The role of iron-hydroxy precipitates in the passivation of chalcopyrite during bioleaching. Miner Eng 13:1117–1127

Third KA, Cord-Ruwisch R, Watling HR (2000) The role of iron-oxidising bacteria in stimulation or inhibition of chalcopyrite bioleaching. Hydrometallurgy 57:225–233

9 Relevance of Cell Physiology and Genetic Adaptability of Biomining Microorganisms to Industrial Processes

Douglas E. Rawlings

9.1 Introduction

The use of microbially based processes in the recovery of metals from ores is now well established and examples of most of these processes are described in this book. The biodiversity of microorganisms that drive these processes is described in Chap. 10 and details of our understanding of the biochemistry and genetics of iron and sulfur oxidation are described in Chap. 14. In this chapter, other properties of microorganisms involved in metal extraction processes as well as the characteristics of heap and tank leaching processes that have made biological processes for mineral processing as robust and efficient as they have proved to be are described.

9.2 Biooxidation of Minerals Is a Marriage Between Chemistry and Biology

There has been a long-standing debate as to the relative contributions of chemistry and biology to the biooxidation of mineral ores. Current understanding is that biooxidation is primarily a chemical process in which ferric iron and protons are responsible for the leaching reactions depending on the type of mineral (Sect. 9.3). The effect of this is that if a mineral (such as chalcopyrite) is recalcitrant to biooxidation at 40°C, the solution to the problem lies primarily in the realm of chemistry rather than biology. For example, an increase in the temperature of the biooxidation process to 80°C will allow the chemical reactions to take place at a much faster rate. However, although the solution to a slow reaction rate is based on chemical considerations, the biology needs to follow the chemistry and microorganisms that are capable of iron and sulfur oxidation at 80°C are required to generate or regenerate the lixiviants (leaching chemicals). With most minerals, separation of the chemistry from the biology into two different processes is unlikely to be as efficient as when both processes are carried out together.

The reason for codependency of chemistry and biology is that the role of the microorganisms is considered to be more than the generation of the ferric iron and acid required for the chemistry of mineral biooxidation.

Biomining
(ed. by Douglas E. Rawlings and D. Barrie Johnson)
© Springer-Verlag Berlin Heidelberg 2007

Contact between the microbes and the mineral is important (though not essential) because the microorganisms produce an exopolysaccharide layer when attached to a mineral (Sand et al. 1995). This serves as a reaction space where mineral dissolution reactions take place more efficiently than in the bulk solution (Fig. 9.1). This increased efficiency is partly because conditions within the exopolysaccharide layer (e.g., concentration of ferric iron, redox potential and pH) are substantially different compared with the those in the bulk solution (Chap. 13).

9.3   General Chemistry of Mineral Biooxidation

Sand and coworkers (Schippers and Sand 1999; Rohwerder et al. 2003) have proposed that from a chemistry of biooxidation perspective, minerals can be divided into two broad categories, acid-insoluble minerals that are solubilized through oxidation by ferric iron (e.g., FeS_2, MoS_2 and WS_2) and acid-soluble minerals that are oxidized by the combined action of ferric iron and protons (e.g., ZnS, PbS, FeAsS, $CuFeS_2$ and MnS_2). In acid-insoluble minerals, the chemical bonds between sulfur and the metal break only after a series of six one-electron removal steps in such a way that thiosulfate is the first free sulfur compound. An example of pyrite oxidation is as follows:

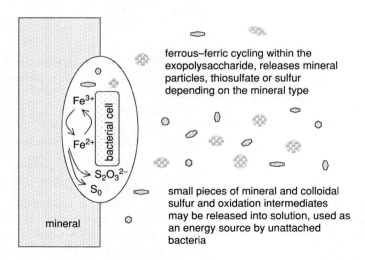

Fig. 9.1. When attached to a mineral, cells produce copious quantities of capsular exopolysaccharide material which serves as a reaction space where mineral leaching takes place most efficiently. The way in which the mineral is degraded depends on the type of mineral and the type of microorganism doing the leaching. (Based on models of Sand et al. 1995 and Tributsch 2001)

$$FeS_2 + 6Fe^{3+} + 3H_2O \rightarrow S_2O_3^{2-} + 7Fe^{2+} + 6H^+. \tag{9.1}$$

Thiosulfate is unstable in acidic liquors and is oxidized by ferric iron, and by sulfur-oxidizing bacteria via tetrathionate and other polythionates to sulfate:

$$S_2O_3^{2-} + 8Fe^{3+} + 5H_2O \rightarrow 2SO_4^{2-} + 8Fe^{2+} + 10H^+. \tag{9.2}$$

In the absence of sulfur-oxidizing microorganisms, rearrangement of the thiosulfate molecules through a variety of intermediates may take place and elemental sulfur may accumulate:

$$8S_2O_3^{2-} \rightarrow S_8 + 8SO_3^{2-}. \tag{9.3}$$

The ferrous iron can be regenerated by iron-oxidizing microorganisms to ferric iron:

$$2Fe^{2+} + 0.5O_2 + 2H^+ \xrightarrow{\text{Iron-oxidizing acidophiles}} 2Fe^{3+} + H_2O. \tag{9.4}$$

With acid soluble minerals the chemical bonds between the metal and sulfur are broken by protons such that after two protons have bound, hydrogen sulfide is released. When ferric iron is present, the sulfur moiety is likely to be oxidized simultaneously and goes though a number of spontaneous rearrangement and oxidation steps via higher polysulfides to elemental sulfur. This is known as the "polysulfide" mechanism. In the absence of microorganisms, chemical oxidation results in more than 90% of the sulfide ending up in the form of elemental sulfur (Schippers and Sand 1999):

$$ZnS + Fe^{3+} + H^+ \rightarrow Zn^{2+} + 0.5H_2S_n + Fe^{2+} \ (n \geq 2), \tag{9.5}$$

$$H_2S_n + 2Fe^{3+} \rightarrow 0.25S_8 + 2Fe^{2+} + 2H^+. \tag{9.6}$$

In the presence of microorganisms the sulfur will be oxidized to sulfuric acid (Eq. 9.7). The polysulfide mechanism can also occur in the absence of ferric iron as a result of the acid produced by sulfur-oxidizing microbes:

$$0.25S_8 + 3O_2 + 2H_2O \xrightarrow{\text{Sulfer-oxidizing acidophiles}} 2SO_4^{2-} + 4H^+. \tag{9.7}$$

9.4 Advantages of Mineral Biooxidation Processes Compared with Many Other Microbe-Dependent Processes

The use of microorganisms for the extraction of metals from mineral ores and concentrates has several substantial advantages that apply to relatively few other industrial processes in which microbes serve as the catalyst. Three of these are described in the following subsections.

9.4.1 There Is a Huge Variety of Iron- and Sulfur-Oxidizing Microorganisms That Are Potentially Useful for Industrial Metal Extraction Processes

The use of reduced iron and sulfur compounds as a source of energy is a natural process that is thought to have been taking place since soon after life on earth was established. There is a creditable theory that the oxidative formation of pyrite (FeS_2) from ferrous sulfide (FeS) and hydrogen sulfide (H_2S or HS^-) was the early source of energy for the fixation of carbon dioxide when life on earth began (Wächtershäuser 1990):

$$FeS+H_2S \rightarrow FeS_2+H_2(\Delta G^\circ = -38.4 \text{ kJ mol}^{-1} \text{ at pH 0 and 25°C}).$$

This reaction is exergonic and in the acidic pH range a temperature increase to 125°C renders the reaction even more exergonic. "Pyrite-pulled reactions" for carbon dioxide fixation are highly exergonic even at the low concentrations of CO_2 (5.7 mmol L^{-1}) and H_2S (7.5 mmol L^{-1}) typical of present day hydrothermal vent waters. An advantage of pyrite formation serving as the aboriginal energy source is that it does not require the necessity for complex molecules such as bacteriochlorophylls or energy couplers, which are themselves seen to be the products of evolution.

There is a general belief that high-temperature, anaerobic, acidic environments were widespread during the early stages of life on earth. The earliest organisms are envisaged as arising in this environment and acquiring their energy by continuously forming and dumping pyrite. Since then, some of these environments have cooled, while others have become more aerobic, giving rise to a wide range of iron- and sulfur-containing environments of different temperatures, degrees of acidity and exposure to oxygen. The occurrence of these different environments has led to the evolution of a variety of mechanisms by which energy may be obtained from reduced iron and sulfur compounds (Chap. 14) plus a huge diversity of microorganisms able to carry out these reactions (Chap. 10). Therefore, should it be advantageous for physical–chemical reasons to carry out a particular process at a certain temperature, pH or redox potential, there is likely to be suitable microbial consortium for the development of the process. If the available microbes do not fulfill the required criteria, there exists the possibility of finding a natural environment somewhere on earth that will serve as a potential source of more suitable microorganisms. Alternately one could attempt to use the most suitable microorganisms available and attempt to "improve" these by placing a selection pressure on the population in such a way that those that cope better with the conditions required for an industrial process are selected (Sect. 9.4.3). There are not many other industrial processes that afford such a large variety of potentially useful microorganisms as mineral biooxidation.

9.4.2 Processes Sterility Is Not Required

In many industrial processes that are dependent on the use of microorganisms it is important that the process is kept free from contamination by undesired organisms. Undesired organisms "waste" substrate and the economic competitiveness of most biological fermentation processes is dependent on keeping them contamination-free. In contrast, the aim of the mineral biooxidation processes is the rapid biodegradation of a mineral or concentrate and one seeks the organisms that are able to do this most effectively. Given the huge volumes of mineral that have to be processed, the relatively low value of the product and the nature of a mining environment, the cost-effective prevention of contamination would be impossible to achieve. Fortunately this is not required. The mineral contains the energy source for the microorganisms and those that use this energy source most effectively are usually those that grow the quickest and dominate the process. As it is not important which microorganisms win the mineral degradation competition under a given set of operating conditions, there is no need to keep the process free from competing microorganisms.

9.4.3 Continuous-Flow, Stirred-Tank Reactors Select for the Most Efficient Organisms

From the description of biomining processes given in Chaps. 1–8 it is clear that "nonsterile" open stirred tanks or heaps exposed to the environment are used. Such processes are susceptible to "contamination" by microorganisms present on the ores, concentrates, inorganic nutrient solutions, water, air, etc. In a continuous-flow process such as that provided by a series of completely mixed leaching tanks, microorganisms in the tanks are continually growing and being washed out. There is therefore a strong positive selection for microbes that grow most effectively on the mineral as those microbes that grow and divide the fastest are subjected to less washout and will dominate the microbial population. There are few biological fermentation processes (sewage treatment processes being another) that have the advantage of continually selecting for the fittest microorganisms.

9.5 Should New Processes Be Inoculated with Established Microbial Consortia?

Frequently, consortia of microbes that are capable of growth at the required temperature and pH are prepared on a suitable mineral before the start-up of a new biooxidation process. These prepared consortia serve as the microbial inoculum during the start-up of a new commercial process. What has not been well established is how readily a population of microorganisms will adapt to a new process and whether any delay in initial process operation

efficiency during the period of adaptation is sufficient to warrant the import
of a previously adapted microbial inoculum. Furthermore, it has not been
established how many of the microbes present in a biooxidation process after
several years of operation are descendants of the microbes that were present
in the initial inoculum. It is possible that many of the microbes that dominate
a process might have been introduced subsequent to start up on nonsterile
minerals, added nutrients, water or air from the environment. Studies to
determine whether strains of bacteria similar to those in the original inoculum are still present in biooxidation aeration tanks many years after start-up
are under way in the author's laboratory. In the case of the Fairview BIOX™
tanks, the microorganisms in the tank inoculum were selected for increased
resistance to arsenic over a period of about 2 years (Sect. 9.1). Unfortunately
none of the original inoculum was maintained, but examination of the
arsenic resistance systems in the Fairview cultures suggested that the arsenic
resistance transposons that were present in the tanks 5 years after start up
(Tuffin et al. 2005, 2006) are still present approximately 15 years later. During
the commissioning of the BIOX™ plant at Tamboraque (Peru), it was decided
not to inoculate with the Fairview culture but to allow the indigenous microorganisms to adapt to the local arsenopyrite concentrate. There is no indication
to suggest that the Tamboraque process has been negatively affected because
it was not inoculated using an established microbial consortium from a similar process. Furthermore, it appears that the arsenic resistance mechanisms
that have been acquired by the microorganisms present in the Tamboraque
biooxidation tanks are closely related (if not identical, unpublished data) to
those present in the tanks in South Africa.

The necessity for inoculation is likely to depend on the type of processes
and the temperature of operation. In stirred-tank processes that operate from
ambient temperatures to about 40°C (possibly as high as 50°C) the advantage
of inoculation over allowing the indigenous microorganisms present on the
mineral or in the local environment to adapt may be small. Inoculation may
well speed up the process during the first few weeks of operation, but it is
quite likely that natural microorganisms present on the mineral heap would
become rapidly established. In contrast, one would predict that microbes
required for processes that operate at 60°C or higher are unlikely to be found
ubiquitously in mineral environments and would therefore need to be delib-
erately introduced. For this reason it has been argued that inoculation of
heaps with organisms, especially those that grow at temperatures of 45°C or
higher, may be a necessity (Chap. 6).

9.6   Types of Organisms

As described in Chap. 10, heap and tank biooxidation processes contain a
variety of microorganisms, with two different organisms usually dominating,
an iron-oxidizer and a sulfur-oxidizer. Conditions in heap reactors are more

varied than in stirred tanks as many more potential ecological niches are present. Regions with very different temperatures, aeration, mineral type, nutrient availability, biofilm formation, etc. exist within a heap and therefore the diversity of microorganisms will be much greater. Nevertheless, the types of microorganisms found in heap-leaching processes are generally similar to those found in stirred-tank processes at similar temperatures although the proportions of microbes present will vary depending on the conditions under which the heaps or tanks are operated.

Microorganisms found to dominate several mineral biooxidation consortia operating at different temperatures are given in Table 10.1. Some additional observations by other researchers are given here. Although *Leptospirillum* and *Acidithiobacillus caldus* have been reported to dominate arsenopyrite BIOX™ tanks operating at 40°C, once methods for distinguishing between leptospirilli were available, the dominant *Leptospirillum* was found to be *L. ferriphilum* rather than what had been reported as *L. ferrooxidans* (Coram and Rawlings 2002). Subsequent studies have confirmed the wide distribution of *L. ferriphilum*. A fluidized-bed reactor operating at 37°C and pH 1.4 was dominated by *L. ferriphilum* with a small proportion of *Ferroplasma*-like archaea (Kinnunen and Puhakka 2004). *L. ferriphilum* like bacteria together with *At. caldus* like and *Sulfobacillus*-like bacteria were found to dominate the first tank in a pilot scale, stirred-tank operation in which three tanks in series were used to treat a polymetallic sulfide ore at 45 C (Okibe et al. 2003). Plumb et al. (Chap. 11) suggest that the slightly greater temperature tolerance of and faster iron oxidation rate by *L. ferriphilum* is the reason it dominates over *L. ferrooxidans* in commercial bioleach tanks and also in heaps that operate at 40°C. Nevertheless, isolates of *L. ferrooxidans* may be dominant in certain commercial processes such as the microbial consortium in the BIOX™ plant at Tamboraque, Lima Peru (Rawlings laboratory, unpublished results).

Whether the type of aeration device affects microbial dominance within a given temperature range is not known.; however, it was suggested that Gram-positive iron- and sulfur-oxidizing bacteria related to *Sulfobacillus thermosulfidooxidans* which are commonly associated with moderately thermophilic processes were favored by bubble-column reactors even though the temperature was only 35°C (Battaglia-Brunet et al. 2002). Similarly, to what extent mineral type may effect the distribution of dominant microorganisms is not yet clear. *At. caldus*, *Sulfobacillus thermosulfidooxidans* and "*Sulfobacillus montserratensis*" like bacteria together with an uncultured thermal soil bacterium were found to dominate the consortium of organisms oxidizing chalcopyrite concentrate at 45°C, with the same bacteria dominating the culture irrespective of whether chalcopyrite, pyrite or an arsenic pyrite concentrate was being oxidized (Dopson and Lindstrom 2004).

What is clear is that consortia of microbes are more robust and efficient that pure cultures. As reported earlier, in a pilot scale, stirred-tank operation in which three tanks in series were used to treat a polymetallic sulfide ore at

45°C, *At. caldus* like, *L. ferriphilum* like and *Sulfobacillus*-like bacteria were found to dominate the first tank (Okibe et al. 2003). When combinations of pure cultures were tested, a mixed culture containing both autotrophic (*Leptospirillum* MT6 and *At. caldus*) and heterotrophic moderate themophiles (*Ferroplasma* MT17) was the most efficient (Okibe and Johnson 2004). The proportions of these bacteria decreased in the second tank, with the numbers of *At. caldus* and a *Ferroplasma*-like archaeon being equally dominant. *Ferroplasma* completely dominated the third tank. The presence of *Ferroplasma*-like organisms is being increasing recognized in bioleaching processes that operate at very low pH (1.4 or less). These archaea appear to be able to oxidize minerals like pyrite in pure culture although not without a small quantity of yeast extract. "Heterotrophically inclined" microbes such as bacteria belonging to the genus *Acidiphilium* (Harrison 1981) or *Ferroplasma*-like archaea (Vásquez et al. 1999; Golyshina et al. 2000) are believed to assist the growth of iron-oxidizing bacteria like *At. ferrooxidans* and the leptospirilli (Hallman et al. 1992; Johnson 1998). This is likely to be due to their ability to provide essential nutrients or to remove toxic metabolic wastes and other substances. How much this symbiosis contributes to the efficiency of a microbial consortium is uncertain (Johnson and Roberto 1997).

9.7   General Physiology of Mineral-Degrading Bacteria

The most important microbes involved in the biooxidation of minerals are those that are responsible for producing the ferric iron and sulfuric acid required for the bioleaching reactions. As described in Chap. 10, a wide variety of these microorganisms may be present. Irrespective of the type of process or the temperatures at which they operate, these microbes have a number of features in common that make them suitable for their role in mineral dissolution. The most important characteristics are as follows: they grow autotrophically by fixing CO_2 from the atmosphere; they obtain their energy by using either ferrous iron or reduced inorganic sulfur compounds (RISCs; some use both) as an electron donor, and generally use oxygen as the electron acceptor; they are acidophiles and grow in low-pH environments (pH 1.4–1.6 is typical); they are remarkably tolerant to a wide range of metal ions (Dopson et al. 2003).

These autotrophic microbes have very modest nutritional requirements, most of which can be met by the aeration of the iron-and/or sulfur-containing mineral suspension in water or the irrigation of a heap. Air provides the source of carbon (CO_2) and oxygen, water the medium for growth and the mineral the energy source and some trace elements. Small quantities of inorganic fertilizer may be added to ensure that nitrogen, phosphate, potassium and trace element limitation does not occur.

9.8 Autotrophy

The fact that the majority of microorganisms responsible for production of the lixiviants (ferric iron and acid) are autotrophic greatly simplifies the mineral bioxidation processes. If it were necessary to feed the microorganisms required for mineral degradation with a carbon source such as molasses, commercial mineral biooxidation processes would almost certainly be unworkable.

Details of the enzymes and pathways used by bacteria such as the acidithiobacilli and leptospirilli have been reviewed recently (Rawlings 2005). The CO_2 concentration present in air is sufficient to avoid carbon limitation when bacteria such as *At. ferrooxidans* are growing on ferrous iron and this bacterium responds to CO_2 limitation by increasing the cellular concentration of CO_2-fixing enzymes like RuBPCase (Codd and Kuenen 1987). *At. ferrooxidans* has been shown to use formic acid as a carbon source provided that the formic acid is provided sufficiently slowly for the concentration to remain low (Pronk et al. 1991). Similarly, genes for a formate hydrogenlyase complex have been located on the genome of *Leptospirillum* type II and it may therefore also grow on formate (Tyson et al. 2004); however, the ability to use formate is unlikely to play a role in commercial processes as natural formate is usually a by-product of anaerobic fermentative metabolism by heterotrophic microorganisms and metabolism of this type is not expected to occur in bioleaching environments.

Not all important bioleaching bacteria are as competent at using CO_2 as *At. ferrooxidans*. Bacteria such as the moderately thermophilic *Sulfobacillus thermosulfidooxidans* and related bacteria require 1% v/v CO_2-enriched air for rapid autotrophic growth (Clark and Norris 1996). This may be partly because the solubility of CO_2 is reduced at 50°C and partly because these bacteria are known to be inefficient at CO_2 uptake. Heterotrophic microorganisms that live off waste products produced by the autotrophs are usually also present and there is evidence that these heterotrophs might provide small quantities of additional carbon for growth of the autotrophs (Johnson and Roberto 1997). In a study of the ability of combinations of different pure cultures of microorganisms to dissolve pyrite, it was concluded that consortia containing heterotrophic microbes were superior to those containing only iron-oxidizers and sulfur-oxidizers (Okibe and Johnson 2004). The heterotrophic microbes would not be expected to contribute directly to the dissolution of the mineral either by the generation of ferric iron or by the removal of sulfur from the surface of the mineral. Their contribution must be due to stimulation of carbon flow between the heterotrophs and autotrophs. In general, not much attention has been paid to the effects of CO_2 limitation in processes that operate up to 50°C.

Hyperthermophilic archaea, like *Sulfolobus*, found in high-temperature leaching processes have also been reported to grow autotrophically but details of the pathway are not known. Work in the laboratory of Norris (1997) has suggested that acetylcoenzyme A (acetyl-CoA) carboxylation may be a key step and that the synthesis of biotin carboxylase and biotincarboxyl carrier protein is increased under conditions of CO_2 limitation. This complex is adjacent to genes encoding a putative propionylcoenzyme A carboxyl transferase and together these observations are in agreement with the suggestion that *Acidianus brierleyi* has a modified 3-hydroxypropionate pathway for CO_2 fixation. Other types of archaea, such as *Ferroplasma*, have the genes necessary to fix CO_2 via the reductive acetyl-CoA pathway (Tyson et al. 2004). Like *Sulfobacillus* spp., autotrophic growth of *Sulfolobus* spp. is enhanced in 1% CO_2-enriched air. As CO_2 solubility is reduced at high temperatures, CO_2 enrichment of air is required for high-temperature processes (Chap. 3). This requirement may be reduced if CO_2 is produced from the decomposition of carbonate in the mineral being processed.

9.9   Nitrogen, Phosphate and Trace Elements

On the basis of dry mass, nitrogen is the second most important element (after carbon) for the synthesis of new cell mass. Exactly how much nitrogen needs to be present in a growth medium will be dependent on the quantity of cell growth to be supported. Ammonia is highly soluble in acid solutions and traces of ammonia present in the air can be readily absorbed into acidic growth media; therefore, it is difficult to estimate the exact nitrogen requirements. Ammonium levels of 0.2 mM have been reported to be sufficient to satisfy the nitrogen requirement of *At. ferrooxidans* (Tuovinen et al. 1971). In commercial operations, inexpensive fertilizer grade ammonium sulfate is usually added to biooxidation tanks or bioleaching heaps to ensure that sufficient nitrogen is available. High concentrations of inorganic or organic nitrogen are inhibitory to iron oxidation.

Biomining microorganisms such as *At. ferrooxidans* (Mackintosh 1978) have the ability to fix nitrogen from the atmosphere and genes for nitrogen fixation have been found in *L. ferrooxidans* (Norris et al. 1995, Parro and Moreno-Paz 2003, 2004) but not in *L. ferriphilum* (Tyson et al. 2004). Nitrogenase enzyme activity is inhibited by oxygen and for this reason it is unlikely that nitrogen fixation would take place in a highly aerated biooxidation tank. However, the aeration of heaps is not homogenous and nitrogen fixation could take place in parts of a heap where oxygen is absent or its concentration sufficiently low. The sensitivity of nitrogenase to oxygen poses a special problem for leptospirilli because they use only iron as the electron donor and are not known to use an electron acceptor other than oxygen. Although the addition of inorganic fertilizer to tanks is relatively efficient and limited quantities are required, this is not the case for heap reactors. The cost of the much less effective addition of nitrogen-containing fertilizer to very extensive heap reactors may warrant the study of conditions under which nitrogen fixation by biomining microorganisms will occur.

Microorganisms cannot obtain phosphate and trace elements from the air as they can carbon or nitrogen. If insufficient quantities of these nutrients are available in the mineral being processed, they need to be added. As for nitrogen, suitable, inexpensive fertilizer-grade chemicals may be used for this purpose.

9.10   Energy Production

As described in Sect. 9.2, the solubilization of minerals is considered to be a chemical process that results from the action of ferric iron and/or acid, typically sulfuric acid. Therefore, irrespective of the temperatures at which they grow, the microorganisms that play the major role in the leaching of metals from minerals are either iron-oxidizers or sulfur-oxidizers.

9.10.1   Iron Oxidation

As described in Sect. 9.3 and several chapters of this volume, ferrous iron is readily oxidized to ferric iron and the electron released can serve as an electron donor to support microbial growth. However, as the Fe^{2+}/Fe^{3+} redox couple has a very positive standard electrode potential in low-pH liquors (Chap. 14.2), only oxygen is able to act as a natural electron acceptor. The oxidation of ferrous iron to produce the ferric iron required for leaching reactions is therefore dependent on the aeration of the mineral. Furthermore, unless the pH is low, ferrous iron will spontaneously oxidize to ferric iron when aerated. Ferric iron must therefore be produced in acid conditions by acidophilic microorganisms. Because the difference in redox potential between the Fe^{2+}/Fe^{3+} and the O_2/H_2O redox couples is small, the oxidation of vast amounts of ferrous iron is required to produce relatively little cell mass.

The molecular biology of ferrous iron oxidation by *At. ferrooxidans*, *Ferroplasma* spp., *Leptospirillum* spp. and *Metallosphera sedula* is discussed in Chap. 14. With the exception of *At. ferrooxidans*, relatively little is known about the mechanism of iron oxidation by acidophilic microorganisms, but it appears that the components of the electron transport chain are very different. This finding has led to the conclusion that the ability to use ferrous iron as an electron donor has probably evolved independently on several occasions (Blake et al. 1993). Some of these differences have important consequences. For example, whereas *At. ferrooxidans* was found to be capable of growth on ferrous iron at redox potentials of up to about +800 mV, *L. ferrooxidans* was capable of oxidation at redox potentials closer to +950 mV (Boon et al. 1998). The effect of this is that although *At. ferrooxidans* can outgrow *L. ferrooxidans* at high ratios of ferrous to ferric iron (as happens during the earlier stages of iron oxidation in batch culture), *L. ferrooxidans* outcompetes *At. ferrooxidans* once the ferric iron concentration becomes high. In continuous-flow aeration tanks such as those used for treating gold-bearing arsenopyrite concentrates, the steady-state concentration of ferric iron is high and ferrous iron oxidation by *At. ferrooxidans* is strongly inhibited. Species of the genus *Leptospirillum* are not as inhibited by the ferric iron and are the dominant iron-oxidizing microorganisms in these tanks (Rawlings et al. 1999).

9.10.2   Sulfur Oxidation

Sulfur oxidation is an important part of the biooxidation of minerals and the acid responsible for the low-pH environment in which mineral oxidation reactions take place is sulfuric acid. This sulfuric acid is produced by the oxidation of RISCs. The RISCs serve as an electron donor, with oxygen serving as the energetically most favorable electron acceptor. Up to seven electrons are available for release when a sulfur atom in pyrite is oxidized to sulfate and eight electrons when sulfide is fully oxidized. The potential amount of energy available in this process is therefore considerable. Naturally occurring RISCs are present wherever sulfide-containing minerals are exposed to the surface. A variety of RISCs are released as a result of the chemical reaction of sulfide minerals with water, oxygen and ferric iron.

There are many challenges to studying the biochemistry and molecular biology of sulfur oxidation. These challenges as well as research progress in biological RISC oxidation by acidophiles are described in Chap. 14. The sulfur oxidation pathways of *At. ferrooxidans* appear to be different from those of most bacteria and archaea but, in general, there appears to be more uniformity in the pathways used by different sulfur-oxidizing organisms than might have been expected.

9.10.3   Other Potential Electron Donors for Acidophilic Microorganisms

In current models, the main role of the microorganisms is to provide the leaching reagents (ferric iron and protons) and the space where these reagents degrade minerals most efficiently (cell exopolysaccharide layer); therefore any other metal oxidation reactions that take place may be considered to be of peripheral significance. Nevertheless, soluble metal ions are frequently present in fairly high concentrations in highly acidic environments. Metal ions which exist in more than one oxidation state and which have redox potentials that are more negative than the O_2/H_2O redox couple have the potential to serve as electron donors for acidophilic bacteria. An *At. ferrooxidans* like bacterium was reported to directly oxidize Cu^+ to Cu^{2+} and U^{4+} to U^{6+}.

Under aerobic conditions these oxidation reactions were coupled to CO_2 fixation (DiSpirito and Tuovinen 1982). However, whenever ferric iron is present, it is difficult to unequivocally demonstrate that biological oxidation of the metal as opposed to chemical oxidation of the metal by ferric iron is responsible. Similarly it has been reported that Mo^{5+} can be oxidized to Mo^{6+} and a molybdenum oxidase has been isolated from cell extracts of *At. ferrooxidans* (Sugio et al. 1992). The potential also exists that the oxidation of oxyanions such as As^{3+} (AsO_2^-) to As^{5+} (AsO_4^{3-}) can serve as an alternate electron donor for acidophilic organisms. An analysis of the *At. ferrooxidans* ATCC23270 genome revealed that as many as 11 cytochromes *c* were present (Yarzábal et al. 2002). One cytochrome *c* was specific for growth on sulfur, three were specific for growth on iron and several were produced on both substrates. The large number of cytochrome *c* molecules might also be a reflection of the versatility of electron donors (and electron acceptors) that the bacterium is capable of using.

Besides metal ions, hydrogen has been identified as a potential electron donor for some acidophilic prokaryotes. In two independent studies, at least four strains of *At. ferrooxidans*, including the ATCC23270 type strain, but no strains of *At. thiooxidans* or *L. ferrooxidans* tested, could use hydrogen as an electron donor to support CO_2 fixation and cell growth with oxygen as an electron acceptor (Drobner et al. 1990, Ohmura et al. 2002). How significant this may be in mineral leaching processes is unknown.

9.10.4. Oxygen and Alternative Electron Acceptors

As described in Sect. 9.10.1, oxygen is essential for the production of ferric iron from ferrous iron in low-pH environments. Oxygen is also the most energetically favorable terminal electron acceptor and is therefore used whenever available. However, as the redox potential of the Fe^{2+}/Fe^{3+} couple is almost as positive as that of the O_2/H_2O couple in acidic liquors, ferric iron is a potentially suitable alternate electron acceptor for most electron donors other than ferrous iron. The oxidation of sulfur and tetrathionate coupled to ferric iron reduction under anaerobic conditions has been shown to occur in the case of *At. ferrooxidans*, although not all isolates of this bacterium could grow by using the H_2- or S^0-coupled reduction of ferric iron (Ohmura et al. 2002). Similarly, ferric iron reduction is widespread amongst Gram-positive mixotrophic iron-oxidizers (Bridge and Johnson 1998).

Ferric iron reduction is not likely to occur in highly aerated tank reactors; however, conditions for ferric iron reduction almost certainly will occur within inadequately aerated regions of heaps and possibly also in microenvironments within well-aerated parts of a heap. Acid production may continue during sulfur oxidation with ferric iron as an electron acceptor, but the removal of ferric iron would reduce the concentration of an important leaching reagent.

Highly acidophilic heterotrophs have an even greater potential to grow by ferric iron respiration since ferric iron reduction can be coupled to the oxidation of many organic compounds. This may be advantageous to heap leaching as it appears that not only soluble but also insoluble amorphous or crystalline ferric iron compounds can be reduced. Some, if not all, *Acidiphilium* species are capable of reductive solublization of a wide range of ferric iron containing minerals such as $Fe(OH)_3$ and jarosite (Bridge and Johnson 1998). In such circumstances, ferric iron respiration has the advantage of mobilizing additional ferrous iron electron donors for the iron-oxidizing obligate autotrophs should aerobic conditions once again prevail.

Besides the ability to use ferric iron, the *At. ferrooxidans* is also able to reduce Mo^{6+}, Cu^{2+} and Co^{2+} when using elemental sulfur as an electron donor (Sugio et al. 1988, 1990). *At. ferrooxidans* and *At. thiooxidans* have been reported to reduce V^{5+} to V^{4+}; however, whether the oxidized vanadium served as an electron acceptor for respiration was unclear as the shake flasks were aerated (Bredberg et al. 2004). The large variety of cytochrome *c* molecules that have been detected in *At. ferrooxidans* might reflect the versatility of this bacterium to use a wide variety of electron acceptors.

9.10.5   Acidophilic Properties

Besides the direct role that protons play in the solubilization of certain minerals (Sect. 9.2), it is essential to bioleaching processes that biomining microorganisms are able to grow at low pH and tolerate high concentrations of acid. There are several reasons for this.

Ferric iron is almost insoluble at neutral pH, whereas in acid solutions (pH<2.5) its solubility is greatly increased. Ferric iron is the principle lixiviant in the solubilization of most minerals and to fulfill this role the ferric iron is required to be in solution. Furthermore, a low-pH environment is required for the microbial iron cycle to take place. As described earlier, ferrous iron serves as an electron donor for cell metabolism under aerobic conditions and ferric iron as an energetically favorable electron acceptor and replacement for oxygen when the concentration of oxygen falls. This iron cycle is likely to contribute to the stability of microbial populations in heap reactors.

Extreme acidophiles such as biomining microbes grow in a low-pH environment (e.g., pH 1.0–2.0), whereas the internal cellular pH is close to neutral (Cox et al. 1979). This difference results in a steep pH gradient across the cell membrane and this pH gradient is important for nutritional purposes, especially when using a weak reductant such as ferrous iron as an electron donor. Autotrophic organisms have a high requirement for compounds such as $NAD(P)H_2$ to reduce CO_2 for the synthesis of the sugars, nucleotides, amino acids and other molecules from which new cell mass is produced (Fig. 9.2). Heterotrophic bacteria do not have as high a demand for $NAD(P)H_2$ as their

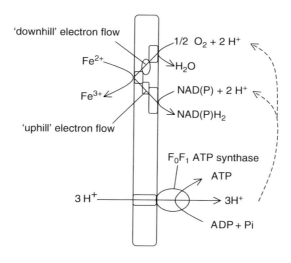

'downhill' electron flow

$1/2\ O_2 + 2\ H^+$

Fe^{2+}

H_2O

Fe^{3+}

$NAD(P) + 2\ H^+$

$NAD(P)H_2$

'uphill' electron flow

F_0F_1 ATP synthase

ATP

$3\ H^+$

$3H^+$

$ADP + Pi$

high proton concentration   cell envelope   lower proton concentration
outside cell (pH 1.5-2.0)   inside cell (±pH 6.5)

Fig. 9.2. The high proton concentration outside the cell serves as the proton motive force for the synthesis of ATP via an F_0F_1 ATP synthase. It also facilitates "uphill" electron flow required for the synthesis of $NAD(P)H_2$ for the reduction of fixed CO_2 to compounds required for the production of cell mass

carbon source is more reduced than CO_2. Hydrogen atoms removed from their source of nutrition may typically be used to satisfy their lower requirements for $NAD(P)H_2$. This option is not available to the chemolithotrophic autotrophs found in biomining processes. Instead, these microbes require a large transmembrane proton gradient to generate the required proton motive force to energize the synthesis of $NAD(P)H_2$. This process is known as reverse electron transport or the "uphill" electron transfer pathway (Elbehti et al. 2000; Brasseur et al. 2002). A way of viewing this is that growth in acid solutions is a nutritional necessity as a large transmembrane pH gradient is required to produce the hydrogen atoms that are assimilated when CO_2 is reduced to cell mass.

9.11   Adaptability of Biomining Microorganisms

The ability of biomining microorganisms to adapt is particularly evident in processes that use biooxidation tanks. After a period of operation, the metabolic capabilities of a population of biomining organisms may improve out of all recognition from the culture originally inoculated into the tanks. Natural populations of microorganisms would be expected to have adapted to survival under the regular feast-or-famine conditions that they experience in nature. For cells to compete in the optimized, controlled conditions of a biooxidation tank, a different set metabolic capabilities would be required. Early experiments on gold biooxidation were carried out in a series of three to four continuous-flow, aerated, stirred tank reactors. The initial rates of arsenopyrite concentrate decomposition were slow because unadapted cultures of biooxidation bacteria were not optimized for rapid growth and because they were sensitive to the arsenic released from the arsenopyrite (Sect. 9.12). The initial arsenopyrite retention time of over 12 days was reduced to less than 7 days after a period of selection of about 2 years in laboratory and pilot plant scale continuous flow processes. During the first 2 years of operation in a full-scale continuous-flow biooxidation plant, the growth rate of the bacteria improved still further such that the retention time could be reduced to 3.5 days. At the same time, the solid concentration in the liquor was increased from 10 to 18% so that the same equipment could be used to treat almost 4 times the amount of concentrate per day as initially.

The remarkable adaptability of *At. ferrooxidans* on exposure to mercury and *At. caldus* on exposure to arsenic is described in Sect. 9.12.

9.12 Metal Tolerance and Resistance

Owing to the action of low pH and ferric iron, many metal ions may be released in solution in the environment in which acidophilic chemolithotrophs grow. It is therefore not unexpected that these microbes should have a high level of general tolerance to metallic and other ions. The levels of resistance of several acidophilic bacteria and archaea to As^{3+}, Cu^{2+}, Zn^{2+}, Cd^{2+} and Ni^+ have recently been reviewed and will not be covered in detail here (Dopson et al. 2003). Levels of resistance within a given species show considerable strain variation, for example, strains of *At. caldus* and *L. ferriphilum* have very different levels of resistance to arsenic (section later). Adaptation to high levels of metal resistance on exposure to a metal is likely to be responsible for much of the strain variation. From a commercial point of view, one may wish to know the maximum level of resistance that can be attained, but this is difficult to predict and will need to be determined empirically for each combination of metal and microbial consortium. Experience suggests that by placing selective pressure on a population of microorganisms for growth in the presence of a metal ion, the level of resistance to most metals can be improved significantly; however, in some cases, this process might take a long time.

Increased resistance to metal ions can arise from two main sources: the occurrence of mutations in genes that are already present in the cell or the acquisition of new genes from other metal-resistant organisms, via the so-called horizontal gene pool. Genome sequencing data on *At. ferrooxidans* and *Leptospirillum* type II (*L. ferriphilum*), plus work from many other groups, suggest that metal resistance is due to a combination of both of these mechanisms. Examples of genes present on the chromosomes of most species of a genus are the efflux genes for arsenic (Butcher et al. 2000), copper, silver, cadmium and several metal cations in *At. ferrooxidans* (genome sequence data; Barreto et al. 2003). Although less well studied, a similar situation occurs in the genome of *Leptospirillum* type II. Another example of a resistance mechanism that might be present in all members of a species because it is associated with general cell physiology is the polyphosphate mechanism for copper resistance of *At. ferrooxidans* (Alverez and Jerez 2004). These workers presented a model whereby the hydrolysis of polyphosphates resulted in the formation of metal–phosphate complexes that are transported out of the cell, enhancing resistance to the metal.

The reason why exposure to a metal ion results in increased resistance as a result of changes internal to a cell is largely untested. It can be predicted that resistance genes for a given metal ion might not have been required for many years and lack of use has resulted in the accumulation of mutations that have inactivated or reduced the activity of the resistance genes. An interesting example of where resistance genes that must already have been present in a cell but which were reactivated when required comes from the work on mercury resistance in *At. ferrooxidans* from the laboratory of Tsuyoshi Sugio. Working with a pure culture, Sugio et al. (2003) were able to take *At ferrooxidans* SUG2.2 cells already resistant to 6 µM Hg^{2+} and adapt them by successive cultivation to produce *At. ferrooxidans* strain MON-1 that was resistant to 20 µM Hg^{2+}. Both strains, SUG2.2 and MON-1, contained similar levels of NADPH-dependent mercury reductase activity, but MON-1 appeared to have acquired an additional mercury resistance mechanism. High-level mercury volatilization in MON-1 was dependent on a mechanism where Fe^{2+} serves as an electron donor (in contrast to NADPH), a mechanism that is so far unique to the species (Inoue et al. 1989). Cytochrome *c* oxidase appeared to deliver electrons directly to mercury via the "uphill" reverse electron transport pathway (Sugio et al. 2003). This property was maintained after several rounds of cultivation on iron in the absence of Hg^{2+}. Interestingly, rusticyanin from mercury-resistant cells enhanced Fe^{2+}-oxidation activity of plasma membranes and activated Fe^{2+}-dependent mercury volatilization activity (Iwahori et al. 2000). As a pure culture of *At. ferrooxidans* had been used in these experiments, the genes for the Fe^{2+}-dependent mercury reduction system must have been present in the cells prior to the start of the experiment. Pressure for growth in higher concentrations of mercury resulted in the selection of mutants in which the previously inactive mercury reduction genes had become active.

Metal resistance genes may also be present on plasmids, transposons or viruses that may move from one microorganism to another and form part of the horizontal gene pool. The need for enhanced metal resistance may result in genes being acquired from this gene pool and lost again when no longer required. For example, when ten *At. ferrooxidans* isolates were screened for Hg^+ resistance, three of the strains contained DNA that hybridized to a transposon Tn*501 mer* gene probe (Shiratori et al. 1989). Bacteria carrying the resistance genes were in general 3–5 times more resistant to Hg^{2+} than strains that did not have *mer* genes. The *mer* genes of the E-15 strain of *At. ferrooxidans* were sequenced and codon usage analysis suggested that the *mer* genes had originated from an organism different from *At. ferrooxidans* (Inoue et al. 1989).

Another example of where resistance genes may be acquired from the horizontal gene pool when needed is the arsenic resistance genes recruited by *At. caldus* (Tuffin et al. 2005) and *L. ferriphilum* (Tuffin et al. 2006). These two bacteria dominate the biooxidation tanks used to treat gold-bearing arsenopyrite concentrate at the Fairview mine and have adapted to become highly resistant to arsenic. Studies on arsenic resistance genes of six strains of isolates of *At. caldus*, three with known exposure to arsenic and three without, have shown that all six strains contain a set of arsenic resistance genes on their chromosomes (Tuffin et al. 2004); however, the three highly arsenic resistant strains contain arsenic resistance genes in addition to those present in all strains. The arsenic resistance genes are present on a transposon belonging to the Tn*21* family that must have been acquired from the horizontal gene pool. All three resistant strains contained a copy of the Tn*AtcArs* transposon and at least one strain had an additional incomplete copy of the transposon. A question to be addressed is, from where did the Tn*AtcArs* acquired by the arsenic resistant strains originate? DNA sequencing data indicated that the closest relative to the *ars* gens is on a transposon present in a heterotrophic bacterium *Alcaligenes faecalis*; however, since the sequence of these genes is sufficiently different *Alcaligenes faecalis* could not have served as the direct source of *ars* genes. The source of the *ars* genes has still to be identified. A similar situation occurs in the case of *L. ferriphilum*. All *L. ferriphilum* strains tested have a set of *ars* genes located on the chromosome, but only *L. ferriphilum* isolated from arsenopyrite tanks at the Fairview mine has an additional set of *ars* genes located in a Tn*21*-family transposon (Tuffin et al. 2006). Interestingly, although the two Tn21-like *ars* transposons from the *At. caldus* and *L. ferriphilum* Fairview isolates are related, they are each more related to transposons found in other bacteria than they are to each other. Therefore, it is likely that the two arsenic resistance transposons were acquired from the horizontal gene pool independently and not passed from one type of bacterium to the other within the biooxidation tanks.

The account of arsenic resistance gene acquisition described here is an illustration of an advantage to be gained by the bioleaching and biooxidation processes being nonsterile, open systems. Exactly how the arsenic resistance transposon entered the BIOX™ microbial population is not known; however, new organisms will continually enter the system and these new strains of iron- and sulfur-oxidizing microbes may be selected. Alternately, microbes already present will have the opportunity of accessing the horizontal gene pool that these organisms contain and that are selected by growth conditions.

9.13   Conclusions

In general, bioleaching processes have proved to be remarkably robust. Where problems have occurred, they have been more frequently of a physical nature or associated with process design and engineering than with the microorganisms involved. This robustness is due to several interrelated factors, many of which have been described in this chapter. Amongst the most important of these is that sterile conditions are not required as whichever organisms decompose the mineral being processed, the quickest will dominate the process and are likely to be the most desirable. Furthermore, processes that use continuous-flow tank reactors and to some extent also heap reactors continually select for the fittest organisms. This may be because the existing microbial consortium has adapted to grow faster by the mutation of inherent genes or by the acquisition of new genes from the horizontal gene pool (e.g., metal resistance genes). Although all life has some limitations, a large pool of different iron- and sulfur-oxidizing microbes exists that provides organisms with overlapping temperature ranges, pH tolerances, redox potential optima, metal resistances and other metabolic characteristics. This enables them to fill the large variety of ecological niches likely to be encountered in mineral biooxidation processes.

References

Alvarez S, Jerez C (2004) Copper ions stimulate polyphosphate degradation and phosphate efflux in *Acidithiobacillus ferrooxidans*. Appl Environ Microbiol 70:5177–5182

Barreto M, Quatrini R, Bueno S, Arriagada C, Valdes J, Silver S, Jedlicki E, Holmes DS (2003) Aspects of the predicted physiology of *Acidithiobacillus ferrooxidans* deduced from an analysis of its partial genome sequence. Hydrometallurgy 71:97–105

Battaglia-Brunet F, Clarens M, d'Hugues P, Godon JJ, Foucher S, Morin D (2002) Monitoring of a pyrite-oxidizing bacterial population using DNA single-strand confirmation polymorphism and microscopic techniques. Appl Microbiol Biotechnol 60:206–211

Blake RC, Schute EA, Greenwood MM, Spencer GM, Ingeldew WJ (1993) Enzymes of aerobic respiration on iron. FEMS Microbiol Rev 11:9–18

Boon M, Brasser HJ, Hansford GS, Heijnen JJ (1998) Comparison of the oxidation kinetics of different pyrites in the prescence of *Thiobacillus ferrooxidans* or *Leptospirillum ferrooxidans*. Hydrometallurgy 53:57–72

Brassuer G, Brusella P, Bonnefoy V, Lemesle-Meunier D (2002) The bc_1 complex of the iron-grown acdiphilic chemolithorophic bacterium *Acidithiobacillus ferrooxidans* functions in the reverse but not in the forward direction. Is there a second bc_1 complex? Biochim Biophys Acta 1555:37–43

Bredberg K, Karlsson HT, Holst O (2004) Reduction of vanadium (V) with *Acidithiobacillus ferrooxidans* and *Acidithiobacillus thiooxidans*. Bioresour Technol 92:93–96

Bridge TAM, Johnson DB (1998) Reduction of soluble iron and reductive dissolution if ferric-iron containing minerals by moderately thermophilic iron-oxidizing bacteria. Appl Environ Microbiol 64:2181–2186

Butcher BG, Deane SM, Rawlings DE (2000) The *Thiobacillus ferrooxidans* chromosomal arsenic resistance genes have an unusual arrangement and confer increased arsenic and antimony resistance to *Escherichia coli*. Appl Environ Microbiol 66:1826–1833

196   Douglas E. Rawlings

Clark DA, Norris PR (1996) *Acidimicrobium ferrooxidans* gen. nov., sp. nov.: mixed-culture ferrous iron oxidation with *Sulfobacillus* species. Microbiology 142:785–790

Codd GA, Kuenen JG (1987) Physiology and biochemistry of autotrophic bacteria. Antonie van Leeuwenhoek 53:3–14

Coram NJ, Rawlings DE (2002) Molecular relationship between two groups of *Leptospirillum* and the finding that *Leptospirillum ferriphilum* sp. nov. dominates South African commercial biooxidation tanks which operate at 40°C. Appl Environ Microbiol 68:838–845

Cox JC, Nicholls DG, Ingledew WJ (1979) Transmembrane electrical potential and transmembrane pH gradient in the acidophile *Thiobacillus ferrooxidans*. Biochem J 178: 195–200

DiSpirito AA, Tuovinen OH (1982) Uranous ion oxidation and carbon dioxide fixation by *Thiobacillus ferrooxidans*. Arch Microbiol 133:28–32

Dopson M, Lindström EB (2004) Analysis of community composition during moderately thermophilic bioleaching of pyrite, arsenical pyrite, and chapcopyrite. Microbial Ecol 48:19–28

Dopson M, Baker-Austin C, Ram Kopponeedi P, Bond P (2003) Growth in sulfidic mineral environments: metal resistance mechanisms in acidophilic micro-organisms. Microbiology 149:1959–1970

Drobner E, Huber H, Stetter KO (1990) *Thiobacillus ferrooxidans*, a facultative hydrogen oxidizer. Appl Environ Microbiol 56:2922–2923

Elbehti A, Brasseur G, Lemesle-Meunier D (2000) First evidence for existence of an uphill electron transfer through the bc_1 and NADH-Q oxidoreductase complexes of the acidophilic obligate chemolithotrophic ferrous ion-oxidizing bacterium *Thiobacillus ferrooxidans*. J Bacteriol 182:3602–3606

Golyshina OV, Pivovarova TA, Karavaiko GI, Kondrat'eva TF, Moore ERB, Abraham WR, Lunsdorf H, Timmis KN, Yakimov MM, Golyshin PN (2000) *Ferroplasma acidiphilum* gen. nov., sp. nov., an acidophilic, autotrophic, ferrous iron-oxidizing, cell-wall-lacking, mesophilic member of the *Ferroplasmacaea* fam. nov., comprising a distinct lineage of the *Archaea*. Int J Sys Evol Microbiol 50:997–1006

Hallmann R, Friedrich A, Koops H-P, Pommerening-Röser A, Rohde K, Zenneck C, Sand W (1992) Physiological characteristics of *Thiobacillus ferrooxidans* and *Leptospirillum ferrooxidans* and physicochemical factors influence microbial metal leaching. 1992, Geomicrobiol J 10:193–206

Harrison AP Jr (1981) *Acidiphilium cryptum* gen. nov., sp. niv., heterotrophic bacterium from acidic mineral environments. Int J Syst Bacteriol 31:327–332

Inoue C, Sugawara K, Shiratori T, Kusano T, Kitawaga Y (1989) Nucleotide sequence of the *Thiobacillus ferrooxidans* chromosomal gene encoding mercury reductase. Gene 84:47–54

Iwahori K, Takeuchi F, Kamimura K, Sugio T (2000) Ferrous iron dependent volatilization of mercury by the plasma membrane of *Thiobacillus ferrooxidans*. Appl Environ Microbiol 66:3823–3827

Johnson DB (1998) Biodiversity and ecology of acidophilic microorganisms. FEMS Microbiol Ecol 27:307–317

Johnson DB, Roberto FF (1997) Heterotrophic acidophiles and their role in the bioleaching of sulfide minerals. In: Rawlings DE (ed) Biomining: theory, microbes and industrial processes. Springer, Berlin Heidelberg New York, pp 259–279

Kunnunen PH-M, Puhakka JA (2004) High-rate ferric sulfate generation by a *Leptospirillum ferriphilum*-dominated biofilm and the role of jarosite in biomass retainment in a fluidized-bed reactor. Biotechnol Bioeng 85:697–705

Mackintosh ME (1978) Nitrogen fixation by *Thiobacillus ferrooxidans*. J Gen Microbiol 105:215–218.

Norris PR (1997) Thermophiles and bioleaching. In: Rawlings DE (ed) Biomining: theory, microbes and industrial processes. Springer, Berlin Heidelberg New York, pp 247–258

Norris PR, Murrel JC, Hinson D (1995) The potential for diazotrophy in iron- and sulfur-oxidizing acidophilic bacteria. Arch Microbiol 164:294–300

Ohmura N, Sasaki K, Matsumoto N, Sakai H (2002) Anaerobic respiration using Fe^{3+}, S^0, and H_2 in the chemoautotrophic bacterium *Acidithiobacillus ferrooxidans*. J Bacteriol 184:2081–2087

Okibe N, Johnson DB (2004) Biooxidation of pyrite by defined mixed cultures of moderately thermophilic acidophiles in pH-controlled bioreactors: significance of microbial interactions. Biotechnol Bioeng 87:574–583

Okibe N, Gericke M, Hallberg KB, Johnson DB (2003) Enumeration and characterization of acidophilic microorganisms isolated from a pilot plant stirred-tank bioleaching operation. Appl Environ Microbiol 69:1936–1043

Parro V, Moreno-Paz M (2003) Gene function analysis in environmental isolates: The *nif* regulon of the strict iron-oxidizing bacterium *Leptospirillum ferrooxidans*. Proc Natl Acad Sci USA 100:7883–7888

Parro V, Moreno-Paz M (2004) Nitrogen fixation in acidophile iron-oxidizing bacteria: The nif regulon of *Leptospirillum ferroxidans*. Res Microbiol 155:703–709

Pronk JT, Meijer WM, Haseu W, van Dijken JP, Bos P, Kuenen JG (1991) Growth of *Thiobacillus ferrooxidans* on formic acid. Appl Environ Microbiol 57:2057–2062

Rawlings DE (2005) Characteristics and adaptability of iron- and sulfur-oxidizing microorganisms used for the recovery of metals from minerals and their concentrates. Microb Cell Fact 4:13

Rawlings DE, Tributsch H, Hansford GS (1999) Reasons why 'Leptospirillum'-like species rather than *Thiobacillus ferrooxidans* are the dominant iron-oxidizing bacteria in many commercial processes for the biooxidation of pyrite and related ores. Microbiology 145:5–13

Rohwerder T, Gehrke T, Kinzler K, Sand W (2003) Bioleaching review part A: progress in bioleaching: fundamentals and mechanisms of bacterial metal sulfide oxidation. Appl Microbiol Biotechnol 63:239–248

Sand W, Gehrke T, Hallmann R, Schippers A (1995) Sulfur chemistry, biofilm, and the (in)direct attack mechanism – critical evaluation of bacterial leaching. App Microbiol Biotechnol 43:961–966

Schippers A, Sand W (1999) Bacterial leaching of metal sulfides proceeds by two indirect mechanisms via thiosulfate or via polysulfides and sulfur. Appl Environ Microbiol 65:319–321

Shiratori T, Inoue C, Sugawara K, Kusano T, Kitawara Y (1989) Cloning and expression of *Thiobacillus ferroxidans* mercury ion resistance genes in *Escherichia coli*. J Bacteriol 171:3458–3464

Sugio T, Tsujita Y, Katagiri T, Inagaki K, Tano T (1988) Reduction of Mo^{6+} with elemental sulfur by *Thiobacillus ferrooxidans*. J Bacteriol 170:5956–5959

Sugio T, Tsujita Y, Inagaki K, Tano T (1990) Reduction of cupric ions with elemental sulfur by *Thiobacillus ferrooxidans*. Appl Environ Microbiol 56:693–696

Sugio T, Hirayama K, Inagaki K (1992) Molybdenum oxidation by *Thiobacillus ferrooxidans*. Appl Environ Microbiol 58:1768–1771

Sugio T, Fujii M, Takeuchi F, Negishi A, Maeda T, Kamimura K (2003) Volatilization by an iron oxidation enzyme system in a highly mercury resistant *Acidithiobacillus ferrooxdians* strain MON-1. 2003 Biosci Biotechnol Biochem 67:1537–1544

Tributsch H (2001) Direct vs indirect bioleaching. Hydrometallurgy 59:177–185

Tuffin IM, de Groot P, Deane SM, Rawlings DE (2004) Multiple sets of arsenic resistance genes are present within highly arsenic resistant industrial strains of the biomining bacterium, *Acidithiobacillus caldus*. Int Congr Ser 1275: 165–172

Tuffin IM, de Groot P, Deane SM, Rawlings DE (2005) An unusual Tn21-like transposon containing an *ars* operon is present in highly arsenic resistant strains of the biomining bacterium *Acidithiobacillus caldus*. Microbiology 151:3027–3039

Tuffin IM, Hector SB, Deane SM, Rawlings DE (2006) The resistance determinants of a highly arsenic resistant strain of *Leptospirillum ferriphilum* isolated from a commercial biooxidation tank. Appl Environ Microbiol 72 (in press)

Tuovinen OH, Niemelä SI, Gyllenberg HG (1971) Effect of mineral nutrients and organic substances on the development of *Thiobacillus ferrooxidans*. Biotechnol Bioeng 13:517–527

Tyson GW, Chapman J, Hugenholtz P, Allen EA, Ram RJ, Richardson PM, Solovyev VV, Rubin EM, Rokhsar DS, Banfield JF (2004) Community structure and metabolism through reconstruction of microbial genomes from the environment. Nature 428:37–43

Vásquez M, Moore ERB, Espejo RT (1999) Detection by polymerase chain reaction-amplification sequencing of an archaeon in a commercial-scale copper bioleaching plant. FEMS Microbiol Lett 173:183–187

Wächtershäuser G (1990) Evolution of the first metabolic cycles. Proc Natl Acad Sci USA 87:200–204

Yarzábal A, Brasseur G, Bonnefoy V (2002) Cytochromes *c* of *Acidithiobacillus ferroxidans*. FEMS Microbiol Lett 209:189–195

10 Acidophile Diversity in Mineral Sulfide Oxidation

PAUL R. NORRIS

10.1 Introduction

The microbial populations in stirred-tank, mineral-processing bioreactors and in ore leaching heaps generally comprise several species. The presence of sulfur and iron as energy sources from most of their mineral substrates and the existence of gradients (chemical and physical) in their environments underlie the diversity in these populations. Many of the important organisms are well known and their relative proportions in the bioreactor populations of some pilot plants and commercial operations have been estimated. Generally, two or three species, specific to a particular temperature range, predominate in the various mixed cultures that exist in the bioreactors. The description of populations in ore leaching heaps has so far been mostly qualitative. Reasons for the coexistence of different species in these mixed cultures or populations are clearer in some cases than in others. Similarly, some of the features that give individual species a competitive advantage over others in mixed cultures have been revealed but others, particularly at elevated temperatures, remain to be defined.

10.2 Acidophiles in Mineral Sulfide Oxidation

The most studied iron- and sulfur-oxidizing acidophile was known as *Thiobacillus ferrooxidans* before it was renamed to reflect the evolutionary separation of acidophilic thiobacilli from the type species of the *Thiobacillus* genus (Kelly and Wood 2000). Now called *Acidithiobacillus ferrooxidans*, it has been the subject of several reviews over many years (e.g., Tuovinen and Kelly 1972; Leduc and Ferroni 1994; Rawlings and Kusano 2001). Other reviews of acidophiles in general have noted their diversity (Norris and Johnson 1998; Hallberg and Johnson 2001), temperature ranges for growth (Norris 1990; Franzmann et al. 2005), taxonomy (Goebel et al. 2000), and ecology and community structures (Johnson 1998; Baker and Banfield 2003). Most of the familiar acidophiles involved in mineral sulfide oxidation can be placed in one of three broad groups on the basis of their temperature ranges for growth and their evolutionary relationships. These are the mesophilic

Biomining
(ed. by Douglas E. Rawlings and D. Barrie Johnson)
© Springer-Verlag Berlin Heidelberg 2007

proteobacteria, Gram-positive moderate thermophiles and thermophilic archaea. This grouping is convenient but not comprehensive or entirely accurate.
Some different organisms in adjacent temperature groups show considerable
overlaps in their temperature ranges for growth, and some phylogenetically
coherent groups have species active over wide temperature ranges.

10.2.1 The Major Species in Laboratory Studies and Industrial Practice

10.2.1.1 Mesophiles

Proteobacteria, particularly the acidithiobacilli and *Leptospirillum* species, are
generally the most active in mineral sulfide oxidation below 45°C. There are
several subgroups of strains within the species *At. ferrooxidans* (Karavaiko
et al. 2003) but the specificity of any characteristics to individual subgroups
and any relationship of these to the capacity for mineral sulfide oxidation are,
as yet, unresolved. Other species of *Acidithiobacillus* oxidize sulfur, but not
iron. The main species are *At. thiooxidans* and *At. caldus*, the latter having
greater thermotolerance and therefore supplanting *At. thiooxidans* in gold-
bearing arsenopyrite-processing bioreactors operating at about 40°C. Some
other sulfur-oxidizing acidithiobacilli have yet to be named, including those
found in marine environments that tolerate higher levels of sodium chloride
than the familiar species (Simmons and Norris 2002). There is also diversity in
the *Leptospirillum* genus. The original isolate was named *Leptospirillum ferrooxidans* (Markosyan 1972). Considerable variety was evident as more
strains were isolated and referred to as *Leptospirillum*-like bacteria (Harrison
and Norris 1985). *Leptospirillum ferriphilum* (Coram and Rawlings 2002) was
subsequently distinguished from *Leptospirillum ferrooxidans* on the basis of
ribosomal RNA (rRNA) gene sequences and copy numbers (two and three
copies respectively). Another species, "*L. ferrodiazotrophum*," was proposed
for a nitrogen-fixing representative of a third rRNA sequence cluster of strains
(Tyson et al. 2005). Some strains of *Leptospirillum* grow up to about 45°C
(Norris 1983; Franzmann et al. 2005) and even higher in the case of a strain
which was named *L. thermoferrooxidans* (Golovacheva et al. 1993), but which
was subsequently lost (Rawlings 2002). The higher-temperature optimum and
tolerance of some strains, for example, *L. ferriphilum*, probably contribute to
their dominance in gold-bearing arsenopyrite processing (Coram and
Rawlings 2002; Franzmann et al. 2005).

10.2.1.2 Moderate Thermophiles

There is not a precise temperature that divides mesophile activity from moder-
ate thermophile activity. Some acidophiles, for example, *At. caldus* and
Acidimicrobium ferrooxidans, are quite active from about 25°C to almost 55°C.
However, for most iron-and/or sulfur-oxidizing acidophiles studied so far,

mesophiles have an optimum temperature about or more often below 40°C and moderate thermophiles have an optimum temperature about or above 45°C.

The *Sulfobacillus* genus was established for a moderate thermophile, *Sb. thermosulfidooxidans* (Golovacheva and Karavaiko 1979). Recent isolates of these Gram-positive spore-forming bacteria include two species from natural geothermal sites that show some growth at almost 65°C and have optimum temperatures about 10°C higher than the 45–50°C optimum temperature of *Sb. thermosulfidooxidans* (Norris, unpublished data). The known temperature range for growth of *Sulfobacillus* was also extended at the other extreme by the isolation of two mesophiles with optimum temperatures of 33 and 37°C (Yahya et al. 1999). A second genus of moderately thermophilic Gram-positive, iron-oxidizing bacteria, not related to *Sulfobacillus*, contains a single species at present, *Acidimicrobium ferrooxidans* (Clark and Norris 1996), although at least one other species remains to be named (Johnson et al. 2003). *Am. ferrooxidans* is widespread in acidic natural geothermal sites and mineral sulfide mine environments (Clark and Norris 1996). It was first found in a commercial copper leach dump (Brierley 1978) but has not been extensively studied in microbial mineral processing (however, see Sect. 10.4.1).

10.2.1.3 Thermophiles

There is an overlap of the temperature ranges for growth of moderately thermophilic iron-oxidizing bacteria with those of some thermoacidophilic archaea; however, the lowest optimum temperatures among the thermophiles studied, for example, about 68°C for *Sulfolobus metallicus*, are well above those of the most thermotolerant *Sulfobacillus* species. Most demonstrations of mineral sulfide oxidation at about 70°C have used *S. metallicus*, although many early studies did not identify it as such (Norris 1997). It may well be the dominant strain in most studies of mineral sulfide oxidation at 65–70°C where the species were not identified (Gericke and Pinches 1999). It is replaced at 75°C by other organisms, such as *Metallosphaera sedula* and various unnamed isolates. Although relatively little work has been published on high-temperature strains and minerals (Norris et al. 2000; Plumb et al. 2002), the number of species known to be directly involved in mineral sulfide oxidation appears to rival that of mesophiles and moderate thermophiles at lower temperatures. Growth on pyrite and chalcopyrite has been observed with novel isolates at 88°C (Norris and Owen 1993) and 90°C (Plumb et al. 2002): in both cases the maximum copper extraction from chalcopyrite was observed at about 85°C.

10.2.2 Less Familiar Iron-Oxidizing Acidophiles

Some of the less well-known iron-oxidizing bacteria have yet to be placed in the context of biomining and could be active in mineral processing operations under conditions not generally used in the laboratory with the more familiar species. Three examples of such conditions – extremes of pH, the presence of organic substrates and of salt – are noted in this section. Some species are influenced by more than one of these conditions. Several species of *Thiomonas*, for example, can oxidize ferrous iron at optimum pH values less acidic than those optimum for the acidithiobacilli and some require organic supplements (Battaglia-Brunet et al. 2006). For some other species, any characteristics that might make them attractive candidates in the bio-mining context remain to be defined: for example, the mesophilic Gram-negative rod "*Thiobacillus* m-1" (Lane et al. 1992) and another related, recently described strain (Karsten et al. 2005), which are quite distinct phylogenetically from *At. ferrooxidans*.

10.2.2.1 Organisms at the Extremes of Acidity

Several strains and/or species of *Ferroplasma* have been isolated in recent years (Golyshina and Timmis 2005). These archaea appear to be favored by low pH and an organic carbon source, although autotrophy has been claimed for one species. Their widespread distribution in natural geothermal environments and in some industrial mineral processing bioreactors has been noted (Edwards et al. 2000; Golyshina et al. 2000; Burton and Norris 2000) and they have also been found in ore leaching heaps (Vásquez et al. 1999; Hawkes et al. 2005). Mesophilic *Ferroplasma acidiphilum* (originally isolated from a pilot plant processing pyrite/arsenopyrite) grows optimally at pH 1.7 (Golyshina et al. 2000), whereas "*Ferroplasma acidarmanus*" grows below pH 1, with an optimum at pH 1.2 (Dopson et al. 2004). A moderately thermophilic species obtained from a commercial, South African arsenopyrite/ pyrite processing bioreactor showed uninhibited growth and iron oxidation at pH 0.8, an optimum temperature of just above 50°C and some growth at 62°C (Norris, unpublished data). The appearance of such organisms in the South African bioreactors coincided with very low pH (0.5) when pH control was removed (Rawlings 2002), and where organic carbon from the primary mineral sulfide-oxidizers was possibly their carbon source. Some mesophilic species of *Sulfobacillus* (Sect. 10.2.1) also appear to tolerate greater acidity than *At. ferrooxidans* and *L. ferrooxidans* (Yahya et al. 1999), in contrast to the more familiar, moderately thermophilic sulfobacilli.

The acidity in some heap leach operations could at times be less than optimum for the well-known acidophiles, and closer to that more usually associated with acid mine drainage environments. Iron-oxidizing bacteria that have been described as moderate acidophiles have been found in mine drainage above pH 3, including species that cluster phylogenetically with *Thiomonas* and *Halothiobacillus* species (Hallberg and Johnson 2003). At higher temperature, the bacterium *Hydrogenobaculum acidophilum* (previously named *Hydrogenobacter acidophilus*) has an optimum pH of 3–4 (Shima and Suzuki 1993). Some strains grow very well on sulfur at 70°C, but are inhibited as the acidity increases towards pH 2 (Norris, unpublished work): iron oxidation and a role in mineral sulfide dissolution have not been reported for these acidophiles.

10.2.2.2 Heterotrophic Acidophiles

The autotrophic nature of the acidophiles which dominate industrial mineral bioprocessing is a key feature of their utility. Obligately heterotrophic, iron-oxidizing acidophiles (Johnson and Roberto 1997), however, have been visible mainly in acid mine drainage environments. The mesophile "*Ferrimicrobium acidiphilum*" is another exception to the simplified subdivision of potential biomining acidophiles (Sect. 10.2) because it is Gram-positive and related to *Am. ferrooxidans* rather than to the proteobacterial (Gram-negative) iron-oxidizing mesophiles. Other Gram-positive, heterotrophic, iron-oxidizing acidophiles (Johnson et al. 2003), both mesophilic and moderately thermophilic, are related to the *Sulfobacillus* genus but more closely to the *Alicyclobacillus* genus, into which some strains have now been incorporated (Karavaiko et al. 2005). Unlike *Sulfobacillus* species, these organisms do not appear to be capable of autotrophic growth as well as heterotrophic growth. They could be active in ore leaching heaps if, as for *Ferroplasma*, some organic carbon source is available. Their survival in heaps, if adverse conditions are encountered, could be favored by their capacity for sporulation.

10.2.2.3 Salt-Tolerant Species

Most acidophiles are sensitive to salt (NaCl), which could restrict their application at some potential biomining sites where only saline water is available. However, several species of halotolerant, iron-oxidizing acidophiles isolated from marine geothermal sites are found in the "*Thiobacillus prosperus*" group (Huber and Stetter 1989). These include moderate thermophiles that are active at 50°C (Norris, unpublished data). Pyrite oxidation by mesophilic species was maintained for months through serial cultures with salt at twice its seawater concentration (Norris and Simmons 2004). Other halotolerant bacteria similar to the mesophilic "*T. prosperus*" strains (Kamimura et al. 1999) and salt-tolerant, iron-oxidizing *Alicyclobacillus*-like bacteria (Crane and Holden 1999; Holden et al. 2001) have isolated from marine harbor sediments.

10.3 Dual Energy Sources: Mineral Dissolution by Iron-Oxidizing and by Sulfur-Oxidizing Bacteria

Commercially important mineral sulfides generally contain both iron and sulfur, which provides scope for microbial diversity in the mineral degradation through the different metabolisms that can sustain growth on each of these oxidizable energy sources. The nature of an individual mineral (its crystal structure) can at least indirectly further determine the type of organism that can extensively degrade it. In pure cultures, species that can oxidize both ferrous iron and sulfur can be involved in degradation of a wide range of mineral sulfides: this range may be restricted in pure cultures of species that oxidize either ferrous iron or sulfur.

In relation to possible mechanisms of dissolution, several mineral sulfides have been placed into two main groups on the basis of their susceptibility to attack by either iron-oxidizing or sulfur-oxidizing bacteria (Kelly et al. 1979), on the basis of their semiconductor electrochemistry (Tributsch and Bennett 1981; Tributsch and Rojas-Chapana 2000; Crundwell 1988) or on the basis of whether their sulfide moiety is oxidized principally through either thiosulfate or polysulfide intermediates (Schippers and Sand 1999): in each case, the relevance of the crystal structures of the minerals was noted.

Sphalerite (ZnS) and galena (PbS) are examples of sulfides whose dissolution can be enhanced by ferric iron but which can be solubilized by sulfur-oxidizing acidophiles in the absence of iron. The dissolution is generally more rapid with *At. thiooxidans* than with *At. ferrooxidans* (Kelly et al. 1979) because the former grows more rapidly on sulfur with faster acid production for the proton attack that can nonoxidatively solubilize these minerals (Crundwell 1988; Sand et al. 2001). In contrast, acidophiles that oxidize iron and sulfur, but not those that oxidize only sulfur, degrade pyrite (FeS_2) and covellite (CuS). Their dissolution appears to require a strong oxidizing agent, such as ferric iron, owing to the nonbonding nature of their valence bands (Crundwell 1988). The involvement of extracellular organic agents and polymers in the process has also been described (Sand et al. 2001). Tributsch and Rojas-Chapana (2000) proposed pyrite dissolution by *At. ferrooxidans* involved cysteine as well as protons and ferric ions in bond disruption, whereas a high redox potential generated with iron in the capsules of *L. ferrooxidans* was sufficient for this species to promote the electrochemical dissolution of the mineral. The growth of *L. ferrooxidans* at higher redox potentials than those tolerated by *At. ferrooxidans* has been implicated in the greater competitiveness of the former in mixed-culture pyrite oxidation (Sect. 10.5).

The relationship of the structures of copper sulfides to their degradation by microbial activity is more complicated than that of the sulfides (e.g., ZnS, PbS) whose acid solubility determines their leaching. *At. thiooxidans* was observed to grow well on sulfur contaminating a covellite sample but copper was still not released (Kelly at al. 1979). In the absence of thiosulfate as the main intermediate in the oxidation of its sulfide moiety, chalcopyrite ($CuFeS_2$) has been grouped with ZnS rather than FeS_2 (Schippers and Sand 1999; Sand et al. 2001). However, extensive dissolution of chalcopyrite concentrates generally requires microbial iron and sulfur oxidation (Norris 1983, 1990). The relationship of $CuFeS_2$ and CuS structures to their dissolution and their passivation behavior has been reviewed (Crundwell 1988). The surface-layer reactions control the dissolution process. The dimerization of sulfur to S_2^{2-} and the reduction of chalcopyritic ferric iron to form a pyritic phase (Klauber 2003) could leave the microbial role in dissolution as regeneration of the mineral-dissolving oxidizing agent, ferric iron, rather than any direct involvement in the solid–liquid interfacial reactions (Parker et al. 2003). If polysulfides or elemental sulfur is not considered to be a passivating factor in chalcopyrite dissolution (Parker et al. 2003; Sandström et al. 2005), the failure of iron-oxidizing *L. ferrooxidans* to degrade chalcopyrite concentrates extensively unless in mixed culture with sulfur-oxidizing bacteria (Norris 1983) could result from the failure to maintain the acidity in a suitable range either for its activity or to prevent excessive jarosite passivation layers.

Given the importance of pyrite and chalcopyrite, and of covellite as an intermediate in chalcocite oxidation, most microbial mineral processing applications depend on the presence of organisms that oxidize iron, or iron and sulfur. However, the competitiveness of sulfur-only-oxidizing strains for the reduced sulfur compounds released or formed during mineral dissolution also ensures their presence under most conditions in the mineral processing operations (Table 10.1; Sect. 10.4.1), where at the very least they play an important part in the acid production that keeps ferric iron in solution.

10.4   Acidophiles in Mineral Processing

10.4.1   Stirred-Tank Bioreactor Cultures

Continuous stirred-tank bioreactors can provide an environment for the long-term selection of stable mixtures of organisms. The conditions in a single tank provide a relatively constant, homogeneous environment, although

Table 10.1. Acidophiles in mineral sulfide concentrate processing at different temperatures. The estimated proportions of species refer to continuous cultures and to primary reactors where several were in series

Mineral concentrate	Temperature (°C)	Major types in populations		
Pyrite/arsenopyrite[a]	40	*Leptospirillum ferrooxidans* (48–57%)	*Acidithiobacillus thiooxidans/ At. caldus* (26–34%)	*At. ferrooxidans* (10–17%)
Mixed sulfides[b]	45	*At. caldus* (65%)	*L. ferrooxidans* (29%)	*Sulfobacillus* sp. (6%)
Nickel concentrate[c]	49	*At. caldus* (63%)	*Acidimicrobium* sp. (32%)	*Sulfobacillus* sp. (4%)
	55	*Sulfobacillus* sp. (93%)	*At. caldus* (5%)	*Acidimicrobium* sp. (2%)
Chalcopyrite[d]	75–78	"*Sulfolobus*" sp. (59%)	*Metallosphaera* sp. (1) (34%)	*Metallosphaera* sp. (2) (5%)

[a]Fairview and São Bento industrial plants (Dew et al. 1997).
[b]Mintek pilot plant (Okibe et al. 2003).
[c]Warwick University laboratory scale (Cleaver and Norris, unpublished data).
[d]Warwick University laboratory scale, HIOX culture (Norris, unpublished data).

one subject to variation in the feed material. Several tanks in series, however, provide different conditions (e.g., solids concentrations and pH) which influence the microbial population. The compositions of some mixed cultures analyzed in relation to commercial or pilot plant operations are summarized in a selective abbreviation of extensive work in each case (Table 10.1). These examples cover the wide range of temperatures at which stirred tanks are operated. Various minerals were involved and different methods were used to examine the cultures: microscope counts of immunofluorescent antibody-reacted bacteria from arsenopyrite oxidation at 40°C; colony-forming-unit plate counts of organisms at 45°C with a mixed sulfide containing $CuFeS_2$, ZnS and FeS_2; fluorescent in situ hybridization with specific 16S rRNA gene probes for moderate thermophiles with a nickel sulfide concentrate; and rRNA gene dot blots and PCR-amplification/clone-bank analyses for a high-temperature culture in chalcopyrite oxidation.

10.4.1.1 Mesophilic Cultures

The populations generally associated with processing minerals rich in pyrite and/or arsenopyrite in stirred-tank reactors at temperatures of about 40°C comprise three (or four) main species: L. ferrooxidans, At. ferrooxidans and At. caldus (and/or At. thiooxidans). This is the case for BIOX™ plants (Dew et al. 1997) and for cobaltiferous pyrite-processing bioreactors (d'Hugues et al. 2003). L. ferrooxidans has generally been reported as the dominant iron-oxidizer in these operations, but most reports predate the taxonomic separation of the species (Sect. 10.2.1), so L. ferriphilum might have been the dominant strain in some cases. Beyond primary tanks, the changing conditions will affect the populations with, for example, an increase in sulfur-oxidizers and fewer iron-oxidizers in BIOX™ downstream tanks (Dew et al. 1997).

10.4.1.2 Thermotolerant and Moderately Thermophilic Cultures

At. caldus can outnumber other bacteria in continuous mineral processing at 45–50°C (Table 10.1), although it would not have the critical role in mineral dissolution (Sect. 10.3). The different temperature optima of the iron-oxidizing bacteria can explain the replacement of Leptospirillum species present at or below 45°C (Okibe et al. 2003; Table 10.1;) by the moderate thermophiles close to 50°C. Fewer studies with moderate thermophiles (compared with the cultures most active at about 40°C) means there is so far less of a recognized "typical" population at 45–50°C. Sulfobacillus species and At. caldus are often dominant (Dopson and Lindström 2004). Sulfobacillus species were present in mineral-processing bioreactors at 45 and 49°C, but seemed to be outcompeted by other iron-oxidizers, Leptospirillum and Acidimicrobium species, respectively, at these temperatures (Table 10.1). The Acidimicrobium species that was the dominant iron-oxidizer in laboratory bioreactors (Table 10.1) that were intended to resemble industrial nickel sulfide processing development (Dew et al. 1999) was closely related to, but distinct from, *Am. ferrooxidans* (Cleaver 2000). *Acidimicrobium* species are thought to be more efficient at scavenging carbon dioxide from air than *Sulfobacillus* species (Clark and Norris 1996) but they surprisingly outcompeted the sulfobacilli even when nickel sulfide processing reactors were gassed with carbon dioxide enriched air (Cleaver and Norris, unpublished data). The *Acidimicrobium* species and *At. caldus* were reduced to minor components of the population when the temperature of a reactor was increased to 55°C (Table 10.1), leaving *Sulfobacillus* species (almost entirely *Sb. thermosulfidooxidans*) to take over and maintain the nickel extraction at the same or a slightly higher level. As with lower-temperature reactors, populations are likely to change through a series of tanks. The population composition in the primary reactor processing a mixed sulfide at 45°C (Table 10.1) changed to contain predominantly *At. caldus* and a *Ferroplasma* species in a secondary reactor, and predominantly *Ferroplasma* in a third reactor in the series (Okibe et al. 2003), where it is possible that the lower pH and organic residues from degraded primary iron- and sulfur-oxidizers favored its growth.

10.4.1.3   High-Temperature Cultures

The enhanced copper extraction that can be obtained from chalcopyrite at high temperatures (Norris 1997) has led to development of bioreactor processes operating at about 78°C: the HIOX process at laboratory pilot plant scale (d'Hugues et al. 2002) and, after extensive laboratory work (Dew et al. 1999), the BioCop process at industrial scale (Batty and Rorke 2005). Mixed cultures have probably evolved in the development of these processes but at least five species were shown to be present in the source culture, including two "*Sulfolobus*"-like strains and species of *Metallosphaera* and *Acidianus*. Of these, only *Metallosphaera* species and one "*Sulfolobus*"-like strain (which has not been named since it has yet to be isolated in pure culture) were present in significant numbers during chalcopyrite oxidation in laboratory reactors over several years of continuous operation (Table 10.1). The factors that affect the balance of species in these cultures include the temperature (between 75 and 80°C), the copper concentration (between 30 and 40 g L^{-1}) and the solids concentration. The characteristics of substrate oxidation by the individual species remain to be described, but different responses of other iron-oxidizing archaeal acidophiles to changing ferrous iron to ferric iron ratios (Sect. 10.5) suggest this could also influence the population composition, as it does in the competition between *At. ferrooxidans* and *L. ferrooxidans* at lower temperatures.

10.4.2   Microbial Populations in Ore Heap Leaching

Various gradients in ore leaching heaps (including those of acidity, temperature, concentrations of oxygen, carbon dioxide and ferrous/ferric iron) should favor the activity of different acidophiles in different regions or niches. Unsurprisingly, a diverse range of organisms have been isolated from heap samples using conventional culturing techniques. More recently, this diversity has been documented using culture-independent molecular biology methods. Some studies have used both traditional and molecular approaches (e.g., Hawkes et al. 2005). Analysis of the relationship of microbial activity, or the presence of particular organisms, to the rate or extent of metal extraction is more difficult with sulfides in ore heaps than in bioreactors. The influence of different temperatures on the activity of mesophiles, moderate thermophiles and thermophilic archaea has been assessed in columns simulating commercial, self-heating biooxidation heaps for refractory gold pretreatment (Brierley 2003). Some correlations of prevalent organism types to factors such as iron and sulfate concentrations have been suggested for large-scale heap tests with copper ore (Demergasso et al. 2005). Most studies have relied on collecting organisms from leach solutions draining from columns or being recycled through heaps. The practical problems of representative sampling of large areas and various depths of heaps and of obtaining efficient, unbiased extraction of organisms (or their nucleic acids) from the minerals and iron salt precipitates require further attention.

10.5   Diversity in Iron Oxidation

As noted in Sect. 10.3, the generation of ferric iron is generally the critical reaction in microbial mineral processing. Surveys of respiratory chain components in various iron-oxidizing acidophiles indicated that a variety of different cytochromes were likely to be involved in transport of electrons from ferrous iron to oxygen in different organisms (Barr et al. 1990; Blake et al. 1992). The characteristics of the cellular components of acidophiles that are involved in the primary interaction with iron probably influence the kinetics of ferrous iron oxidation and its inhibition by the end-product ferric iron.

Nongrowing cell suspensions have often been used, for convenience, to gauge the responses of different species to different ferrous iron and ferric iron concentrations and some relevance of the observed effects to growing cells has been found. However, various experimental factors influence the measured parameters, including the range of substrate ferrous iron and product ferric iron concentrations examined, the concentration of cells in suspensions, the pH of the suspensions and the method used, for example, following oxygen uptake with Warburg respirometers and oxygen electrodes, or electric current change in electrochemical cells. The kinetics aspects of iron and organism interactions have been reviewed, principally with reference to *At. ferrooxidans* (Nemati et al. 1998) but also to *L. ferrooxidans* and the competition between these two species (Hansford 1997). Several models have been proposed to describe the interactions, and these may incorporate other factors as well as the ferrous iron/ferric iron concentrations (e.g., the redox potential; Meruane et al. 2002). However, in order to compare data for a wide range of acidophiles, the affinity and inhibition constants collected here were obtained using the same apparatus (Clark-type oxygen electrode cells) and conditions, except for temperature (Table 10.2).

10.5.1 Mesophiles

A similar behavior of *L. ferrooxidans* and *At. ferrooxidans* with regard to iron oxidation kinetics was reported by Eccleston et al. (1985); however, there was a generally higher affinity for the substrate and less sensitivity to

Table 10.2. Michaelis constant (K_m) and competitive inhibition constant (K_i) values for ferrous iron oxidation by cell suspensions of acidophiles in oxygen electrode cells. All cultures were grown on ferrous iron with thermophiles supplemented with CO_2 or yeast extract (*YE*) as indicated

Organism	Temperature (°C)	K_m (mM Fe^{2+})	K_i (mM Fe^{3+})	Reference
At. ferrooxidans (DSMZ 583)	30	0.7	8–10	Kelly and Jones (1978)
At. ferrooxidans (DSMZ 583)	30	1.3	3.1	Norris et al. (1988)
L. ferrooxidans (DSMZ 2705)	30	0.5	33	Eccleston et al. (1985)
L. ferrooxidans (DSMZ 2705)	30	0.3	43	Norris et al. (1988)
Sulfobacillus thermosulfidooxidans (strain TH1) (YE)	50	1.0	2.7	Norris et al. (1988)
Sb thermosulfidooxidans (strain BC1) (YE)	45	0.7	3.6	Clark (1995)
Sb thermosulfidooxidans (strain BC1) (CO_2)	45	0.3	5.7	Clark (1995)
Sb acidophilus (strain ALV) (CO_2)	50	3.0	1.1	Norris et al. (1988)
Acidimicrobium ferrooxidans (strain TH3) (YE)	50	0.5	1.9	Norris et al. (1988)
Am. ferrooxidans (str. ICP, DSMZ 10331) (YE)	45	0.2	1.4	Clark (1995)
Am. ferrooxidans (str. ICP, DSMZ 10331) (CO_2)	45	0.5	0.4	Clark (1995)
Sulfolobus metallicus (strain LM) (CO_2)	65	0.4	1.7	Norris (1992)
Metallosphaera sedula (DSMZ 5348) (CO_2)	65	1.0	1.0	Norris (1992)
Acidianus brierleyi (DSMZ 1651) (CO_2)	65	0.4	1.9	Norris (1992)

the ferric product with *L. ferrooxidans* compared with *At. ferrooxidans* in earlier work (Kelly and Jones 1978; Table 10.2). The differences in the oxidation constants with these species were more noticeable in subsequent work (Norris et al. 1988; Table 10.2). In addition, the different sensitivities to copper (*L. ferrooxidans* was more sensitive) and nitrate (*At. ferrooxidans* was more sensitive), which were indicated in the earlier work (Eccleston et al. 1985), were used to examine competition between the species in ferrous-iron-limited chemostat cultures. At dilution rates that allowed coexistence of the two species, *L. ferrooxidans* was dominant at about 95% of the cells (Norris et al. 1988). The respective responses of these mesophiles to substrate and product concentrations were suggested to explain the selection for *L. ferrooxidans*. This influence of the ferrous iron to ferric iron ratio on these bacteria has been extensively analyzed in terms of redox potential and the activity of the organisms in pyrite dissolution (Hansford 1997; Rawlings et al. 1999). It was suggested that *At. ferrooxidans* was initially favored in mixtures on pyrite because it can rapidly attack the mineral, whereas *L. ferrooxidans* was adapted to the very positive redox potentials that develop as its degradation proceeds. However, the initial rates of pyrite dissolution by each organism were similar when batch cultures were inoculated (Norris and Kelly 1982). The pyrite dissolution that was limited sooner in the *At. ferrooxidans* culture, as the unfavorable ferrous iron to ferric iron ratio developed, could also have been restricted by the increasing acidity. *At. ferrooxidans* has been shown to be more sensitive than *L. ferrooxidans* to increasing acidity during their growth on pyrite (Norris 1983; Helle and Onken 1988).

The influence of metal ions other than ferric iron on kinetics aspects of iron oxidation by acidophiles have been relatively little studied. Competitive inhibition constants of 230 mM copper (Norris 1992) and 343 mM copper (Kupka and Kupsáková 1999) have been obtained for ferrous iron oxidation by *At. ferrooxidans*. In contrast, uncompetitive inhibition by copper was observed with *L. ferrooxidans* as the affinity for ferrous iron and the maximum rate of iron oxidation were both reduced (Norris et al. 1988).

10.5.2   Thermophiles

Sb. acidophilus showed a relatively low affinity for ferrous iron and high sensitivity to ferric iron in comparison with *Sb. thermosulfidooxidans* (Table 10.2). This indication from nongrowing cell suspensions that *Sb. acidophilus* might be relatively poorly adapted for growth on ferrous iron was confirmed with observations of its relatively limited growth on pyrite, its greater sensitivity to ferric iron during batch culture growth, and a failure to maintain it in continuous culture without much of the substrate ferrous iron remaining in excess (Norris et al. 1988). It remains to be seen whether similar correlations hold between cell-suspension oxidation kinetics and mineral dissolution capacities for other potentially competing organisms, such as *S. metallicus* and *M. sedula*, where the observed parameters would appear to favor *S. metallicus* (Table 10.2). The similar ratio of the Michaelis constant to the competitive inhibition constant seen with *S. metallicus* and *At. ferrooxidans* (and therefore similar relationships to iron concentrations) might be reflected in the prevalence of these species in mineral sulfide enrichment cultures at their respective optimum temperatures from widespread samples; however, a comparison of the mesophile and thermophile using an electrochemical cell indicated far less favorable kinetics for the thermophile (Meruane et al. 2004). There has been insufficient comparative work with different organisms and methods to establish a consensus in this context and several aspects of these data remain to be investigated further. For example, the apparently greater sensitivity to ferric iron of autotrophically grown *Am. ferrooxidans* in comparison with that of lithoheterotrophically grown cells (Table 10.2) could be reflected in the difficulty of growing this organism in the absence of yeast extract in pure culture on pyrite. The basis and the reproducibility of the observed influence of the carbon source on the apparent kinetics constants seen with the *Acidimicrobium* and *Sulfobacillus* species (Table 10.2) is also unknown.

10.6 Summary

The involvement of acidophiles in mineral sulfide dissolution through their oxidation of iron and sulfur has been related to the solubilities and crystal structures of minerals and, more recently, to interfacial electrochemical events where the local environment might be more aggressive than the bulk solution in promotion of degradation. However, the bulk solution redox potential, and specifically the ferrous iron to ferric iron ratio, also at least indirectly influences selectively the activity of different microbial species. The discussion of mineral dissolution mechanisms at the molecular level is probably more advanced in the minerals electrochemistry field than in the area of the microbe–metal interactions, particularly for organisms other than *At. ferrooxidans* and *L. ferrooxidans*. The known diversity of acidophilic microorganisms that can interact with mineral sulfides is steadily increasing, while relatively few acidophilic microorganisms appear to predominate under the conditions generally used in industrial processes. Where maintenance of acidity and ferric iron in solution is possible with any one of a variety of species, the actual composition of the population might be unimportant, but as any parameter (such as the temperature, ferric iron or salt concentration) becomes more selective, more attention is required in the selection and development of a culture for application.

References

Baker BJ, Banfield JF (2003) Microbial communities in acid mine drainage. FEMS Microbiol Ecol 44:139–152

Barr DW, Ingledew WJ, Norris PR (1990) Respiratory components of iron-oxidizing, acidophilic bacteria. FEMS Microbiol Lett 70:85–90

Battaglia-Brunet F, Joulian C, Garrido F, Dictor M-C, Morin D, Coupland K, Johnson DB, Hallberg KB, Baranger P (2006) Oxidation of arsenite by *Thiomonas* strains and charcaterization of *Thiomonas arsenovorans* sp. nov. Antonie van Leeuwenhoek (in press)

Batty JD, Rorke GV (2005) Development and commercial demonstration of the BioCOP™ thermophile process. In: Harrison STL, Rawlings DE, Petersen J (eds) Proceedings of the 16th international biohydrometallurgy symposium, pp 153–161

Blake RC, Shute EA, Waskovsky J, Harrison AP (1992) Respiratory components in acidophilic bacteria that respire on iron. Geomicrobiology J 10:173–192

Brierley JA (1978) Thermophilic iron-oxidizing bacteria found in copper leaching dumps. Appl Environ Microbiol 36:523–525

Brierley JA (2003) Response of microbial systems to thermal stress in biooxidation-heap pretreatment of refractory gold ores. Hydrometallurgy 71:13–19

Burton NP, Norris PR (2000) Microbiology of acidic, geothermal springs of Montserrat: environmental rDNA analysis. Extremophiles 4:315–320

Clark DA (1995) The study of acidophilic, moderately thermophilic iron-oxidizing bacteria. PhD thesis, University of Warwick

Clark DA, Norris PR (1996) *Acidimicrobium ferrooxidans* gen. nov. sp. nov.: mixed culture ferrous iron oxidation with *Sulfobacillus* species. Microbiology 141:785–790

Cleaver A (2000) Mineral sulphide oxidation, mixed cultures in bioreactors. PhD thesis, University of Warwick

Coram NJ, Rawlings DE (2002) Molecular relationship between two groups of the genus *Leptospirillum* and the finding that *Leptospirillum ferriphilum* sp. nov. dominates in South African commercial biooxidation tanks that operate at 40°C. Appl Environ Microbiol 68:838–845

Crane AG, Holden PJ (1999) Leaching of harbour sediments by estuarine iron-oxidising bacteria. In: Amils R, Ballester A (eds) Biohydrometallurgy and the environment toward the mining of the 21st century, part A. Elsevier, Amsterdam, pp 347–356

Crundwell FK (1988) The influence of the electronic structure of solids on the anodic dissolution and leaching of semiconducting sulphide minerals. Hydrometallurgy 21:155–190

Demergasso CS, Galleguillos P. PA, Escudero G. LV, Zepeda A. VJ, Castillo D, Casamayor EO (2005) Molecular characterization of microbial populations in a low-grade copper ore bioleaching test heap. Hydrometallurgy 80:241–253

Dew DW, Lawson EN, Broadhurst JL (1997) The BIOX™ process for biooxidation of gold-bearing ores or concentrates. In: Rawlings DE (ed) Biomining. Springer, Berlin Heidelberg New York, pp 45–80

Dew DW, van Buuren C, McEwan K, Bowker C (1999) Bioleaching of base metal sulphide concentrates: a comparison of mesophile and thermophile bacterial cultures. In: Amils R, Ballester A (eds) Biohydrometallurgy and the environment toward the mining of the 21st century, part A. Elsevier, Amsterdam, pp 229–238

d'Hugues P, Foucher S, Gallé-Cavalloni, Morin D (2002) Continuous bioleaching of chalcopyrite using a novel extremely thermophilic mixed culture. Int J Miner Process 66:107–119

d'Hugues P, Battaglia-Brunet F, Clarens M, Morin D (2003) Microbial diversity of various metal-sulphides bioleaching cultures grown under different operating conditions using 16S-rDNA analysis. In: Tsezos M, Hatzikioseyian A, Remoundaki E (eds) Biohydrometallurgy: a sustainable technology in evolution, part II. National Technical University of Athens, pp 1313–1323

Dopson M, Lindström EB (2004) Analysis of community composition during moderately thermophilic bioleaching of pyrite, arsenical pyrite, and chalcopyrite. Microb Ecol 48:19–28

Dopson M, Baker-Austin C, Hind A, Bowman JP, Bond PL (2004) Characterization of *Ferroplasma* isolates and *Ferroplasma acidarmanus* sp. nov., extreme acidophiles from acid mine drainage and industrial bioleaching environments. Appl Environ Microbiol 70:2079–2088

Eccleston M, Kelly DP, Wood AP (1985) Autotrophic growth and iron oxidation and inhibition kinetics of *Leptospirillum ferrooxidans*. In: Caldwell DE, Brierley JA, Brierley CL (eds) Planetary ecology, Van Norstrand Reinhold, New York, pp 263–272

Edwards KJ, Bond PL, Gihring TM, Banfield JF (2000) An archaeal iron-oxidizing extreme acidophile important in acid mine drainage. Science 287:1796–1799

Franzmann PD, Haddad CM, Hawkes RB, Robertson WJ, Plumb JJ (2005) Effects of temperature on the rates of iron and sulfur oxidation by selected *Bacteria* and *Archaea*: application of the Ratkowsky equation. Miner Eng 18:1304–1314

Gericke M, Pinches A (1999) Bioleaching of copper sulphide concentrate using extremely thermophilic bacteria. Miner Eng 12:893–904

Goebel BM, Norris PR, Burton NP (2000) Acidophiles in biomining. In: Priest FG, Goodfellow M (eds) Applied microbial systematics. Kluwer, Dordrecht, pp 293–314

Golovacheva RS, Karavaiko GI (1979) A new genus of thermophilic spore-forming bacteria, *Sulfobacillus*. Microbiology 48:658–665

Golovacheva RS, Golyshina OV, Karavaiko GI, Dorofeev AG, Pivovarova TA, Chernykh NA (1993) A new iron-oxidizing bacterium, *Leptospirillum thermoferrooxidans* sp. nov. Microbiology 61:744–750

Golyshina OV, Timmis KN (2005) *Ferroplasma* and relatives, recently discovered cell wall-lacking archaea making a living in extremely acid, heavy metal-rich environments. Environ Microbiol 7:1277–1288

Golyshina OV, Pivovarova TA, Karavaiko GI, Kondrat'eva TF, Moore ERB, Abraham W-R, Lunsdorf H, Timmis KN, Yakimov MM, Golyshin PN (2000) *Ferroplasma acidiphilum* gen. nov. sp. nov, an acidophilic, autotrophic, ferrous-iron-oxidizing, cell-wall-lacking, mesophilic member of the *Ferroplasmaceae* fam. nov., comprising a distinct lineage of the *Archaea*. Int J Syst Evol Microbiol 50:997–1006

Hallberg KB, Johnson DB (2001) Biodiversity of acidophilic prokaryotes. Adv Appl Microbiol 49:37–84

Hallberg KB, Johnson DB (2003) Novel acidophiles isolated from moderately acidophilic mine drainage waters. Hydrometallurgy 71:139–148

Hansford GS (1997) Recent developments in modeling and the kinetics of bioleaching. In: Rawlings DE (ed) Biomining. Springer, Berlin Heidelberg New York, pp 153–175

Harrison AP Jr, Norris PR (1985) *Leptospirillum ferrooxidans* and similar bacteria: some characteristics and genomic diversity. FEMS Microbiol Lett 30:99–102

Hawkes RB, Franzmann PD, Plumb JJ (2005) Moderate thermophiles including '*Ferroplasma cyprexacervatum*' sp. nov. dominate an industrial-scale chalcocite heap bioleaching operation. In: Harrison STL, Rawlings DE, Petersen J (eds) Proceedings of the 16th international biohydrometallurgy symposium, pp 657–666

Helle U, Onken U (1988) Continuous microbial leaching of a pyritic concentrate by *Leptospirillum*-like bacteria. Appl Microbiol Biotechnol 28:553–558

Holden PJ, Foster LJ, Neilan BA, Berra G, Vu QM (2001) Characterisation of novel salt tolerant iron-oxidising bacteria. In: Ciminelli VST, Garcia O Jr (eds) Biohydrometallurgy: fundamentals, technology and sustainable development, part A. Elsevier, Amsterdam, pp 283–290

Huber H, Stetter KO (1989) *Thiobacillus prosperus* sp. nov., represents a new group of halotolerant metal-mobilizing bacteria isolated from a marine geothermal field. Arch Microbiol 151:479–485

Johnson DB (1998) Biodiversity and ecology of acidophilic microorganisms. FEMS Microbiol Rev 27:307–317

Johnson DB, Roberto FF (1997) Heterotrophic acidophiles and their roles in bioleaching of sulfide minerals. In: Rawlings DE (ed) Biomining. Springer, Berlin Heidelberg New York, pp 259–279

Johnson DB, Okibe N, Roberto FF (2003) Novel thermo-acidophilic bacteria isolated from geothermal sites in Yellowstone National Park: physiological and phylogenetic considerations. Arch Microbiol 180:60–68

Kamimura K, Kunomuraugio K, Sugio T (1999) Isolation and characterization of a marine iron-oxidizing bacterium requiring NaCl for growth. In: Amils R, Ballester A (eds) Biohydrometallurgy and the environment toward the mining of the 21st century, oart A, Elsevier, Amsterdam, pp 741–746

Karavaiko GI, Tourova TP, Kondrat'eva TF, Lysenko AM, Kolganova TV, Ageeva SN, Muntyan LN, Pivovarova TA (2003) Phylogenetic heterogeneity of the species Acidithiobacillus ferrooxidans. Int J Syst Evol Microbiol 53:113–119

Karavaiko GI, Bogdanova TI, Tourova TP, Kondrat'eva TF, Tsaplina IA, Egorova MA, Krasil'nikova EN, Zakharchuk LM (2005) Reclassification of 'Sulfobacillus thermosulfidooxidans subsp. thermotolerans' strain K1 as Alicyclobacillus tolerans sp. nov. and Sulfobacillus disulfidooxidans Dufresne et al. 1996 as Alicyclobacillus disulfidooxidans comb. nov., and emended description of the genus Alicyclobacillus. Int J Syst Evol Microbiol 55:941–947

Karsten C, Harneit K, Sand W, Stackebrandt E, Schumann P (2005) Characterization of novel iron-oxidizing bacteria. In: Harrison STL, Rawlings DE, Petersen J (eds) Proceedings of the 16th international biohydrometallurgy symposium, pp 729–735

Kelly DP, Jones CA (1978) Factors affecting metabolism and ferrous iron oxidation in suspensions and batch cultures of Thiobacillus ferrooxidans: relevance to ferric iron leach solution regeneration. In: Murr LE, Torma AE, Brierley JA (eds) Metallurgical applications of bacterial leaching and related microbiological phenomena. Academic, New York, pp 19–44

Kelly DP, Wood AP (2000) Reclassification of some species of Thiobacillus to the newly designated genera Acidithiobacillus gen. nov., Halothiobacillus gen. nov. and Thermithiobacillus gen. nov. Int J Syst Evol Microbiol 50:511–516

Kelly DP, Norris PR, Brierley CL (1979) Microbiological methods for the extraction and recovery of metals. In: Bull AT, Ellwood DC, Ratledge C (eds) Microbial technology: current state, future prospects. Society for General Microbiology symposium 29. Cambridge University Press, Cambridge, pp 263–307

Klauber C (2003) Fracture-induced reconstruction of a chalcopyrite (CuFeS2) surface. Surf Interface Anal 35:415–428

Kupka D, Kupsáková I (1999) Iron (II) oxidation kinetics in Thiobacillus ferrooxidans in the presence of heavy metals. In: Amils R, Ballester A (eds) Biohydrometallurgy and the environment toward the mining of the 21st century, part A. Elsevier, Amsterdam, pp 387–396

Lane DJ, Harrison AP Jr, Stahl D, Pace B, Giovannoni SJ, Olsen GJ, Pace N (1992) Evolutionary relationships among the sulfur- and iron-oxidizing eubacteria. J Bacteriol 174:269–278

Leduc LG, Ferroni GD (1994) The chemolithotrophic bacterium Thiobacillus ferrooxidans. FEMS Microbiol Rev 14:103–120

Markosyan GE (1972) A new acidophilic iron bacterium, Leptospirillum ferrooxidans. Biol Zh Arm 25:26

Meruane G, Salhe C, Wiertz J, Vargas T (2002) Novel electrochemical-enzymatic model which quantifies the effect of the solution Eh on the kinetics of ferrous iron oxidation with Acidithiobacillus ferrooxidans. Biotechnol Bioeng 80:280–288

Meruane G, Cárcamo C, Vargas T (2004) Kinetics of ferrous iron oxidation with Sulfolobus metallicus at 70°C. In: Tsezos M, Hatzikioseyian A, Remoundaki E (eds) Biohydrometallurgy: a sustainable technology in evolution, part I. National Technical University of Athens, pp 277–283

Nemati M, Harrison STL, Hansford GS, Webb C (1998) Biological oxidation of ferrous sulphate by Thiobacillus ferrooxidans: a review on the kinetic aspects. Biochem Eng J 1:171–190

Norris PR (1983) Iron and mineral oxidation studies with Leptospirillum-like bacteria. In: Rossi G, Torma AE (eds) Recent progress in biohydrometallurgy. Associazione Mineraria Sarda, Iglesias, pp 83–96

Norris PR (1990) Acidophilic bacteria and their activity in mineral sulfide oxidation. In: Ehrlich HL, Brierley CL (eds) Microbial mineral recovery. McGraw-Hill, New York, pp 3–27

Norris PR (1992) Thermophilic archaebacteria: potential applications. In: Danson MJ, Hough DW, Lunt GG (eds) The archaebacteria: biochemistry and biotechnology. Biochemical Society symposium 58. Portland, London, pp 171–180

Norris PR (1997) Thermophiles in bioleaching. In: Rawlings DE (ed) Biomining. Springer, Berlin Heidelberg New York, pp 247–258

Norris PR, Johnson DB (1998) Acidophilic microorganisms. In: Horikoshi K, Grant WD (eds) Extremophiles: life in extreme environments. Wiley, New York, pp 133–154

Norris PR, Kelly DP (1982) The use of mixed microbial cultures in metal recovery. In: Bull AT, Slater JH (eds) Microbial interactions and communities. Academic, London, pp 443–474

Norris PR, Owen JP (1993) Mineral sulphide oxidation by enrichment cultures of novel thermoacidophilic bacteria. FEMS Microbiol Rev 11:51–56

Norris PR, Simmons S (2004) Pyrite oxidation by halotolerant, acidophilic bacteria. In: Tsezos M, Hatzikioseyian A, Remoundaki E (eds) Biohydrometallurgy: a sustainable technology in evolution, part II. National Technical University of Athens, pp 1347–1351

Norris PR, Barr DW, Hinson D (1988) Iron and mineral oxidation by acidophilic bacteria: affinities for iron and attachment to pyrite. In: Norris PR, Kelly DP (eds) Biohydrometallurgy. Science and Technology Letters, Kew, pp 43–59

Norris PR, Burton NP, Foulis NAM (2000) Acidophiles in bioreactor mineral processing. Extremophiles 4:71–76

Okibe N, Gericke M, Hallberg KB, Johnson DB (2003) Enumeration and characterization of acidophilic microorganisms isolated from a pilot plant stirred-tank bioleaching operation. Appl Environ Microbiol 69:1936–1943

Parker A, Klauber C, Kougianos A, Watling HR, van Bronswijk (2003) An X-ray photoelectron spectroscopy study of the mechanism of oxidative dissolution of chalcopyrite. Hydrometallurgy 71:265–276

Plumb JJ, Gibbs B, Stott MB, Robertson WJ, Gibson JAE, Nichols PD, Watling HR, Franzmann PD (2002) Enrichment and characterisation of thermophilic acidophiles for the bioleaching of mineral sulphides. Miner Eng 15:787–794

Rawlings DE (2002) Heavy metal mining using microbes. Ann Rev Microbiol 56:65–91

Rawlings DE, Kusano T (2001) Molecular genetics of *Thiobacillus ferrooxidans*. Microbiol Rev 58:39–55

Rawlings DE, Tributsch H, Hansford GS (1999) Reasons why '*Leptospirillum*'-like species rather than *Thiobacillus ferrooxidans* are the dominant iron-oxidizing bacteria in many commercial processes for the biooxidation of pyrite and related ores. Microbiology 145:5–13

Sand W, Gehrke T, Jozsa P-G, Schippers A (2001) (Bio)chemistry of bacterial leaching – direct vs. indirect bioleaching. Hydrometallurgy 59:159–175

Sandström Å, Shchukarev A, Paul J (2005) XPS characterisation of chalcopyrite chemically and bio-leached at high and low redox potential. Miner Eng 18:505–515

Schippers A, Sand W (1999) Bacterial leaching of metal sulfides proceeds by two indirect mechanisms via thiosulfate or via polysulfides and sulfur. Appl Environ Microbiol 65:319–321

Shima S, Suzuki K-I (1993) *Hydrogenobacter acidophilus* sp. nov., a thermoacidophilic, aerobic, hydrogen-oxidizing bacterium requiring sulfur for growth. Int J Syst Bacteriol 43:703–708

Simmons S, Norris PR (2002) Acidophiles of saline water at thermal vents of Vulcano, Italy. Extremophiles 6:201–207

Tributsch H, Bennett, JC (1981) Semiconductor-electrochemical aspects of bacterial leaching. 1. Oxidation of metal sulphides with large energy gaps. J Chem Technol Biotechnol 31:565–577

Tributsch H, Rojas-Chapana JA (2000) Metal sulfide semiconductor electrochemical mechanisms induced by bacterial activity. Electrochimica Acta 45:4705–4716

Tuovinen OH, Kelly DP (1972) Biology of *Thiobacillus ferrooxidans* in relation to the microbiological leaching of sulphide ores. Z Allg Mikrobiol 12:311–346

Tyson GW, Lo I, Baker BJ, Allen EE, Hugenholtz P, Banfield JF (2005) Genome-directed isolation of the key nitrogen fixer *Leptospirillum ferrodiazotrophum* sp. Nov. from an acidophilic microbial community. Appl Environ Microbiol 71: 6319–6324.

Vásquez M, Moore ERB, Espejo RT (1999) Detection by polymerase chain reaction-amplification and sequencing of an archaeon in a commercial-scale copper bioleaching plant. FEMS Microbiol Lett 173:183–187

Yahya A, Roberto FF, Johnson DB (1999) Novel mineral-oxidizing bacteria from Montserrat (W.I.): physiological and phylogenetic characteristics. In: Amils R, Ballester A (eds) Biohydrometallurgy and the environment toward the mining of the 21st century, part A. Elsevier, Amsterdam, pp 729–739

11   The Microbiology of Moderately Thermophilic and Transiently Thermophilic Ore Heaps

JASON J. PLUMB, REBECCA B. HAWKES, PETER D. FRANZMANN

11.1   Introduction

As early as 1966, leaching experiments were reported that described the effects of temperature on the oxidation of mineral sulfide ores. In these experiments, a fivefold greater amount of copper dissolution from chalcopyrite ore was achieved at 65°C than at 35°C. The exothermic nature of oxidation reactions in mineral sulfide ore heap and dump operations was documented in the following year (Beck 1967). In his summary of the role of bacteria in copper mining operations, Beck described briefly the generation of temperatures between 60 and 80°C in operations containing low-grade sulfide mineral ores. It was considered that chemical rather than biological oxidation was the dominant process at these elevated temperatures, owing to the limited knowledge at that time of bioleaching microorganisms other than mesophiles such as *Acidithiobacillus ferrooxidans*. Since these early reports, a greater understanding of the microorganisms involved in bioleaching has been achieved, and the relationship between temperature and the rate of mineral sulfide dissolution well defined (Norris 1990; Stott et al. 2003). The discovery of moderately thermophilic and extremely thermophilic acidophiles has increased significantly our appreciation of the number of available bioleaching acidophiles, and created new processing options for bioprocessing of mineral sulfide ores and concentrates. For refractory minerals such as chalcopyrite, the increased Cu leaching rates achieved at higher temperatures (above 60°C) are of critical importance for successful industrial-scale processing of this sulfide. In the application of heap bioleaching for the processing of low-grade chalcopyrite ore, the successful operation of heaps at elevated temperatures in order to achieve optimal leaching rates and Cu recovery poses significant challenges. These include the design and engineering of heaps in order to generate and sustain high temperatures and favorable growth conditions for the sustained growth of thermophilic microorganisms. In addition, strategies to successfully inoculate the heap environment are needed to provide the appropriate population of thermophilic microorganisms for maximal copper bioleaching. For successful heap leaching of mineral sulfide ores other than chalcopyrite (e.g., chalcocite and refractory gold), operation at high temperature is not essential; however, depending on the ore content and grade, elevated temperatures are often achieved. To aid development

Biomining
(ed. by Douglas E. Rawlings and D. Barrie Johnson)
© Springer-Verlag Berlin Heidelberg 2007

of new technologies or strategies for thermophilic heap bioleaching, it is useful to summarize what is currently known about the microbiology of moderately thermophilic or transiently thermophilic heap bioleaching operations. This chapter summarizes some recent findings, discusses some of the fundamental aspects of the effect of temperature on the growth of bioleaching organisms and describes the microbiology of past and current moderately thermophilic to thermophilic heap bioleaching operations.

11.2   Heat Generation Within Bioleaching Heaps

Reactions involved in the biological oxidation of mineral sulfides are energy-producing or exothermic reactions. For example, the complete oxidation of pyrite and chalcopyrite yields 2,578 and 2,883 kJ mol^{-1}, respectively (Amend and Shock 2001). The amount of free energy produced by these oxidation reactions provides favorable thermodynamics for the growth of microorganisms and results in the generation of heat. The generation of elevated temperatures in a heap bioleaching environment is dependent on the local climate, ambient temperature, sulfide oxidation rate, irrigation strategy and rate, aeration rate, and the evaporation rate of solution from the system (Schnell 1997; Dixon 2000). Also, mineralogy type and grade influences the potential for generation of heat in bioleaching operations. This has been well studied, and the influences of these processes have been incorporated into computer models that describe heap bioleaching of mineral sulfides (Pantelis and Ritchie 1993; Dixon 2000; Leahy et al. 2005).

In reactor bioleaching processes that treat high-grade pyritic or arsenopyritic concentrates, such as the BIOX™ process, the heat generated during operation of the reactors led to a requirement for cooling of the reactors in order to provide favorable operating temperatures for the mesophilic and moderately thermophilic microorganisms involved (Dew et al. 1997). Similarly, in heap or dump leaching operations, the generation of heat from mineral sulfide oxidation has led to numerous observations of elevated temperatures. Testwork conducted in 1995 by Kennecott Utah Copper as part of the Bingham Canyon Heap Leach Program demonstrated the generation of heat from a 960,000-t heap containing low-grade (0.27% Cu) chalcopyrite ore with about 4% pyrite content (Ream and Schlitt 1997). Over almost 20 years of operation, heap exhaust air temperature (approximately 30–50°C) was maintained at about 30°C above ambient air temperature (approximately 0–20°C). Internal heap temperatures up to 66°C were measured at depths between 6 and 12 m beneath the heap surface. These observations are interesting for two reasons. Firstly, sufficient heat generation achieved and maintained temperatures suited to moderately thermophilic and thermophilic microorganisms in a heap situated at 2,070 m above sea level, where ambient temperatures in winter were on average close to 0°C.

Secondly, it is noteworthy that elevated temperatures could be produced by ore of such low grade.

Significantly greater temperatures were generated from a mineral sulfide ore of similar grade in the Newmont Mining Corporation's 800,000-t biooxidation pretreatment demonstration heap facility in Nevada (Shutey-McCann et al. 1997). Heap biooxidation of gold-bearing ore with a sulfide-S content of between 3.3 and 3.8% generated an average internal heap temperature of 52°C. In parts of the heap, temperatures in the range 60–75°C were measured. The generation of heap temperatures greater than 60°C provided ideal temperatures for the growth of thermophilic microorganisms, although no iron-oxidizing bacteria capable of growth at 65°C were detected. In a commercial biooxidation heap commissioned in 1999 by Newmont Mining Corporation, heap temperatures of up to 81°C were achieved through the processing of an ore with a sulfide-S content of 1.4–1.8% (Tempel 2003). The ability to generate high temperatures in these refractory gold-bearing ore bioleaching heaps of relatively low grade was due to the pyrite content of the ore and the establishment of active bioleaching microbial populations in the heap aided by a successful inoculation process.

Similar temperatures were achieved in an 85,000-t test heap at the Straits Resources Nifty Copper Operation in Australia (Readett et al. 2003). This test heap contained a blend of 60% chalcocite ore and 30% transitional ore comprising small amounts of chalcopyrite and pyritic material. During construction of the heap, thermocouples were installed within the heap at depths ranging from 0.5 to 2.5 m to measure changes in temperature. Figure 11.1

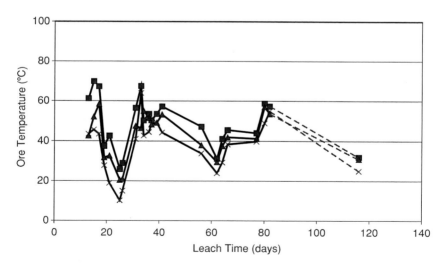

Fig. 11.1. Heap temperature profile of a chalcocite test heap operated by Straits Resources at the Nifty Copper Operation in Western Australia (reprinted from Readett et al. 2003, copyright 2003, with permission from the Minerals, Metals and Materials Society)

shows the heap temperatures measured during the first 120 days of operation. Heat generation in the heap occurred quite rapidly, with temperatures increasing to 60°C or greater within the first 20 days of operation. After reaching close to 70°C in some parts of the heap, temperatures decreased rapidly from day 19 onwards owing to failure of the aeration fan. Upon resumption of aeration, heap temperatures increased to levels almost as high as the initial peak temperature. Temperatures fluctuated gradually over time between 30 and 60°C until around day 120, after which further temperature measurements could not be made because of failure of the thermocouples owing to corrosion. The dependence of heat generation on aeration is implied given the marked decrease in heap temperature following cessation of aeration at day 19; however, the failure to sustain heap temperatures above 60°C may indicate dependence on factors other than adequate aeration for generation of elevated temperatures. One important factor could have been the depletion of readily oxidizable mineral after the initial period of elevated temperature. Analysis of the mineralogy of the ore in the Nifty test heap showed a pyrite content of 3% with some of the pyrite associated with pyrobitumen. It is thought that rapid oxidation of the pyrite/pyrobitumen component of the ore led to the generation of elevated temperatures in the heap. Presumably, as this material was depleted over time, heat generation declined, leading to lower heap temperatures. Another possible reason for the failure to sustain heap temperatures above 60°C was the absence of suitable microorganisms to maintain iron and sulfur oxidation at these elevated temperatures.

High temperatures can also occur in bioleaching heaps that contain low-grade ore without the use of forced aeration, such as occurs at the Myanmar Ivanhoe Copper Company Monywa Project (unpublished data). The Monywa Project is a successful heap leach solvent extraction–electrowinning operation that treats a low-grade (0.4% Cu mainly as chalcocite) ore with a pyrite content of up to 4%. The Monywa bioleaching heaps contain either crushed and agglomerated ore or run-of-mine ore. No aeration is provided, but the lifts are excavated routinely using a backhoe to improve solution permeability and overcome some of the physical barriers to leaching. Although the heaps are not instrumented with probes, freshly excavated material from a depth of between 2 and 5 m from within the heap showed temperatures of at least 46°C. As rapid cooling occurred upon exposure to the air, temperatures higher than 46°C undoubtedly occur within the heaps. Microbial populations in the Monywa heaps contain abundant moderately thermophilic strains, indicating that high temperatures were achieved and an active bioleaching population was established and maintained without forced aeration.

These examples show that sufficient heat generation to achieve temperatures suitable for thermophilic microorganisms in bioleaching heaps occurs even with relatively low grade ores; however, the maintenance and control of heap temperatures within an optimal range for bioleaching by thermophiles is more difficult. This is due largely to the problems associated with managing bioleaching heaps at large scale, and the limited process variables that can be manipulated in order to control temperature. The main process parameters used to control temperature in heaps are the rates of aeration and irrigation, although the use of permeable "shade" cloth covers to minimize heat loss through evaporation was attempted at a heap operation in Quebrada Blanca (Schnell 1997). An optimal combination of aeration and irrigation provides sufficient oxygen to achieve nonlimiting conditions for the oxidation of mineral sulfides, without quenching the heat generated through cooling by overirrigation. A more detailed analysis of these effects was provided by Dixon (2000). Optimization of heap temperature control will be required in order to exploit the known benefits of high-temperature bioleaching of chalcopyrite.

11.3   Effect of Temperature on Bioleaching Microorganisms

Increased chemical reaction rate with increased temperature is often described using the Arrhenius equation. Unlike chemical reactions, the effects of temperature on microbial growth rates do not obey typical Arrhenius kinetics. In the early 1980s, Ratkowsky et al. (1982, 1983) defined the relationship between temperature and bacterial growth rate, and published an equation that described bacterial growth rate throughout the entire biokinetic temperature range for any single strain. The Ratkowsky equation is as follows:

$$\sqrt{1/t} = b \, (T - T_{min}) \, (1 - e^{c \, (T - T_{max})}), \qquad (11.1)$$

where T is the temperature, t is generally the generation time or the time taken to reach a specific condition (e.g., the time taken for a culture to increase its optical density to 0.3 absorbance units), T_{min} is the theoretical extrapolated minimum temperature for growth, T_{max} is the theoretical extrapolated maximal temperature for growth, and b and c are fitting parameters. A study that compared available models for describing bacterial growth as a function of temperature concluded that modified forms of the Ratkowsky model were the most suitable (Zwietering et al. 1991). Although the Ratkowsky model is very useful for modeling the temperature dependence of microbial growth, conceptual hypotheses for the underlying biochemistry that defines the response of microorganisms to temperature are only beginning to be developed. Recently, Ratkowsky et al. (2005) concluded that the mechanisms that apply to denaturation of globular proteins at high and low temperatures probably affect microbial growth rate in a similar manner, as the 3D structures of enzymes change as temperatures approach the T_{min} and T_{max} of the enzyme. The status of biophysical structures of biomolecules such as enzymes involved in microbial growth at different temperatures determines the growth response of a microorganism to temperature.

Fluctuations in temperature can have a pronounced effect on the growth of individual microorganisms and also on the makeup of mixed microbial populations. Microorganisms can be grouped according to the temperature range over which they grow. Generally, the temperature-based groupings are based on the range of temperatures in which their T_{opt} occurs and are as follows: psychrophiles (T_{opt}<15°C), mesophiles (T_{opt}=15–40°C), moderate thermophiles (T_{opt}=40–60°C), thermophiles (T_{opt}=60–80°C) and hyperthermophiles (T_{opt}>80°C). To date, no truly psychrophilic bioleaching acidophiles have been described. Bioleaching by the hyperthermophilic *Sulfolobus solfataricus*-like strain JP3 was reported (Plumb et al. 2002); however, most known bioleaching microorganisms are mesophiles, moderate thermophiles or thermophiles. As temperature increases, a succession from psychrophiles and mesophiles, to moderate thermophiles, and finally to thermophiles and hyperthermophiles can occur. Whether or not this succession will occur in a mineral sulfide ore heap is dependent on an available source of these microorganisms indigenous to the ore, or as an introduced inoculum. It was suggested that bacterial succession in bioleaching heaps, in addition to changing in response to changing temperatures, may occur in response to changes in redox, pH, oxygen content and solution chemistry (Brierley 1999). With the exception of temperature, the response of bioleaching microorganisms to changes in these physical and chemical variables is relatively poorly understood. The response of pure cultures of bioleaching microorganisms to temperature when grown under ideal laboratory conditions is relatively well understood; however, more studies are required to define the microbial response to temperature in heap environments.

Franzmann et al. (2005) described the effect of temperature on the rates of iron and sulfur oxidation by selected bioleaching microorganisms. In this study the Ratkowsky model was used to determine the temperature "operating window" for 12 species of *Bacteria* and *Archaea*, mesophiles, moderate thermophiles and thermophiles, commonly associated with tank or heap bioleaching processes. Using the Ratkowksy model, theoretical values for the so-called cardinal temperatures, minimum temperature (T_{min}), optimum temperature (T_{opt}) and maximum temperature (T_{max}) for oxidation of iron and/or sulfur by each strain were determined. The results of this study are summarized in Figs. 11.2 and 11.3, and Table 11.1. Figure 11.2 shows a comparison of the temperature "operating windows" for five selected iron-oxidizing bioleaching microorganisms, with the square root of the reciprocal of the time to double ferric iron concentration plotted against temperature. The data show that the strains tested oxidized ferrous iron over a broad range of temperatures, although at temperatures lower than 20°C iron oxidation rates were relatively slow, and at around 60°C, there was limited overlap between the moderate thermophiles and the thermophile *Acidianus brierleyi*. Other thermophiles capable of oxidizing ferrous iron optimally at temperatures around 60°C await discovery. Not all strains tested were moderate thermophiles; however, each strain was capable of oxidizing ferrous iron

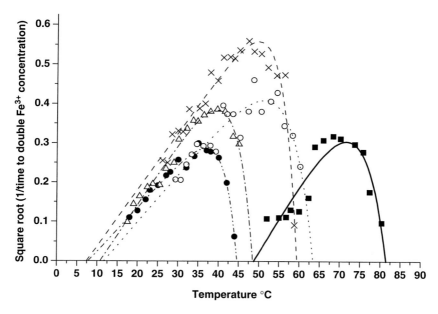

Fig. 11.2. Ratkowsky plots showing the relationship of temperature to the oxidation of iron for a range of common bioleaching organisms: *Leptospirillum ferrooxidans* (*closed circles*), *L. ferriphilum* (*triangles*), *Acidimicrobium ferrooxidans* (*crosses*), *Sulfobacillus thermosulfidooxidans* (*open circles*), *Acidianus brierleyi* (*squares*). (Reprinted from Franzmann et al. 2005, copyright 2005, with permission from Elsevier)

within the moderately thermophilic temperature range (40–60°C), although *At. ferrooxidans*, which is a true mesophile, was not tested for its relationship of temperature to ferrous iron oxidation rate. In a heap environment, the transition from mesophilic temperatures to thermophilic temperatures would probably involve contributions from mesophiles, moderate thermophiles and thermophiles for the generation of ever-increasing temperatures, and this emphasizes the importance of microbial succession within the heap.

Figure 11.3 shows good rates of sulfur oxidation over a broad range of temperatures for five selected sulfur-oxidizing bioleaching microorganisms, with the square root of the reciprocal of the time to increase sulfate concentration to 3,000 mg L^{-1} plotted against temperature. T_{min} values for each strain estimated by the fitting of the Ratkowsky equation to the data for each strain, especially for *At. ferrooxidans* and *At. caldus*, which gave large error values for T_{min} estimations (Table 11.1), are much lower than expected. Franzmann et al. (2005) suggested that this was caused by the fitting of the equation to sulfur oxidation rate data that showed a more rapid drop-off of the oxidation rate at temperatures above T_{opt} than is usually observed with growth data, and because for most of the strains, no data points were available for temperatures near T_{min}. Even with the poor estimation of T_{min} values, reasonable

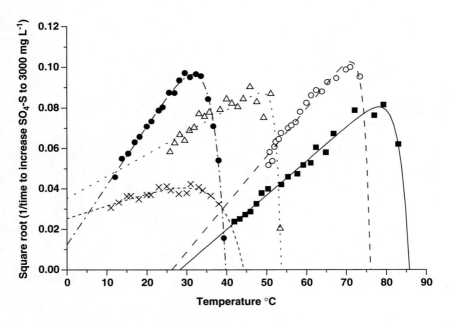

Fig. 11.3. Ratkowsky plots showing the relationship of temperature to the oxidation of elemental sulfur for a range of common bioleaching organisms: *Acidithiobacillus ferrooxidans* (*crosses*), *At. thiooxidans* (*closed circles*), *At. caldus* (*triangles*), *Sulfolobus metallicus* (*open circles*), *Ad. brierleyi* (*squares*). Note, the T_{min} values for sulfur oxidation are lower than expected because there are few data points near T_{min}, and the rapid drop-off of the sulfur oxidation rate above the T_{opt} for sulfur oxidation forces the curve to fit to a lower T_{min} value. (Reprinted from Franzmann et al. 2005, copyright 2005, with permission from Elsevier)

rates of sulfur oxidation were measured at temperatures lower than 20°C. *S. metallicus*, showed good rates of sulfur oxidation at temperatures around 60°C, which provided good coverage of the gap between the T_{opt} values for the moderate thermophiles and the thermophile *Ad. brierleyi*. All strains with the exception of *At. thiooxidans*, were capable of oxidizing sulfur within the moderately thermophilic temperature range. As was the case for ferrous iron oxidation, this indicates the potential role of mesophiles, moderate thermophiles and thermophiles in the transition from mesophilic to thermophilic temperatures in a heap environment.

The cardinal temperatures in Table 11.1 provide a description of the temperature "operating window" for each strain with an unprecedented level of precision. The equation derived for each species can be used to support observations made about the performance of each species in a bioleaching environment. For example, the slightly greater temperature tolerance of and faster iron oxidation rate by *Leptospirillum ferriphilum* explain why this ferrous iron oxidizer dominates over *L. ferrooxidans* in commercial bioleach tanks that operate at 40°C (Coram and Rawlings 2002) or in natural acid mine

Table 11.1. Extrapolated values of the cardinal temperatures for the activity of common bioleaching organisms for ferrous iron or sulfur oxidation derived from application of the Ratkowsky equation

Organism	Substrate	Cardinal temperatures (°C)		
		T_{min}[a]	T_{opt}	T_{max}
L. ferrooxidans	Fe^{2+}	7.8±1.4	36.7±0.6	44.6±0.2
L. ferriphilum	Fe^{2+}	10.7±1.1	38.6±1.5	48.5±1.0
Fp. acidiphilum	Fe^{2+}	12.7±6.1	39.6±0.9	47.2±0.7
Am. ferrooxidans	Fe^{2+}	7.4±3.0	48.8±1.1	59.5±0.3
"Fp. cupricumulans"	Fe^{2+}	22.6±0.8	53.6±0.5	63.2±0.2
Sb. thermosulfidooxidans	Fe^{2+}	11.7±5.3	51.2±3.1	63.5±1.4
Ad. brierleyi	Fe^{2+}	48.7±2.6	71.5±2.2	81.5±0.9
At. ferrooxidans	S^0	−40.4±19.1	29.6±5.5	44.3±2.6
At. thiooxidans	S^0	−4.2±0.8	32.8±0.3	39.7±0.1
At. caldus	S^0	−32.4±11.0	48.8±1.1	53.6±0.1
S. metallicus	S^0	26.3±1.6	71.0±4.4	76.0±0.1
Ad. brierleyi	S^0	28.2±1.2	78.1±1.8	85.8±0.9

[a] The T_{min} values for sulphur oxidation are lower than expected because there were few data points near T_{min}, and the rapid drop off of sulphur oxidation rate above the T_{opt} for sulphur oxidation forces the curve to fit to a lower T_{min} value.

drainage biofilms growing at 42°C (Tyson et al. 2004). A comparison of ferric iron production at 40°C calculated for each strain from its Ratkowsky equation shows that the "ferric iron doubling time" for *L. ferriphilum* was 6.7 h compared with 13.6 h for *L. ferrooxidans*. This indicates that *L. ferriphilum* should rapidly outcompete *L. ferrooxidans* in a continuous process at 40°C. Also, in a bioleaching heap operating at moderately thermophilic temperatures, *L. ferriphilum* would be expected to dominate over *L. ferrooxidans* as the temperature increased above 40°C owing to its ability to grow and oxidize ferrous iron at slightly higher temperatures. It should be noted that cardinal temperatures may vary for different strains of the same species. Coram and Rawlings (2002) showed that three out of nine strains of *L. ferriphilum* did not grow at 45°C.

Fitting the Ratkowsky model shows that oxidation rates for each microorganism increase with temperature until T_{opt} is exceeded. Beyond T_{opt} the oxidation rate decreases until it ceases at T_{max}. To incorporate this into a mathematical heap bioleaching model, Ritchie (1997) used the terms T_{sick} and T_{kill}, which define the temperatures at which the oxidation rate starts to decrease or ceases, respectively. These terms correspond well to T_{opt} and T_{max}, although by definition T_{sick} is slightly greater than T_{opt}, as only after T_{opt} is exceeded does the oxidation rate start to decrease. At temperatures lower than T_{sick} it was assumed that the oxidation rate was constant. Application of

the Ratkowsky model shows this is not the case. The square root of the reciprocal of the doubling time of the reaction product shows linearity between
T_{min} and T_{opt}. For a narrow temperature range below T_{opt} the effect of temperature on the reaction rate shows a typical Arrhenius response (Franzmann
et al. 2005). Incorporation of the Ratkowsky equation into heap leaching
models would be expected to enable better prediction of the effect of temperature on the behavior of bioleaching microorganisms.

In an ideal mixed microbial community, high oxidation rates will be maintained through microbial succession, assuming that microorganisms capable
of growing at higher temperature are present. Succession from mesophiles to
moderate thermophiles appears readily achievable either with or without
inoculation of heaps. This is due largely to the ubiquity of these bioleaching
acidophiles in mineral sulfide environments and their ability to rapidly colo-
nize a bioleaching heap solution circuit. In addition, most mineral sulfide
deposits from which heap bioleach pads are constructed reach temperatures
above the T_{min} of the moderate thermophiles, so populations of moderate thermophiles should be able to develop in such deposits, at least if temperature
was the only constraint on growth. Thermophilic bioleaching acidophiles are
less ubiquitous and are found mostly in volcanically or geothermally active
environments. Some exceptions to this were the discoveries of thermophilic
acidophiles in self-heating waste piles at a coal mine in Wales and an abandoned uranium mine in Germany, and also in the low-pH solutions draining
from a copper mine in Rio Tinto, Spain (Marsh and Norris 1983; Fuchs et al.
1995; Gómez et al. 1996). Despite these three discoveries of thermophiles in
nongeothermally active sites, which were not newly constructed environments, there are no definitive reports of the colonization of bioleaching heaps
by thermophilic acidophiles unless inoculation was undertaken. This poses a
problem for the operation of heaps at high temperatures as microbial succession from moderate thermophiles to thermophiles is unlikely without the
addition of thermophilic acidophiles to heaps via inoculation. Unless the
source material regularly reaches temperatures above the T_{min} of the thermophiles, 48°C in the case of *Ad. brierleyi*, there will be no opportunity for
thermophiles to grow within the deposit for which the heap was constructed,
at least in the shorter term. For economic recovery of metals from heaps, ideally, full recovery of metals would be achieved in less than 1 year.

11.4 Microbial Populations of Moderately Thermophilic or Transiently Thermophilic Commercial Bioleaching Heaps

The number of studies of the microbial populations associated with commercial bioleaching operations is continuing to expand. Characterization of
the microbial populations involved in bioleaching processes increases the
potential to optimize the biological phase of the process and hence improve

process performance. In most commercial bioleaching processes, operators rely on the application of largely unknown mixed populations of bioleaching microorganisms to their processes, although some examples where well-defined microbial populations have been used to inoculate heaps have been reported. The following text provides a summary of reports that describe the microbiology of inoculated and uninoculated moderately thermophilic and transiently thermophilic heap bioleaching operations.

11.4.1 Newmont Biooxidation Heaps

As described in Chap. 6, the Newmont Mining Corporation commissioned a commercial-scale biooxidation heap for the processing of gold-bearing mineral sulfide ore. This heap was inoculated using a patented process that involved the addition of a mixed inoculum of iron- and sulfur-oxidizing microorganisms to crushed and agglomerated ore particles (Brierley and Hill 1993). The inoculum was sprayed onto the ore as it was transported on a conveyor belt. The inoculum had the ability to initiate oxidation of the ore prior to commencement of the heap operation. The inoculated ore was then stacked into a heap for processing. The initial inoculum contained a mix of mesophilic and moderately thermophilic bacteria. Owing to the generation of significant heat, the design operating temperature of 60°C was exceeded, so after 6 months of operation thermophilic archaea were inoculated into the biosolution pond associated with the heap. The inoculation with thermophiles proved successful. Analysis of heap solutions showed that both mesophiles and thermophiles were maintained at 10^6–10^8 cells mL^{-1} (Tempel 2003). In contrast, the numbers of moderate thermophiles decreased to 10^2 cells mL^{-1}. An explanation for the decline in the numbers of moderate thermophiles was the establishment of two different growth environments within the biooxidation heap process that each had a temperature that was unsuitable for optimal growth by moderate thermophiles. Firstly, the hot parts (above 60°C) of the heap would have exceeded the T_{max} of the moderate thermophiles, leading to their death, or to the formation of heat-resistant spores. The high temperature favored the growth of thermophiles. Secondly, the average heap effluent temperatures of 40°C or lower indicate that the biosolution pond and the other parts of the solution circuit would have favored the growth of mesophilic microorganisms.

The outcomes of the Newmont biooxidation heap operations have provided useful insights into the way microbial populations respond to a changing heap environment. The generation of heat to achieve temperatures suited to the growth of thermophiles, and the subsequent establishment of an active population of thermophilic microorganisms is the first reported example of a successfully operated thermophilic bioheap for processing ore. This outcome provided useful know-how for other operators in their endeavors to operate bioheaps under similar conditions.

11.4.2   Nifty Copper Operation Heap Bioleaching

The chalcocite test heap constructed by Straits Resources at the Nifty Copper Operation in Australia was described earlier as another ore heap operation that achieved moderately thermophilic and transiently thermophilic heap temperatures (Readett et al. 2003). Unlike the Newmont heap, the test heap at Nifty was not inoculated with a mix of known bioleaching microorganisms. The ore was inoculated during agglomeration and irrigation using an unknown microbial population that had naturally developed at the site. The source of the inoculum was the existing solution circuit used for heap irrigation and the native microbial population associated with the ore. Solution and solid samples were collected periodically for microbiological analyses. Direct cell counts performed using a phase-contrast microscope showed that heap leachate collected over the duration of the test program contained 10^6–10^7 cells mL^{-1}, usually dominated by rod-shaped cells typical of species of the genus *Acidithiobacillus*. Viable cell counts using a most probable number (MPN) method showed that the heap leachate contained relatively low numbers of mesophiles (10^2–10^4 cells mL^{-1}) and usually fewer moderate thermophiles (10^3 cells mL^{-1} or fewer). Pure strains isolated from selected MPN cultures were identified as species of *Sulfobacillus* using 16S ribosomal RNA (rRNA) gene sequence analysis (unpublished data). Similar numbers of mesophiles and moderate thermophiles were detected in the intermediate leach solution used to irrigate the test heap. It was considered surprising that only low numbers of moderate thermophiles occurred in the leachate given that heap temperatures above 40°C were achieved. No thermophilic microorganisms were cultured from any of the samples obtained from the Nifty heap during this or subsequent work (Readett et al. 2003; Keeling et al. 2005). In subsequent analysis of leachate samples collected after decommissioning of the test heap, *Ferroplasma acidiphilum* like organisms were obtained in MPN cultures performed at 35°C using growth media containing yeast extract (unpublished data). The 16S rRNA gene sequences from these organisms showed 99% sequence homology with sequence data from the type strain of *Fp. acidiphilum*. Inclusion of yeast extract in growth media is generally considered essential to achieve good growth of members of this genus (Golyshina et al. 2000; Dopson et al. 2004). The MPN culture medium used to test for growth of mesophiles in the original samples collected from the test heap did not contain yeast extract, so it is possible that mesophilic *Ferroplasma* spp. were present in the test heap but were not cultured.

Culture-independent methods were applied also to describe the microbiology of the Nifty test heap. Microbial phospholipid fatty acid (PLFA) analysis was used to provide estimates of biomass and to characterize microbial diversity in solid samples obtained from the test heap. Analysis of solid samples collected from "cool," "warm" and "hot" parts of the test heap provided a biomass estimate of between 5×10^6 and 6×10^7 cells (g dry wt)$^{-1}$, using the conversion factor where 1 µg PLFA equates to 2.2×10^8 bacterial cells (Virtue et al. 1996). According to PLFA analysis, the heap solid samples contained relatively low amounts of biomass. This finding was consistent with the other quantitative data obtained using direct cell counts and MPN viable counts which showed relatively low cell numbers. The only specific PLFA signature detected in solid samples from the test heap was that of *At. ferrooxidans*. No phospholipids characteristic of Archaea were detected in any of the samples analyzed.

Analysis of heap solid samples was also performed using a molecular biology based method that used the polymerase chain reaction (PCR) to amplify specific fragments of the 16S rRNA gene from genomic DNA extracted from the solid sample. The amplified gene fragments were then separated using denaturing gradient gel electrophoresis (DGGE) and individual DNA fragments were sequenced to provide the identity of the source microorganism. 16S rRNA gene sequences from species of the genus *Sulfobacillus* were detected in the Nifty mine samples. The detection of members of this (mainly) moderately thermophilic genus using PCR-DGGE was consistent with the growth of *Sulfobacillus* spp. in MPN cultures at 50°C.

The Nifty test heap contained a microbial population of low biomass and limited diversity. It is conceivable that the decrease in heap temperature from greater than 60°C to less than 60°C was due to lack of a microbial population capable of iron and sulfur oxidation at thermophilic temperatures in order to sustain the high temperatures. The inoculation and successful colonization of the heap with thermophiles would be expected to increase mineral sulfide oxidation rates in the absence of naturally occurring thermophiles.

11.4.3   Myanmar Ivanhoe Copper Company Monywa Project

The Monywa project operated by the Myanmar Ivanhoe Copper Company described earlier is a successful chalcocite heap leach solvent extraction--electrowinning operation in Myanmar. Heaps of crushed and agglomerated ore or run-of-mine ore have been constructed on three different leach pads and are operated using standard heap leaching practices, except that the heaps are not forced-aerated. The heap solution chemistry is extremely acidic, with a solution pH of usually less than 1.5 and in some cases less than 1.0. This high acidity, thought to be due to the pyrite content in the ore, promotes high solution concentrations of ferric iron (up to 18 g L^{-1}), which creates favorable conditions for leaching. As mentioned earlier, heap temperatures of at least 46°C were measured in samples obtained immediately after excavation of parts of the heaps. Hawkes et al. (2005) described the microbiology of the Monywa Project heaps using a range of culture-dependent and culture-independent techniques to analyze heap leachate and heap solids samples.

Direct cell counts using a phase-contrast microscope showed that all leachate solutions contained cells, with cell numbers generally about 2×10^7

cells mL^{-1}. MPN cultures incubated at 35°C were dominated by small motile rods and showed very little ferrous iron oxidation after 2 weeks' incubation. Similar cell morphotypes were noted in cultures inoculated with either heap solids or heap leachates. Sulfur-oxidizing-bacteria with similar cell morphology were later isolated at 35°C and identified as strains of *At. caldus*. MPN counts obtained at 35°C were between 10^4 and 10^6 cells mL^{-1} for all samples tested. MPN cultures incubated at 50°C contained large nonmotile rods typical of *Sulfobacillus* spp. and small irregular-shaped cocci, with the nonmotile rods observed only in the lowest dilutions. All MPN cultures incubated at 50°C showed ferrous iron oxidation, with almost complete ferrous iron oxidation after 2 weeks' incubation. Viable cells were detected in all of the samples obtained from freshly excavated heaps. MPN counts obtained at 50°C were between 10^3 and 10^5 cells mL^{-1} for all samples tested. Compared with MPN counts obtained with samples from the chalcocite test heap at the Nifty Copper Operation, the Monywa heap MPN counts were generally at least 2 orders of magnitude greater.

Six different pure strains were isolated from the Monywa heap solids and leachate samples. These microorganisms were enriched from samples incubated at 35, 44 and 55°C. No growth was observed in enrichment cultures performed at 65°C. Strains BH1 and BH2 were isolated from cultures incubated at 55°C. Strain BH1 was isolated from the highest-dilution culture showing growth in a MPN viable count using serial decimal dilutions to extinction at pH 1.2. Strain BH2 was isolated by serial decimal dilutions to extinction from a separate enrichment using medium at pH 0.8. BH1 and BH2 were small nonmotile pleomorphic cocci capable of oxidizing iron mixotrophically when provided with yeast extract, and were identified as members of the archaeal genus *Ferroplasma* using 16S rRNA gene sequence analysis. The closest matching reference sequence was that of *Fp. acidiphilum* which was isolated from a bioleaching reactor in Kazakhstan (Golyshina et al. 2000). On the basis of only 95% sequence homology similarity with *Fp. acidiphilum* and subsequent phylogenetic analysis, strains BH1 and BH2 were defined as a new species of the genus *Ferroplasma*. Hawkes et al. (2005) proposed the name *"Fp. cyprexacervatum"*, later to become *"Ferroplasma cupricumulans"* (Hawkes et al. in press; (*cu.pri.cu'mu.lans. N.L. net. n. cuprum copper, L. part. pres. cumulans heaping up*) for this new moderately thermophilic species of the genus Ferroplasma. In addition to different 16S rRNA gene sequence, BH1 and BH2 differed physiologically from *Fp. acidiphilum*. The Ratkowsky model was used to fit data obtained using a temperature-gradient incubator and estimates of T_{min}, T_{opt} and T_{max} of 15, 55.2 and 63°C, respectively, were obtained (Fig. 11.4). According to the description of *Fp. acidiphilum* this species grew between 15 and 45°C with an optimum growth temperature of 35°C (Golyshina et al. 2000). The optimal pH for growth of strain BH2 was between pH 1.0 and 1.2, whereas *Fp. acidiphilum* grew optimally at pH 1.7 (Golyshina et al. 2000).

Two other strains were isolated from enrichment cultures incubated at 35°C. Their cells were motile short rods and were isolated on solid medium

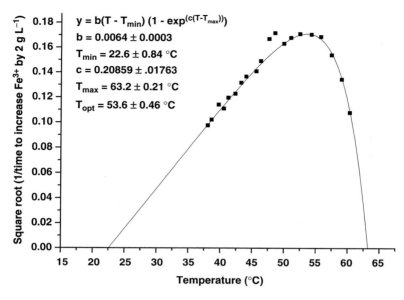

Fig. 11.4. Ratkowsky plot showing the effect of temperature on the oxidation of ferrous iron by "*Ferroplasma cyprexacervatum*" (strain BH2) and estimates of T_{min}, T_{opt} and T_{max} of 15, 55.2 and 63°C, respectively

containing 20 g L^{-1} ferrous sulfate and 2 g L^{-1} sodium tetrathionate. No ferrous iron oxidation occurred in association with the growth of these strains. They were identified as strains of *At. caldus* using 16S rRNA gene sequence analysis. The final two strains were isolated from enrichments at 44°C on autotrophic medium containing 20 g L^{-1} ferrous sulfate. Both strains were highly motile spirilla capable of ferrous iron oxidation and were identified as strains of *L. ferriphilum* using 16S rRNA gene sequence analysis. The rate of ferric iron production by one of the strains was used to determine its temperature and pH optima. Estimates of T_{min}, T_{opt} and T_{max} were 7.5, 41.3 and 50.9°C, respectively. The optimal pH range for growth was between 1.1 and 1.5, slightly lower than the optimal pH range for the type strain of *L. ferriphilum* of between 1.4 and 1.8 (Coram and Rawlings 2002) and that of another strain isolated from a stirred-tank bioleaching reactor (Okibe et al. 2003).

Culture-independent analysis of heap solids, heap leachates and primary enrichment cultures using PCR-DGGE analysis of 16S rRNA genes detected sequences similar to those of *At. caldus* and *L. ferriphilum* from ore and leachate samples, respectively. Sequences identical to that of "*Fp. cupricumulans*" were detected in heap leachate samples only, and sequences similar to those of *Sulfobacillus* spp. were detected in the primary enrichment cultures. Sequences similar to those of species of *Proteobacteria* unrelated to known bioleaching acidophiles were also detected in ore samples.

The combination of culture-dependent and culture-independent methods provided a good description of the microbial populations in the Monywa heap bioleaching operation. In general, the results of the culture-dependent analyses agreed with the results obtained using the culture-independent methods. The study showed that the low-pH environment in the nonaerated Monywa heaps supported a comparatively large microbial population of limited diversity that was characterized by the ability to grow at moderately thermophilic temperatures and extremely low pH. The microbial population in the Monywa heaps was remarkably similar to that found in a stirred-tank bioleaching reactor processing a mixed sulfide concentrate at about 45°C (Okibe et al. 2003). Although not determined, it seems possible that sufficient heat was generated within parts of the Monywa heaps to support the growth of thermophiles. The failure to detect any thermophiles in the Monywa heaps again supports the hypothesis of a restricted distribution of thermophiles in nongeothermally active areas.

11.5 Summary

The generation of heat from the exothermic oxidation of mineral sulfides is an important phenomenon associated with mineral sulfide bioleaching. When this heat generation occurs in large-scale ore heaps it is often unpredictable and difficult to control. Fluctuations in heap temperature can have significant effects on the microbial populations in the heap environment as evidenced by the examples from selected heap operations summarized in this chapter. Application of the Ratkowsky model to define the cardinal temperatures (T_{min}, T_{opt} and T_{max}) provides a precise means of describing the effect of temperature on microbial growth and activity. Depending on the microbial populations in the heap, increased temperature may promote greater microbial oxidation rates as the temperature approaches the T_{opt} of the microorganisms, or alternatively, it max lead to destruction of the microbial populations as the T_{max} of the organisms is exceeded. For the successful leaching of mineral sulfides, such as chalcopyrite, that require elevated temperature to leach rapidly, maintaining temperatures at or near T_{opt} will influence greatly the bioleaching performance.

The monitoring of microbial populations in heaps is crucial for evaluating bioleaching performance, as it provides an assessment of how microbial populations change in response to temperature and other heap conditions. A better understanding of microbial succession due to increases in temperature will enable better management of heaps in order to improve performance. The detection and monitoring of specific microorganisms, such as thermophilic microorganisms, will help the prediction of bioleaching performance as heap temperature exceeds moderately thermophilic temperatures, and will enable operators to decide if inoculation of the heap with thermophiles is required.

Monitoring heap microbial populations is not trivial and requires a combination of culture-dependent and culture-independent methods. The microbiological analysis of heap solids is difficult and advances in analytical techniques will provide improved monitoring methods for describing microorganisms associated with ore particles. Further studies of the microorganisms that develop in heap environments, especially moderately thermophilic or transiently thermophilic heaps, will no doubt lead to the discovery of novel bioleaching microorganisms.

Examination of moderately thermophilic and transiently thermophilic bioleaching environments in both tank and heap operations that have not been inoculated has revealed microbial populations of similar composition. The dominant microorganisms seem to consistently be strains of *Sulfobacillus thermosulfidooxidans*, *At. caldus*, *Ferroplasma* spp. and *L. ferriphilum*. Greater understanding of the physiology of these and other microorganisms will improve their utility in applications that require a transition from mesophilic to thermophilic heap bioleaching, such as is anticipated to be required for the economic bioleaching of chalcopyrite ores and concentrates.

References

Amend JP, Shock EL (2001) Energetics of overall metabolic reactions of thermophilic and hyperthermophilic *Archaea* and *Bacteria*. FEMS Microbiol Rev 25:175–243

Beck JV (1967) The role of bacteria in copper mining operations. Biotechnol Bioeng 9:487–497

Brierley CL (1999) Bacterial succession in bioheap leaching. In: Amils R, Ballester A (eds) Biohydrometallurgy and the environment toward the mining of the 21st century. International biohydrometallurgy symposium. Elsevier, Amsterdam, pp 91–97

Brierley JA, Hill D (1993) Biooxidation process for recovery of gold from heaps of low-grade sulfidic and carbonaceous sulfidic ore materials. US Patent 5,246,486

Coram NJ, Rawlings DE (2002) Molecular relationship between two groups of the genus *Leptospirillum* and the finding that *Leptospirillum ferriphilum* sp. nov. dominates in South African commercial biooxidation tanks that operate at 40°C. Appl Environ Microbiol 68:838–845

Dew DW, Lawson EN, Broadhurst JL (1997) The BIOX™ process for biooxidation of gold-bearing ores or concentrates. In: Rawlings DE (ed) Biomining. Springer, Berlin Heidelberg New York, pp 46–80

Dixon DG (2000) Analysis of heat conservation during copper sulfide heap leaching. Hydrometallurgy 58:27–41

Dopson M, Baker-Austin C, Hind A, Bowman JP, Bond PL (2004) Characterisation of *Ferroplasma* isolates and *Ferroplasma acidarmanus* sp. nov., extreme acidophiles from acid mine drainage and industrial bioleaching environments. Appl Environ Microbiol 70:2079–2088

Franzmann PD, Haddad CM, Hawkes RB, Robertson WJ, Plumb JJ (2005) Effects of temperature on the rates of iron and sulphur oxidation by selected bioleaching *Bacteria* and *Archaea*: application of the Ratkowsky equation. Miner Eng 18:1304–1314

Fuchs T, Huber H, Teiner K, Burggraf S, Stetter KO (1995) *Metallosphaera prunae*, sp. nov., a novel metal-mobilising, thermacidophilic *Archaeum*, isolated from a uranium mine in Germany. Syst Appl Microbiol 18:560–566

Golyshina OV, Pivovarova TA, Karavaiko GI, Kondrat'eva TF, Moore ERB, Abraham W, Lunsdorf H, Timmis KN, Yakimov MM, Golyshin PN (2000) *Ferroplasma acidiphilum* gen. nov., sp. nov., an acidophilic, autotrophic, ferrous-iron-oxidising, cell-wall-lacking, mesophilic member of the *Ferroplasmaceae* fam. nov., comprising a distinct lineage of the archaea. Int J Syst Evol Microbiol 50:997–1006

Gómez E, Blázquez ML, Ballester A, González F (1996) Study by SEM and EDS of chalcopyrite bioleaching using a new thermophilic bacteria. Miner Eng 9:985–999

Hawkes RB, Franzmann PD, O'Hara G, Plumb JJ *Ferroplasma cupricumulans* sp. nov., a novel moderately thermophilic, acidophilic archaeon isolated from an industrial-scale chalcocite bioleach heap. In press.

Hawkes RB, Franzmann PD, Plumb JJ (2005) Moderate thermophiles including *Ferroplasma cyprexacervatum* sp. nov. dominate an industrial-scale chalcocite heap bioleaching operation. In: Harrison STL, Rawlings DE, Petersen J (eds) Proceedings of the 16th international biohydrometallurgy symposium, pp 657–666

Keeling SE, Palmer M-L, Caracatsanis FC, Johnson JA, Watling HR (2005) Leaching of chalcopyrite and sphalerite using bacteria enriched from a spent chalcocite heap. Miner Eng 18:1289–1296

Leahy MJ, Davidson MR, Schwarz MP (2005) A model for heap bioleaching of chalcocite with heat balance: bacterial temperature dependence. Miner Eng 18:1239–1252

Marsh RM, Norris PR (1983) The isolation of some thermophilic, autotrophic, iron- and sulfur-oxidising bacteria. FEMS Microbiol Lett 17:311–315

Norris PR (1990) Acidophilic bacteria and their activity in mineral sulfide oxidation. In: Ehrlich HL, Brierley CL (eds) Microbial mineral recovery. McGraw-Hill, New York, pp 3–27

Okibe N, Gericke M, Hallberg KB, Johnson DB (2003) Enumeration and characterisation of acidophilic microorganisms isolated from a pilot plant stirred-tank bioleaching operation. Appl Environ Microbiol 69:1936–1943

Pantelis G, Ritchie AIM (1993) Optimising oxidation rates in heaps of pyritic material. In: Torma E, Wey JE, Lackshmanan, (eds) Biohydrometallurgy technologies, bioleaching processes, vol 1. The Minerals, Metals and Materials Society, Warrendale, pp 731–738

Plumb JJ, Gibbs B, Stott MB, Robertson WJ, Gibson JAE, Nichols PD, Watling HR, Franzmann PD (2002) Enrichment and characterization of thermophilic acidophiles for the bioleaching of mineral sulfides. Miner Eng 15:787–794

Ratkowsky DA, Olley J, McMeekin TA, Ball A (1982) Relationship between temperature and growth rate of bacterial cultures. J Bacteriol 149:1–5

Ratkowsky DA, Lowry RK, McMeekin TA, Stokes AN, Chandler R.E (1983) Model of bacterial culture growth rate throughout the entire biokinetic temperature range. J Bacteriol 154:1222–1226

Ratkowsky DA, Olley J, Ross T (2005) Unifying temperature effects on the growth rate of bacteria and the stability of globular proteins. J Theor Biol 233:351–362

Readett D, Sylwestrzak L, Franzmann PD, Plumb JJ, Robertson WR, Gibson JAE, Watling H (2003) The life cycle of a chalcocite heap bioleach system. In: Young CA, Alfantazi AM, Anderson CG, Dreisinger DB, Harris B, James A (eds) Hydrometallurgy 2003 – 5th international conference in honour of Professor Ian Ritchie, vol 1. Leaching and solution purification. The Minerals, Metals and Materials Society, Warrendale, pp 365–374

Ream BP, Schlitt WJ (1997) Kennecott's Bingham Canyon heap leach program, part 1: the test heap and SX-EW pilot plant. paper presented at ALTA 1997, copper hydrometallurgy forum, Brisbane

Ritchie AIM (1997) Optimization of biooxidation heaps. In: Rawlings DE (ed) Biomining. Springer, Berlin Heidelberg New York, pp 201–226

Schnell HA (1997) Bioleaching of copper. In: Rawlings DE (ed) Biomining. Springer, Berlin Heidelberg New York, pp 21–43

Shutey-McCann ML, Sawyer FP, Logan T, Schindler AJ, Perry RM (1997) Operation of Newmont's biooxidation demonstration facility. In: Hausen DM (ed) Global exploitation of heap leachable gold deposits. The Minerals, Metals and Materials Society, Warrendale, pp 75–82

Stott MB, Sutton DC, Watling HR, Franzmann PD (2003) Comparative leaching of chalcopyrite by selected acidophilic *Bacteria* and *Archaea*. Geomicrobiol J 20:215–230

Tempel K (2003) Commercial biooxidation challenges at Newmont's Nevada operations. In: 2003 SME annual meeting, preprint 03-067, Society of Mining, Metallurgy and Exploration, Littleton

Tyson GW, Chapman J, Hugenholtz P, Allen EE, Ram RJ, Richardson PM, Solovyev VV, Rubin EM, Rokhsar DS, Banfield JF (2004) Community structure and metabolism through reconstruction of microbial genomes from the environment. Nature 428:37–43

Virtue P, Nichols PD, Boon PI (1996) Simultaneous estimation of microbial phospholipid fatty acids and diether lipids by capillary gas chromatography. J Microbiol Methods 25:177–185

Zwietering MH, de Koos JT, Hasenack BE, de Wit JC, van't Riet K (1991) Modelling of bacterial growth as a function of temperature. Appl Environ Microbiol 57:1094–1101

12 Techniques for Detecting and Identifying Acidophilic Mineral-Oxidizing Microorganisms

D. Barrie Johnson, Kevin B. Hallberg

12.1 Biodiversity of Acidophilic Microorganisms That Have Direct and Secondary Roles in Mineral Dissolution

The majority of microorganisms that have the potential to accelerate the dissolution of minerals in acidic milieu are prokaryotic, though some species of fungi can also solubilize minerals, chiefly as a result of their production of metal-chelating organic acids such as citric and oxalic (Ehrlich 2002). The minerals that have been most widely studied in this context are sulfides, which may be found within igneous, and (nonoxidized) sedimentary and metamorphic rocks. Sulfide minerals may be categorized into those that are "acid-soluble" and others that are "acid-insoluble" (Rohwerder et al. 2003). The former (which includes chalcopyrite, $CuFeS_2$, and sphalerite, ZnS) are solubilized by protons (or soluble ferric iron), while the latter (which includes pyrite, FeS_2, and molybdenite, MoS_2) are attacked by chemical oxidants, of which ferric iron is the primary reagent in biomining operations and in most environmental situations. Since ferric iron is highly insoluble above approximately pH 2.5, it follows that it is only effective at oxidizing sulfides in acidic liquors. Two classes of acidophilic microorganisms are therefore important primary agents in accelerating the dissolution of sulfide minerals at low pH: (1) those that generate (sulfuric) acid by oxidizing sulfur, sulfide and reduced inorganic sulfur compounds (e.g., thiosulfate and tetrathionate) – the "sulfur-oxidizing prokaryotes," – and (2) those that oxidize iron(II) (ferrous iron) to iron(III) (ferric iron) – the "iron-oxidizing prokaryotes."

In addition to those acidophiles that have direct roles in accelerating mineral dissolution in acidic environments, there is a second group that can have a major impact in the overall process owing to their positive (or negative) interactions with the primary mineral sulfide-oxidizing prokaryotes. These "secondary" microorganisms are mostly heterotrophic, and include eukaryotes (some protozoa, microalgae and fungi) as well as bacteria and archaea. Some of these microorganisms may also contribute directly to mineral oxidation. For example, many *Acidiphilium* spp. can oxidize reduced forms of sulfur, but since characterized species (with the exception of *Acidiphilium acidophilum*) require organic carbon, these acidophiles only have important

Biomining
(ed. by Douglas E. Rawlings and D. Barrie Johnson)
© Springer-Verlag Berlin Heidelberg 2007

roles in mineral oxidation when they are members of consortia that include autotrophic prokaryotes.

The biodiversity of mineral-oxidizing acidophiles was described in Chap. 10; however, it is important to note, when considering the various approaches that are available for their detection and identification, that these microorganisms can have very different physiological characteristics. For example, amongst iron- and sulfur-oxidizing acidophiles, there are some, such as *Acidithiobacillus* spp. and *Leptospirillum* spp., that are autotrophic and fix inorganic carbon (CO_2), and others (obligate heterotrophs) that have an absolute requirement for organic carbon. An added complexity is another group can use either organic or inorganic carbon as a carbon source. These prokaryotes have sometimes been described as "mixotrophs," though this term has also been used to describe prokaryotes that use organic carbon as a carbon source and inorganic electron donors. A second, frequently used means of differentiating acidophiles is on the basis of their response to temperature. Although there are no fixed rules for delineating temperature cutoffs, mesophilic acidophiles are usually considered as microorganisms that grow optimally between 20 and 40°C, while thermo-acidophiles grow optimally at higher temperatures – moderate thermophiles at approximately 40–60°C and extreme thermophiles at above 60°C. It is important to note that these are temperature optima, and that some mesophilic acidophiles (such as some *Leptospirillum* spp.), will grow at above 40°C and some moderate thermophiles (such as *Acidicaldus organivorans*) at above 60°C. Thermal characteristics need to be considered when targeting specific mineral-oxidizing prokaryotes, by using suitable incubation temperatures. Another important physicochemical variable is pH. Acidophiles vary greatly in the degree to which they tolerate acidity. Some archaea, such as "*Ferroplasma acidarmanus*" (an iron-oxidizer) and *Picrophilus* spp. (moderately thermophilic heterotrophs) can grow at pH 0, while most mineral-oxidizing bacteria do not grow below pH 1. The terms "extreme acidophiles" and "moderate acidophiles" have been used to describe those microorganisms that grow optimally at pH<3 or pH 3–5, respectively. Recently, it has been shown that some novel species of iron-oxidizers (*Thiomonas* spp. and *Frateuria*-like γ-proteobacteria) are moderately acidophilic. Although it is unlikely that such bacteria are important in commercial heap-leaching and stirred-tank biomining operations, there is increasing evidence for their widespread distribution in mine drainage waters (Hallberg and Johnson 2003).

12.2 General Techniques for Detecting and Quantifying Microbial Life in Mineral-Oxidizing Environments

12.2.1 Microscopy-Based Approaches

Mineral-oxidizing microorganisms may be observed using light and electron microscopes. However, a significant drawback to most microscopic techniques, with the notable exceptions of those involving the use of specific oligonucleotide probes or specific antigen/antibody systems described in Sect. 12.5, is that they do not differentiate or identify different species of acidophilic prokaryotes. Phase contrast microscopy allows the visualization of live microorganisms, and acidophiles that occur as highly motile cells with distinctive morphologies (such as vibrioid and spirilla *Leptospirillum* spp.) are readily recognized. This is also the case with sporulating bacteria, such as *Sulfobacillus* spp.; endospores appear as bright, phase-dense objects within cells, though these can sometimes be confused with other cellular inclusions, such as sulfur granules. Counting chambers (such as Thoma cells) can be used to enumerate bacteria, but these require a minimum of more than 10^6 cells per milliliter to obtain meaningful counts. For more accurate and sensitive enumeration, microorganisms present in liquid samples may be concentrated by filtering through membrane filters (pore size 0.1–0.2 μm), and may be stained with a general dye. There are a number of alternative stains that may be used, but 4′,6-diamidino-2-phenylindole (DAPI) and SYBR Green (both of which bind to DNA, and also RNA in the case of SYBR Green) are particularly effective at detecting living (though not necessarily metabolically active) cells, and have greater sensitivity than other fluorescent dyes such as acridine orange.

One of the major problems in using microscopy to assess microbial populations in mineral-leaching environments is to differentiate living organisms from particulate matter. Liquid samples are usually much more easily processed than solid materials, though large amounts of suspended colloidal materials can be problematic. Ferric iron compounds are often amongst the most frequently encountered materials that interfere with microscopic detection of microorganisms in mineral-leaching environments. Oxalic acid may be used in a pretreatment to remove amorphous and poorly crystalline ferric iron minerals, such as schwertmannite and ferrihydrite (Ramsay et al. 1988), though these acids can also destroy the bacteria being examined. When dealing with solid materials, such as sulfidic minerals, the major problem often encountered in detecting and enumerating acidophiles is dislodging them from mineral surfaces. Many bacteria have a great propensity to attach to surfaces, and this is certainly the case for mineral-oxidizing acidophiles, many of which appear to attach selectively to certain minerals (e.g., sulfides, rather than to gangue minerals) and locations on the minerals (e.g., fracture planes, in preference to smooth surfaces; Tributsch and Rojas-Chapana 2004). Detaching bacteria and archaea from solid surfaces is often difficult, particularly those that synthesize copious amounts of exopolymeric materials, which increases the strength of their attachment (Harneit et al. 2005).

12.2.2   Biomass Measurements

There are a number of different techniques available for measuring total microbial biomass in environmental samples and most of these may be used, in theory, in acidic mineral-leaching situations. Modifications of standard protocols are often necessary owing to the high concentrations of protons, metals and other solutes that are characteristic of mineral leachates. However, a major drawback of many of these techniques is, again, their inability to differentiate between different types of microorganisms present in a particular system.

One compound that is ubiquitous in all living microorganisms is adenosine triphosphate (ATP), which is often considered to be the "energy currency" of life. Concentrations of ATP may be readily measured in laboratory and environmental samples. The most widely used method involves the use of an enzyme (luciferase, which is usually sourced from the tails of fireflies) and a long-chain aldehyde substrate (luciferin). In the presence of ATP, luciferase oxidizes luciferin, releasing photons which may be detected in a luminometer. This is a highly sensitive technique, readily detecting femtomolar concentrations of ATP, and commercial reagents and equipment for this purpose are widely available. A "typical" acidophilic mineral-oxidizing bacterium contains about 10^{-21} mol ATP, though this depends both on the size and on the metabolic activity of the organism. However, because it is an enzymatic assay, careful handling of mineral leachates and similar test materials is required to ensure that their elevated concentrations of dissolved metals and protons do not inhibit measurements of ATP.

Other cellular components of microorganisms may also be quantified to determine the total microbial biomass present in an environmental (or industrial) sample (White et al. 1997), though not all of these have been used in the context of biomining. These include proteins (Ramsay et al. 1988; Karan et al. 1996), muramic acid (for Gram-positive bacteria) and other cell wall components, and lipids (White et al. 1997). Most polar lipids in microorganisms are phospholipids, and may be determined as lipid phosphates or phospholipid ester-linked fatty acids (PFLA). PFLA patterns may be used to indicate spatial and temporal variations in microbial community compositions (White et al. 1997).

12.2.3   Measurements of Activity

Measurements of microbial activity are important in mineral-leaching environments. Because of the importance of redox transformations of iron and sulfur in these situations, measurements of rates of oxidation and reduction of these two elements are particularly meaningful. The oxidation of iron is readily assessed by monitoring changes is ferrous iron concentrations (e.g., using the "ferrozine" assay; Lovley and Phillips 1987), while sulfur oxidation can be determined by measuring changes in sulfate concentrations (Kolmert et al. 2000). Alternatively, an oxygen electrode (e.g., a Clarke electrode) can be used to determine rates of oxygen consumption associated with these reactions. Rates of ferric iron or sulfate reduction can also be determined by measuring concentrations of ferrous iron or sulfate, though in such cases test materials need to be incubated under zero or limiting oxygen concentrations. By coupling measurements of redox transformation of iron or sulfur with those of biomass (e.g., protein concentrations) it is possible to determine specific rates of transformations of these elements. Activities of heterotrophic acidophiles (apart from those involved in iron and sulfate reduction) may be assessed by monitoring changes in an organic substrate/electron donor, or by measuring rates of oxygen consumption, which is more appropriate when a complex organic substrate, such as yeast extract, is used.

The heat generated through the biologically catalyzed oxidation of pyrite and other sulfidic minerals may be used to quantify the activity of these acidophiles (Rohwerder et al. 1998). Since the amount of heat output in a typical sample is very small, a specialized instrument (a microcalorimeter) that can accurately record microwatt outputs of thermal energy is required. With this technique, a representative sample of solid material (typically 2–20 g) is placed into a glass container, leaving sufficient air space (50–70%) so that the reaction does not become oxygen-limited. The glass vessel is securely sealed and placed in the microcalorimeter. Following an equilibration period (typically 30 min) at a nominated temperature (generally the average temperature of the site from where the sample was taken), heat output is measured over a 2–4-h period. If the dominant sulfide mineral present is pyrite, its rate of oxidation (as milligrams per kilogram per hour) in the sample can be evaluated by using the ΔH value (1,546 kJ mol^{-1} pyrite) that has been calculated for the complete oxidation of this mineral to ferric iron and sulfate (Schippers and Bosecker 2005).

12.3   Cultivation-Dependent Approaches

Cultivation-based techniques, as the term implies, have the prerequisite that the microorganism(s) in question can be grown under defined conditions in the laboratory. While there is the possibility, or even probability, that mineral-leaching environments contain microorganisms that may not be cultivated *in vitro*, currently most known acidophiles can be cultivated in synthetic liquid and also on solid media.

12.3.1   Enrichment Media

Liquid media for enriching populations of mineral-leaching and other acidophilic prokaryotes may be variously formulated to encourage the growth of target microorganisms. Also important is the temperature, pH and oxygen status under which enrichment cultures are incubated, in order to select for specific groups of acidophiles (thermophiles, anaerobes, etc.). Even so, care needs to be taken when interpreting data from enrichment cultures. For example, the widespread use of ferrous sulfate liquid media to enrich for iron-oxidizing chemolithotrophs invariably favors the growth of *Acidithiobacillus ferrooxidans* rather than *Leptospirillum* spp., as the former has a faster growth rate on ferrous iron. On the other hand, with use of suitably formulated enrichment cultures together with selective solid media, it is possible to both select for and isolate most characterized mineral-oxidizing and other acidophilic microorganisms (Table 12.1). Alternatively, enrichment cultures may be diluted to extinction in order to obtain the dominant microorganism present; the highest dilution displaying positive growth is assumed to be a pure culture of the most numerous organism in the original enrichment culture.

12.3.2   Most Probable Number Counts

Acidophilic microorganisms can be enumerated using a statistical approach whereby samples are diluted and inoculated into a series of tubes containing growth media. Following incubation, the tubes are examined and scored for positive or negative growth, and the results are compared with standard tables to determine the most probable number (MPN) of microorganisms present (Schippers and Bosecker 2005). By varying the growth medium and

Table 12.1. Possible routes for isolating target mesophilic and moderately thermophilic acidophilic mineral-oxidizing prokaryotes, based on initial growth in liquid enrichment cultures followed by isolation on solid media

Target acidophile	Liquid medium for enrichment (pH; temperature, °C)	Streak to plate[a]
Acidithiobacillus ferrooxidans	Ferrous sulfate (2.0; 30)	Fe<u>o</u>
At. thiooxidans	Elemental S (2.5; 30)	FeS<u>o</u>
At. caldus	Elemental S (2.5; 45)	FeS<u>o</u>
Leptospirillum spp.[b]	Pyrite (1.5; 37)	iFe<u>o</u>
"*Ferrimicrobium acidiphilum*"	Ferrous sulfate/yeast extract (2.0; 30)	Fe<u>o</u>
Sulfobacillus thermosulfidooxidans	Ferrous sulfate/yeast extract (1.8; 45)	FeS<u>o</u>
Sb. acidophilus	Ferrous sulfate/yeast extract (1.8; 45)	FeS<u>o</u>
"*Sb. montserratensis*"	Ferrous sulfate/yeast extract (1.8; 30)	FeS<u>o</u>
Acidimicrobium ferrooxidans	Ferrous sulfate/yeast extract (1.8; 45)	FeS<u>o</u>
Ferroplasma spp.	Ferrous sulfate/yeast extract (1.5; 37)	FeS<u>o</u>
Thiomonas spp.	Ferrous sulfate/thiosulfate/ yeast extract (4.0, 30)	FeT<u>o</u>

[a]See Table 12.2
[b]Protracted incubation (10–20 days) is recommended in order to obtain greater numbers of *Leptospirillum* spp. than *At. ferrooxidans* in liquid pyrite-containing medium.

incubation conditions, it is possible to enumerate different physiological groups of acidophiles. Again, care is required to correctly interpret the data obtained. For example, 9K liquid medium (Silverman and Lundgren 1959) has frequently been used to determine MPN counts of iron-oxidizing acidophiles, yet neither heterotrophic iron-oxidizers such as *Ferroplasma* and "*Ferrimicrobium*," nor *Thiomonas*-like moderate acidophiles grow in 9K medium, owing to the absence of organic carbon and inappropriate pH, respectively. Therefore, MPN counts using 9K or other highly acidic, organic substrate-free liquid media are likely to underestimate total numbers of iron-oxidizing prokaryotes present in a sample.

12.3.3   Cultivation on Solid Media and on Membrane Filters

Solid media are used routinely in microbiology to cultivate neutrophilic microorganisms. In contrast, growth of acidophiles on solid media, in particular iron-oxidizers and sulfur-oxidizers, was, for many years, reported to be nonreproducible or, in the case of *Leptospirillum* spp., nonexistent. A number of different media formulations have been published, mostly designed to improve the plating efficiency of *At. ferrooxidans* (described in Johnson 1995). A radically different approach, referred to as the "overlay technique," has been used to facilitate the growth of the entire range of known moderately thermophilic and mesophilic acidophiles (Johnson and McGinness 1991; Johnson 1995; Hallberg and Johnson 2003; Johnson et al. 2005). With this, toxic organic materials, invariably present in agar-based gelling agents and also produced during plate incubation as a result of on-going acid hydrolysis of the polysaccharide, are removed by a heterotrophic acidophile which is incorporated into the lower layer of a two-layered gel. A number of different overlay plate formulations have been used to select for different groups of mineral-oxidizing and other acidophiles (Table 12.2). In most cases, the heterotroph used in the gel underlayer is an *Acidiphilium cryptum* like bacterium (strain SJH), though an "*Acidocella aromatica*" like isolate (strain PFBC) is superior for plating heterotrophic acidophiles, as this bacterium metabolizes the major toxins (organic acids) present in the solid media, but is unable to utilize materials such as yeast extract and glucose, which serve as carbon/energy sources for heterotrophs present in the inoculum.

There are a number of inherent advantages in using the overlay technique for detecting and enumerating mineral-oxidizing and other acidophilic bacteria, though a major detraction is the time required (approximately 3–20 days) for microorganisms to develop on solid media. Firstly, direct plating of samples eliminates the problems of bias (and potential elimination of poorly competitive strains) associated with enrichment cultures, allowing a more accurate picture of indigenous biodiversity of samples to be assessed. Secondly, colonies of acidophilic microorganisms generally display very different morphological characteristics, which allows them to be differentiated

Table 12.2. Overlay solid media for isolating and cultivating mineral-oxidizing and other acidophilic microorganisms

Medium code	Energy sources	pH	Target isolates
Acidiphilium SJH in the gel underlayer			
iFeo	Ferrous iron	~2.5	Fastidious iron-oxidizers
Feo	Ferrous iron/TSB	~2.5	Iron-oxidizers (also some heterotrophs)
FeSo	Ferrous iron/tetrathionate/TSB	~2.5	Iron-oxidizers and sulfur-oxidizers (also some heterotrophs)
FeTo	Ferrous iron/thiosulfate/TSB	~4.0	Moderately acidophilic iron-oxidizers, sulfur-oxidizers and heterotrophs
Acidocella PFBC in the gel underlayer			
YE3o	Yeast extract	~3.0	Extremely acidophilic heterotrophs
YE4o	Yeast extract	~4.0	Moderately acidophilic heterotrophs

All of the media are gelled with agarose (e.g., Sigma type I) at a final concentration of 0.5% (w/v). In most cases, three separate solutions are prepared: (1) basal salts/tryptone soya broth (*TSB*) (or yeast extract)/(tetrathionate or thiosulfate), acidified to pH 2.5 (pH 6.5 in the case of FeTo medium) and heat-sterilized; (2) ferrous sulfate (a 1 M acidified (pH 2.0) and filter-sterilized stock solution); (3) a concentrated (2%) heat-sterilized agarose solution. On cooling to about 45°C, the three solutions are combined to give the desired final concentrations. In the case of ferrous iron, these are 25 mM (iFeo, Feo and FeSo media), 5 mM (FeTo medium) and 500 μM (yeast extract overlay media). The final concentration of tetrathionate in FeSo medium is 2.5 mM, and that of thiosulfate (in FeTo medium) is 10 mM. TSB is added to a final concentration of 0.025% (w/v) in Feo, FeSo and FeTo media, and yeast extract is added to a final concentration of 0.02%. Combined media are inoculated with either *Acidiphilium* SJH or *Acidocella* PFBC (grown in corresponding liquid media) and poured as thin gels in sterile Petri plates. Once these have solidified, a top layer of sterile medium is added.

and aids their preliminary identification (Johnson et al. 2005). For example, colonies of iron-oxidizing isolates are readily identified from their deposition of ferric compounds, causing them to be encrusted in rust-like materials. The sizes and detailed morphologies of colonies of iron-oxidizers differ between species and often strains, though a single strain will usually produce uniform colonies on any particular medium formulation. Confirmation of the identities of plate isolates is most accurately made by using a biomolecular approach (generally by analysis of their 16S ribosomal RNA, rRNA, genes; Johnson et al. 2005).

A significant limitation of agar(ose)-gelled solid media is their use at elevated temperatures (above approximately 50°C), so they are generally limited to cultivating mesophilic and moderately thermophilic acidophiles. Gelrite is an alternative polysaccharidic gelling agent produced from bacteria, in contrast to agar, which is derived from algae. Solid media produced using Gelrite are far more thermostable than agar gels, their strength and stability being determined by the concentrations of divalent cations present. Lindström and Sehlin (1989) used a double-layered Gelrite-gelled solid medium to grow colonies of the thermo-acidophile *Sulfolobus acidocaldarius* strain BC65 (now known to be *Sulfolobus metallicus*). The inoculated plates were incubated at 65°C, and small colonies were evident after about 4 days of incubation. The efficiency of the plating technique was essentially 100%.

A different approach that circumvents the problems of the toxicity and thermostability of gelling agents is to grow acidophiles on membrane filters. Tuovinen and Kelly (1973) pioneered this work with *At. ferrooxidans*, filtering liquid cultures through membrane filters (which varied in their effectiveness) that were then placed onto agar-gelled solid media, as avoiding direct contact of the cells with agar was considered to be very important. Later, de Bruyn et al. (1990) eliminated the need for a gelled medium altogether, floating inoculated polycarbonate membranes on low-pH (1.6) ferrous sulfate liquid medium. Colonies of a fastidious *At. ferrooxidans* like acidophile were evident after about 5 days of incubation at 30°C.

12.4 Polymerase Chain Reaction (PCR)-Based Microbial Identification and Community Analysis

Advances in molecular biological tools have transformed the field of microbial ecology, allowing for the analysis of microbial communities without prior cultivation and revealing a remarkable diversity of microorganisms (Hugenholtz et al. 1998), including acidophiles. The most commonly used approaches in molecular microbial ecology focus on the small subunit of the rRNA, SSU rRNA, or its gene, and these have been described in detail elsewhere (Amann et al. 1995; Head et al. 1998). In general, the biodiversity of microorganisms in a particular ecological niche can be assessed by extraction of nucleic acid from that environment followed by amplification of the SSU rRNA gene by the polymerase chain reaction (PCR) using universal or strain-specific primers. The resulting PCR products are then subjected to a variety of analyses to assess microbial identification, community diversity and composition and, with some techniques, quantitative analysis of specific microorganisms within a community. While these molecular approaches are useful for understanding microbial communities without the reliance on cultivation techniques, they are not always capable of providing information concerning the physiology of unknown microorganisms. Therefore, the combined use of biomolecular tools and cultivation techniques is recommended (Fig. 12.1), as this approach not only provides information on the diversity of microorganisms in samples being analyzed, but also results in the isolation of indigenous microflora that can be then subjected to further studies in the laboratory.

The focus of most molecular ecological studies, the SSU rRNA, is common to all organisms, though the molecule is somewhat smaller (16S) in prokaryotes than in eukaryotes (18S). It is a highly conserved molecule (e.g., it shares a high degree of sequence identity among all organisms) that is made up of regions of near identity interspersed with regions of high sequence variability.

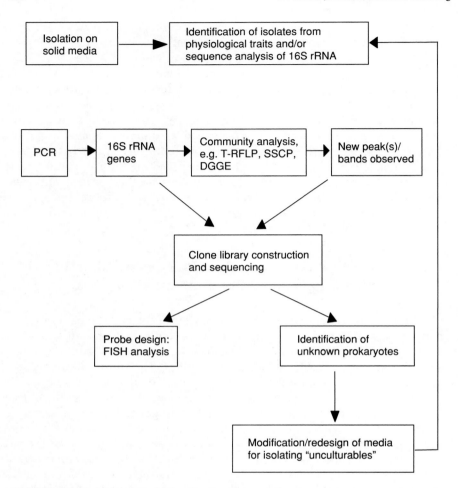

Fig. 12.1. Scheme for analysis of acidophilic communities using the dual approach of cultivation and molecular microbial ecology. Enrichment cultures may be used ahead of plate isolation or polymerase chain reaction (*PCR*) based analyses, though the resulting diversity of microorganisms is likely to be greatly diminished, compared with that for non-enriched samples. *rRNA* ribosomal RNA, *T-RFLP* terminal restriction fragment length polymorphism, *SSCP* single-stranded DNA conformation polymorphism, *DGGE* denaturing gradient gel electrophoresis, *FISH* fluorescent *in situ* hybridization

The regions of the 16S rRNA that are nearly identical in all organisms are targets for PCR primers, to facilitate the amplification (formation of multiple copies) of the gene from all microorganisms – the so-called universal primers. Universal primer pairs are available to amplify nearly the entire 16S rRNA gene from bacteria and archaea, as well as internal portions of the gene (Table 12.3). In contrast, the regions of high variability can serve as targets for species-specific PCR primers.

Table 12.3. Target sites, sequences and specificity of oligonucleotides used for amplification of 16S ribosomal RNA (*rRNA*) genes from bacteria and archaea, including those used to amplify variable regions of the gene for community analysis by denaturing gradient gel electrophoresis or single-stranded DNA conformation polymorphism (*SSCP*)

Primer[a]	Target site[b]	Sequence (5' to 3')[c]	Specificity	Reference
27F	8–27	AGAGTTTGATCMTGGCTCAG	Bacteria	Lane (1991)
1387R	1404–1387	GGGCGGWGTGTACAAGGC	Bacteria	Marchesi et al. (1998)
20F	3–20	TCCGGTTGATCCYGCCRG	Archaea	Orphan et al. (2000)
1392R	1407–1392	GACGGGCGGTGTGTRC	Universal	Lane (1991)
1492R	1513–1492	TACGGYTACCTTGTTACGACTT	Universal	Lane (1991)
357F-GC[d]	341–357	CCTACGGGAGGCAGCAG	Bacteria	Lane (1991)
907R	926–907	CCGTCAATTCMTTTGAGTTT	Universal	Lane (1991)
ARC363F-GC	344–363	ACGGGGYGCAGCAGGCGCGA	Archaea	Raskin et al. (1994)
ARC915R	934–915	GTGCTCCCCCGCCAATTCCT	Archaea	Stahl and Amann (1991)
w49[e] (F)	331–350	ACGGTCCAGACTCCTACGGG	Bacteria	Zumstein et al. (2000)
w34[e] (R)	533–515	TTACCGCGGCTGCTGGCAC	Universal	Zumstein et al. (2000)
w36[f] (F)	333–348	TCCAGGCCCTACGGGG	Archaea	Leclerc et al. (2001)

[a]Numbers corresponds to the position of the 16S rRNA to which the 3' end of the primer anneals. *F* and *R* are forward and reverse primers, respectively.
[b]*Escherichia coli* numbering of Brosius et al. (1981)
[c]*M* is A or C, *Y* is C or T, *R* is A or G, *W* is A or T
[d]GC is a 40 nucleotide GC-rich sequence attached to the 5' end of the primer for use with DGGE, and has the sequence 5'-CGC CCG CCG CGC CCC GCG CCC GTC CGC CGC CCC CCG CCC G-3' (Muyzer et al. 1996).
[e]w49 is the forward primer and w34 is the reverse primer for amplification of the V3 variable region of bacterial 16S rRNA genes for SSCP analysis.
[f]w36 is the forward primer used in conjunction with the w34 primer to amplify the V3 region of archaeal 16S rRNA genes for SSCP analysis.

12.4.1   Rapid Identification and Detection of Specific Acidophiles in Communities

In the simplest application of molecular biology to identify microorganisms, these primers are used to amplify the 16S rRNA gene from an isolate for sequence analysis. Following amplification, the gene can be cloned prior to sequencing; however, given that the Taq polymerase used to replicate the DNA during PCR is prone to making errors (i.e., inserts incorrect nucleotides), it is often better to sequence using the PCR product as the template. A cloned gene will contain any misincorporated nucleotides that have been inserted into that molecule and may affect sequence comparison, while the PCR product will contain a mixture of faithfully replicated genes in addition to the few mistakes. Sequencing with the latter as a template will most likely assure that the correct sequence of the gene is obtained. With use of this gene sequence, a phylogenetic comparison (Sect. 12.4.4) with those of known microorganisms can yield a presumptive identification of that isolate.

Biomolecular approaches other than cloning and sequencing have also been used for the rapid analyses of mineral-leaching populations. These include restriction enzyme mapping of the 16S rRNA genes amplified from bioleachate liquors (Rawlings 1995; Rawlings et al. 1999; Johnson et al. 2005), often referred to as amplified ribosomal DNA restriction enzyme analysis (ARDRA or ARDREA). For use of this technique, however, 16S rRNA gene sequence information is required. Another genetic marker that can be amplified with universal PCR primers and used to rapidly study populations of acidophiles is the spacer region between the 16S and 23S rRNA genes which can be distinguished on the basis of its size in different prokaryotes (Pizarro et al. 1996; Bergamo et al. 2004). While these regions can vary in size even within a species, partial sequencing of the 16S rRNA gene adjacent to a spacer region can be used to confirm which organism is represented by a particular amplified spacer region. Species-specific primers, on the other hand, can be used to determine the presence of microorganisms of interest in any given environment. Such an approach has been used to detect iron- and sulfur-oxidizing acidophiles in laboratory-scale bioreactors (De Wulf-Durand et al. 1997).

12.4.2 Techniques for Microbial Community Analysis

In contrast to the amplification of a 16S rRNA gene from an individual microorganism, community analysis based on 16S rRNA genes requires a means to separate genes that have different sequences, as these represent distinct microbial species. This can be accomplished by constructing a 16S rRNA gene library in a host bacterium, usually *Escherichia coli*. These libraries are constructed by cloning the individual PCR products into plasmids and transforming *E. coli* with these plasmids. Each colony of *E. coli* arises from a single cell that has been transformed with one plasmid (i.e., has an individual cloned 16S rRNA gene). Genes that are contained within these colonies are then screened to determine their uniqueness, and to determine the number of each unique gene cloned. Screening is typically done by restriction enzyme fragment length polymorphism (RFLP) following PCR amplification using PCR primers that target the plasmid (to avoid amplifying the *E. coli* gene). By plotting the number of unique clones identified against the total number of clones screened, important information on the comprehensiveness of the clone library (in terms of accounting for the entire microbial community under investigation) can be obtained. Once unique genes ("clones") have been identified by RFLP analysis, they can be sequenced to determine the phylo-genetic relationship of the microorganism represented by that gene.

While gene libraries are powerful tools for microbial community analysis, their construction and analysis is a time consuming and expensive process and thus they are not usually used to compare spatial or temporal variation of microbial communities. Rapid methods for such comparisons exist, and are often referred to as gene "fingerprinting." Two of these methods are based on analysis of a portion of the gene amplified by PCR using primers that target internal regions of the RNA gene, while the third can be applied to the nearly completely amplified genes using the universal primers. In the first of the techniques (Muyzer and Smalla 1998), portions of the 16S rRNA gene containing the variable regions are amplified using archaeal or bacterial primers with a "GC clamp" (Table 12.3), and the resulting products are sepa-rated by denaturing gradient gel electrophoresis (DGGE). The denaturant typically used is a mixture of urea and formamide, or else a temperature gra-dient is used (often referred to as TGGE); both serve to separate (melt) the DNA into single strands that are prevented from completely denaturing by the GC clamp. The double stranded genes migrate through the gel up to the point at which they are partially denatured, which is sequence-dependent. In this way, genes with different sequences will separate from each other. The resulting bands can be quantified by densitometry to give a measure of the relative abundance of individual microorganisms within a population. Further work, however, is required to assign bands to specific microorgan-isms, such as band extraction and gene sequencing, or to perform DGGE analysis on clones from a clone library made in parallel with the DGGE analy-sis. Care must be taken with the latter approach, however, as two different sets of primers are used for the respective PCR, which may yield different results in terms of relative abundances of microorganisms.

Another fingerprinting method, single-stranded DNA conformation poly-morphism (SSCP), is also based on the amplification of a small, variable region of the 16S rRNA gene, and also relies on sequence-specific secondary structure for the separation of the resulting PCR products (Orita et al. 1989). In this case, however, the small product is denatured and allowed to renature during electrophoresis, which affects the mobility of the products through the gel. The renaturation occurs in a sequence-dependent manner, thus allowing the separation of unique genes within a community. In SSCP, the PCR prod-ucts are labeled with a fluorescent molecule during PCR (using a primer syn-thesized with a fluorochrome attached to the 5′ end), and are standardized (e.g., given a "size") by comparison with known size standards labeled with a different fluorochrome. This allows for comparison of SSCP products with those of known acidophiles, or with products from clones, to identify microorganisms within a community. In addition, the relative abundance of the products can be determined by measuring the fluorescence intensity of each product relative to the total fluorescence of all products. As with DGGE/TGGE, though, comparison of relative abundances of microorganisms determined by clone libraries and SSCP may differ owing to the different primer sets used.

A third fingerprinting method commonly employed for microbial ecological studies is terminal RFLP (T-RFLP; Marsh 1999). Here, nearly complete 16S rRNA genes in a sample are amplified using a fluorescently labeled primer to yield a mixture of labeled 16S rRNA genes. These amplification products are digested with restriction enzymes to produce labeled terminal restriction enzyme fragments (T-RFs), which are then denatured, and the single-stranded T-RFs are separated by electrophoresis under denaturing conditions (e.g., in the presence of urea). Comparison of the migration time of the T-RFs to internal standards, labeled with a different fluorochrome, allows accurate sizing of the fragments to within ±1 nucleotide. Ideally, each T-RF represents a single microorganism, though in practice microorganisms of different species often share one T-RF. Therefore, digestion with up to three different restriction enzymes is usually necessary to accurately identify a microorganism on the basis of T-RF size. As in SSCP, the relative abundance of microorganisms represented by a T-RF can be determined by measuring the fluorescence of each T-RF relative to the sum of the fluorescence. An advantage of T-RFLP over the other fingerprinting methods is that computer analysis of gene sequences can be carried out to determine the theoretical T-RFs of known microorganisms for comparison with T-RFs obtained from environmental samples. In this way, identification of microorganisms in a sample can be determined rapidly. Alternatively, a database of experimentally derived T-RFs from known microorganisms can be created to allow for environmental microorganism identification. Lastly, by using the same primers as those typically used to make gene clone libraries, comparative information on microorganism abundance can be obtained, as well as identification of unknown T-RFs.

12.4.3 PCR Amplification from Community RNA for Identification of Active Microorganisms

Each of the previous approaches is based on amplification of the 16S rRNA gene. While this is a rapid and generally reliable approach, the detection of DNA is only an indication of the presence of the microorganism containing that gene. To get information on the active microorganisms of a sample, RNA-based analyses should be performed. Two such approaches that do not involve PCR are described later in this chapter (Sect. 12.5.2). To detect RNA by PCR methods, and thus employ the community analysis techniques described before, a viral enzyme is employed to transcribe RNA into DNA (referred to as copy DNA or cDNA) in a process known as reverse transcription (or transcriptase) PCR (RT-PCR). Following conversion of RNA isolated from a sample into cDNA, PCR amplification is carried out as normal. This product can then be subjected to further analysis by any of the previously described techniques to investigate the active microbial community. RT-PCR can also be employed on genes other than the 16S rRNA genes, but much greater care needs to be taken as messenger RNA is notoriously unstable relative to rRNA.

12.4.4 Phylogenetic Analysis of Amplified Genes for Microbial Identification

A final key step to identifying mineral-oxidizing microorganisms is phylogenetic (relationships based on molecular similarity) analysis of the 16S rRNA gene (or any other gene) sequences obtained by any of the methods already described. A new gene sequence can be compared with those from classified microorganisms, isolates obtained in other studies, or other sequences obtained by molecular methods, to provide information on the identity of the microorganism in question. Gene sequences can be found in a host of databases, the largest of which include GenBank (http://www.ncbi.nlm.nih.gov/), EMBL (http://www.embl.de/) and the DNA Databank of Japan (http://www. ddbj.nig.ac.jp/). The Ribosomal Database Project (http://rdp.cme.msu.edu/ index.jsp) is a specialist database containing only rRNA sequences and a variety of tools for sequence analysis. A new gene sequence can be compared with any of these databases, often using the basic local alignment search tool (BLAST) at GenBank, to find those genes that are most identical. The information gained from such a search can be quite informative, but caution is required when analyzing such results. There is no firm rule that relates the degree of 16S rRNA gene sequence identity to taxonomical relationships (those based on physiological and phenotypic traits) but, as a rule of thumb, gene sequences that share greater than 95% identity are from two microorganisms that probably belong to the same genus, while identities above 98% indicate a high probability that the two microorganisms belong to the same species (Stackebrandt and Goebel 1994). Currently, the only accepted definition that two microorganisms belong to the same species is if they share greater that 70% similarity of their entire genomes. Also, it is important to remember when determining phylogenetic relationships that caution should be used when attempting to assign phenotypic traits to microorganisms discovered only by molecular means. A good example of this is the well-known acidophiles *At. ferrooxidans* and *At. thiooxidans*, which both share 16S rRNA gene sequence identity of approximately 97%, and yet the latter is an obligately aerobic sulfur-oxidizing microorganism, while the former is also capable of oxidizing iron and is a facultative anaerobe.

More detailed phylogenetic analysis requires specialized computer software. Once related microorganisms have been identified by homology searches, the gene sequences of all the microorganisms are aligned. Several alignment programs exist, but the commonly used ones are ClustalX (Thompson et al. 1997) and ARB (http://www.arb-home.de/). ClustalX runs on Windows, while the ARB package operates on UNIX or Linux, and both are available at no cost on the Internet. These gene sequence alignments are then used for the construction of phylogenetic trees, again using a variety of software packages available on the Internet; ARB also includes tools for the production of trees. To assure that the phylogenetic relationships are accurate, different algorithms should be used for tree production, giving a final consensus tree. In addition, statistical analysis of the topology of the tree (the phylogenetic relationships) can be performed by bootstrap analysis (Felsenstein 1985). Trees that are thus formed can easily be viewed with the freely available Treeview (Page 1996). Great care must be taken when performing true phylogenetic analysis, especially if microorganisms that are very distantly related to any others are discovered, but this process can be rapid when one simply wants to make a tree showing the relationship of newly discovered microorganisms to known microorganisms.

12.4.5   Other Genes Useful for Microbial Identification and Community Analysis

While most of the molecular microbial ecology methods discussed here focus on the 16S rRNA gene, it is by no means the only one that is useful in this context. Other genes that are highly conserved can also be used in the previously described methods, including a range of functional genes that encode for growth and maintenance functions in bacteria and archaea (so-called housekeeping genes) as well as for enzymes responsible for different modes of growth. While there is a range of universal primers to amplify genes encoding for functional enzymes, those that may be more relevant to acidophiles include the genes encoding for sulfate reduction (in the context of mine effluents), rusticyanin and those that target the large subunit of type I and type II ribulose-1,5-bisphosphate carboxylase/oxygenase (*cbbL* or *cbbM*), enzymes involved in the fixation of carbon dioxide. Primers have been designed to target two different enzymes in the sulfate reduction pathway, and include the adenosine 5'-phosphosulfate reductase subunit A gene (*apsA*) (Friedrich 2002) and the dissimilatory sulfite reductase subunits A and B genes (*dsrAB*) (Wagner et al. 1998). These genes have been amplified from nearly all classified sulfate-reducing prokaryotes, including the archaeon *Archaeoglobus profundus*, and an extensive database of these gene sequences exists for comparative analysis and potential microbial identification. Primers have also been designed to amplify the genes encoding two different isoforms of the blue-copper protein rustacyanin from strains of *At. ferrooxidans* (Sasaki et al. 2003). The latter primer pair is more restricted in nature than those used for sulfate-reducing prokaryote analysis, but nevertheless it provides a means to target a specific species of mineral-oxidizing bacteria, since the only acidophile known to contain rusticyanin is *At. ferrooxidans*.

12.5   PCR-Independent Molecular Detection and Identification of Acidophiles

Although the approaches described in Sect. 12.4 are useful in assessing environmental biodiversity, they are at best only semiquantitative. They all rely on the use of PCR to amplify target genes, a process that has been shown to be subject to various biases (von Wintzingerode et al. 1997). In addition, although PCR-based techniques yield useful information concerning population sizes and diversity, they are often time-consuming and require specialist skills and equipment, and more rapid and simpler approaches for enumerating and identifying microorganisms are more useful in an industrial situation, such as biomining.

12.5.1   Immunological Detection and Identification of Acidophiles

Immunoassay is a PCR-independent approach, providing rapid, yet accurate enumeration of specific microorganisms. In the immunoassay, cells are immobilized on nitrocellulose membranes and are detected by antibodies that recognize target microorganisms. Antibodies have been produced that target different acidophiles, including iron- and sulfur-oxidizing bacteria and the archaeon *Sulfolobus metallicus* (Apel et al. 1976; Arredondo and Jerez 1989; Jerez and Arredondo 1991; Amaro et al. 1994). However, the existence of multiple serotypes among Gram-negative acidophiles (Koppe and Harms 1994; Hallberg and Lindström 1996) limits the use of this approach. Lipopolysaccharides (LPS), a major constituent of the outer membrane of Gram-negative bacteria, are potent antigens, and differences in the LPS among bacteria of the same species are recognized by different antibodies (a method of determining serotypes). To use immunoassays effectively, a thorough knowledge of microorganisms present in a leaching operation is needed, as is a judicious choice of antibodies that target cellular components other than the variable LPS molecules.

As with all enumeration techniques, immunoassays are far less effective for detecting microorganisms attached to solids. Aside from using methods to detach the microorganisms, such as vigorous shaking in dilute detergent solutions, an innovative modification of immunoassays has been described to enumerate attached acidophiles. Enzyme-linked immunofiltration assay (ELIFA) makes use of specially adapted 96-well plates that have filters in the wells (Dziurla et al. 1998). These filters have small enough pores to retain both planktonic microorganisms and those attached to mineral particles. Following a washing step, the bacteria can be detected with antibodies that have been cross-linked to an enzyme, which produces a colored product or fluorescence. The amount of product is directly related to the total number of bacteria contained in the well, and the solids do not interfere with detection of the product. Providing care is taken, ELIFA is reported to be able to allow detection of all microorganisms on the mineral particles that are applied to the wells.

12.5.2 Detection and Enumeration of Acidophiles by RNA-Targeting Methods

A different approach to characterizing populations of acidophiles (a variant of DGGE; Stoner et al. 1996) makes use of the differential migration patterns of 5S rRNA in acrylamide gels run with a gradient of urea. In this method, RNA is purified directly from microbial populations and separated by DGGE. The 5S rRNA from each different species migrates in a unique manner in these gels, and therefore the presence of a specific band in an electrophoresis gel can be used to confirm the presence of a particular microorganism in a mixed population. Since rRNA is the target molecule, this method gives information on the presence of active cells in the population, as only active cells contain appreciable amounts of RNA. Quantification of the microorganisms present can be performed by measuring the relative intensities of each band, but this is difficult to relate directly to cell number as each microorganism may have a different number of rRNA molecules, which will also vary depending on the growth rate of that microorganism. A drawback to this method is that the composition of the culture examined needs to be well known, or else a large number of pure cultures of acidophiles must be tested in this manner in order to assign 5S rRNA bands to specific microorganisms.

Fluorescent *in situ* hybridization (FISH) is a cultivation-independent technique that can yield highly accurate counts of different groups or species of acidophiles in a particular sample (Amann et al. 1990b). As the target molecule in FISH is typically 16S rRNA, this technique gives quantitative data on the active microorganisms in a population. FISH is a multistep technique that starts with the fixation of samples of acidophiles using ethanol or formaldehyde, preserving the structure of the microbial population at the moment of sampling (the *in situ* population). Next a gene probe (an oligonucleotide coupled to a fluorescent dye) is applied to the fixed sample, which binds to complementary nucleotide sequences in the 16S rRNA molecule (not the gene) present in the target microorganism(s). After washing out unbound probes from the sample, a fluorescence microscope is used to observe and count the number of target cells present in the original mixed population, although other methods of counting, such as flow cytometry, can also be used (Amann et al. 1990a). Given the nature of the 16S rRNA molecule, probes can be designed to target regions of the RNA that are shared amongst many microorganisms (e.g., those that target *Bacteria* or *Archaea*), those that are shared by subgroups of microorganisms such as the α-, β- and γ-*Proteobacteria*, etc., or species-specific regions.

Probe design to target specific microorganisms requires knowledge of the 16S rRNA sequence, which is usually obtained from gene sequence analysis

(Sect. 12.4). Once a short region of 18–20 nucleotides unique to the targeted acidophile has been found, an oligonucleotide complementary to this region can be synthesized (there are a number of companies that offer this service). It has been shown that some regions of the 16S rRNA molecule are more accessible to probes than others (Fuchs et al. 1998), though by using "helper oligonucleotides" (unlabeled oligos that bind to the RNA on either side of the probe) this problem can be overcome (Fuchs et al. 2000). When novel species of acidophiles are discovered and targeted by FISH probes, great care has to be taken to ensure the stringency of new probes. It is important to ensure that they detect target cells, but also that the oligonucleotide sequence of the probe and the hybridization conditions are such that all other microorganisms that may be present do not also retain the fluorescent probe. This entails a considerable amount of careful practical work, though this is aided by the fact that extremely acidic environments tend to contain few if any active neutrophilic microorganisms, reducing the likelihood that new probes will detect nontarget microorganisms.

A number of species-specific oligonucleotide probes that target more familiar and recently discovered acidophiles have been described (Table 12.4). These probes have been used to study bacteria and archaea found in mine water and acid streamer growths (Bond et al. 2000b; Bond and Banfield 2001; Hallberg et al. 2006). In other studies, FISH has been used to assess the planktonic microbial community of the Rio Tinto (Gonzalez-Toril et al. 2003), and in laboratory-scale bioreactors with pyrite as a growth substrate or sediment samples from a mine tailings pile (Peccia et al. 2000). A problem with the use of FISH for some environmental samples, especially solids from heaps, is that the low activity of the microorganisms results in low fluorescence against a high background of nonspecific fluorescence. This has been overcome by the adaptation of catalyzed reporter deposition FISH (CARD-FISH), where an enzyme that enhances the fluorescence signal is conjugated to the probe (Pernthaler et al. 2002). CARD-FISH has been successfully applied to a heap-leaching operation (Demergasso et al. 2005).

12.6   Future Perspectives on Molecular Techniques for Detection and Identification of Acidophiles

Although a number of molecular techniques for detecting and identifying mineral-oxidizing and other acidophiles have been described in this chapter, rapid advances continue to be made in tools used for molecular microbial ecology. One important advance is the use of oligonucleotide arrays to detect microbes (Zhou 2003). Here, the probes to detect microorganisms are immobilized on a solid surface, which can hold many thousands of such probes, and are then hybridized with labeled RNA extracted from the community. Microarray technology, therefore, allows for the potential rapid detection and

Table 12.4. Oligonucleotide probes that target the 16S rRNA molecule of commonly encountered mineral-oxidizing acidophiles

Probe Name	Target Organism	Sequence (5' to 3')	Reference
TH1187	Thermoplasmales	GTACTGACCTGCCGTCGAC	Bond and Banfield (2001)
FER656	Ferroplasma	CGTTTAACCTCACCGATC	Edwards et al. (2000b)
LF581	Leptospirillum groups I and II[a]	CGGCCTTTCACCAAAGAC	Schrenk et al. (1998)
LF1252	Leptospirillum group III[a]	TTACGGGCTCGCCTCCGT	Bond and Banfield (2001)
LF655	Leptospirillum groups I, II and III[a]	CGCTTCCCTCTCCCAGCCT	Bond and Banfield (2001)
ACM732	Acidimicrobium and relatives[b]	GTACCGGCCCAGATCGCTG	Bond and Banfield (2001)
ACM995	Am. ferrooxidans	CTCTGCGGCTTTTCCCTCCATG	Paul Norris, University of Warwick, UK
SUL228	Sb. thermosulfidooxidans[c]	TAATGGGCCGGCGAGCTCCC	Bond and Banfield (2001)
ACD840	Acidiphilium spp.	CGACACTGAAGTGCTAAGC	Bond and Banfield (2001)
Acdp821	Acidiphilium spp.	AGCACCCCAACATCCAGCACACAT	Peccia et al. (2000)
Thio820	Acidithiobacillus spp.[d]	ACCAAACATCTAGTATTCATCG	Peccia et al. (2000)
TF539	At. ferrooxidans	CAGACCTAACGTACCGCC	Bond and Banfield (2001)
ATT0223	At. thiooxidans	AGACGTAGGCTCCTCTTC	Hallberg et al. (2006)
THC642	At. caldus	CATACTCCAGTCAGCCCGT	Edwards et al. (2000a)

[a] The *Leptospirillum* grouping is based on phylogenetic analysis of environmental clones from the Iron Mountain site (Bond et al. 2000a).
[b] This probe targets *Am. ferrooxidans* as well as related clones from Iron Mountain and "*Frm. acidiphilum*," but not the environmental clones IMBA84 and TRA2-10 from Iron Mountain.
[c] This probe also targets *Sb. acidophilus*, but with a one-base mismatch in the target region, and thus it may not be specific for *Sb. thermosulfidooxidans* under certain hybridization conditions.
[d] *At. thiooxidans*, *At. ferrooxidans* and *Acidithiobacillus* sp. DSM612 (previously known as *Thiobacillus thiooxidans* DSM612) are the targets for this probe, but *At. caldus* is not a target.

identification of a great many more microorganisms in a community than has previously been possible. Given that the number of known acidophiles is vastly limited compared with the number of environments, such as active sewage sludge, it is possible that a carefully designed array could be made to detect all known acidophiles. A powerful extension of microarrays is the coupling of arrays with the labeling of macromolecules such as DNA or RNA with isotopes of a substrate (e.g., $^{13}CO_2$). Separation of the macromolecules that incorporated the isotope from those that remained unlabeled, followed by hybridization to a microarray allows for the identification of those microorganisms in a community that are metabolically active with that substrate (Adamczyk et al. 2003).

Other advances include the recent sequencing of the dominant genomes in an abandoned pyrite mine, which has allowed a hypothetical reconstruction of the community interactions (Tyson et al. 2004). This work identified microorganisms that were important to the oxidation of pyrite, leading to the formation of acid mine drainage, and showed that a key microbe in the process was a novel, nitrogen-fixing *Leptospirillum* spp. (Tyson et al. 2005). Such a genome-based approach to studying microbial communities is further enhanced by a recently described approach to study gene expression in acidophiles (Parro and Moreno-Paz 2003). The application of these new techniques will further advance our understanding of the roles of acidophilic microorganisms in mineral processing operations and the wider environment.

References

Adamczyk J, Hesselsoe M, Iversen N, Horn M, Lehner A, Nielsen PH, Schloter M, Roslev P, Wagner M (2003) The isotope array, a new tool that employs substrate-mediated labeling of rRNA for determination of microbial community structure and function. Appl Environ Microbiol 69:6875–6887

Amann R, Binder BJ, Olson RJ, Chisholm SW, Devereux R, Stahl DA (1990a) Combination of 16S rRNA-targeted oligonucleotide probes with flow cytometry for analyzing mixed microbial populations. Appl Environ Microbiol 56:1919–1925

Amann RI, Krumholz L, Stahl DA (1990b) Fluorescent-oligonucleotide probing of whole cells for determinative, phylogenetic, and environmental studies in microbiology. J Bacteriol 172:762–770

Amann RI, Ludwig W, Schleifer KH (1995) Phylogenetic identification and *in situ* detection of individual microbial cells without cultivation. Microbiol Rev 59:143–169

Amaro AM, Hallberg KB, Lindström EB, Jerez CA (1994) An immunological assay for detection and enumeration of thermophilic biomining microorganisms. Appl Environ Microbiol 60:3470–3473

Apel WA, Dugan PR, Filppi JA, Rheins MS (1976) Detection of *Thiobacillus ferrooxidans* in acid mine environments by indirect fluorescent-antibody staining. Appl Environ Microbiol 32:159–165

Arredondo R, Jerez CA (1989) Specific dot-immunobinding assay for detection and enumeration of *Thiobacillus ferrooxidans*. Appl Environ Microbiol 55:2025–2029

Bergamo RF, Novo MTM, Verissimo RV, Paulino LC, Stoppe NC, Sato MIZ, Manfio GP, Prado PI, Garcia O, Ottoboni LMM (2004) Differentiation of *Acidithiobacillus ferrooxidans* and

A. thiooxidans strains based on 16S-23S rDNA spacer polymorphism analysis. Res Microbiol 155:559–567

Bond PL, Banfield JF (2001) Design and performance of rRNA targeted oligonucleotide probes for *in situ* detection and phylogenetic identification of microorganisms inhabiting acid mine drainage environments. Microb Ecol 41:149–161

Bond PL, Smriga SP, Banfield JF (2000a) Phylogeny of microorganisms populating a thick, subaerial, predominantly lithotrophic biofilm at an extreme acid mine drainage site. Appl Environ Microbiol 66:3842–3849

Bond PL, Druschel GK, Banfield JF (2000b) Comparison of acid mine drainage microbial communities in physically and geochemically distinct ecosystems. Appl Environ Microbiol 66:4962–4971

Brosius J, Dull TJ, Sleeter DD, Noller HF (1981) Gene organization and primary structure of a ribosomal RNA operon from *Escherichia coli*. J Mol Biol 148:107–127

de Bruyn JC, Boogerd FC, Bos P, Kuenen JG (1990) Floating filters, a novel technique for isolation and enumeration of fastidious, acidophilic, iron-oxidizing, autotrophic bacteria. Appl Environ Microbiol 56:2891–2894

Demergasso C, Echeverria A, Escudero L, Galleguillos P, Zepeda V, Castillo D (2005) Comparison of fluorescent in situ hybridization (FISH) and catalyzed reporter deposition (CARD-FISH) for visualization and enumeration of archaea and bacteria ratio in industrial heap bioleaching operations. In: Harrison STL, Rawlings DE, Petersen J (eds) Proceedings of the 16th international biohydrometallurgy symposium. 16th International Biohydrometallurgy Symposium, Cape Town, pp 843–851

De Wulf-Durand P, Bryant LJ, Sly LI (1997) PCR-mediated detection of acidophilic, bioleaching-associated bacteria. Appl Environ Microbiol 63:2944–2948

Dziurla M-A, Achouak W, Lam B-T, Heulin T, Berthelin J (1998) Enzyme-linked immunofiltration assay to estimate attachment of thiobacilli to pyrite. Appl Environ Microbiol 64:2937–40

Edwards KJ, Bond PL, Banfield JF (2000a) Characteristics of attachment and growth of *Thiobacillus caldus* on sulphide minerals: a chemotactic response to sulphur minerals? Environ Microbiol 2:324–332

Edwards KJ, Bond PL, Gihring TM, Banfield JF (2000b) An archaeal iron-oxidizing extreme acidophile important in acid mine drainage. Science 287:1796–1799

Ehrlich HL (2002) Geomicrobiology, 4th edn. Taylor and Francis, New York

Felsenstein J (1985) Confidence-limits on phylogenies – an approach using the bootstrap. Evolution 39:783–791

Friedrich MW (2002) Phylogenetic analysis reveals multiple lateral transfers of adenosine-5′-phosphosulfate reductase genes among sulfate-reducing microorganisms. J Bacteriol 184:278–289

Fuchs BM, Wallner G, Beisker W, Schwippl I, Ludwig W, Amann R (1998) Flow cytometric analysis of the *in situ* accessibility of *Escherichia coli* 16S rRNA for fluorescently labeled oligonucleotide probes. Appl Environ Microbiol 64:4973–4982

Fuchs BM, Glockner FO, Wulf J, Amann R (2000) Unlabeled helper oligonucleotides increase the *in situ* accessibility to 16S rRNA of fluorescently labeled oligonucleotide probes. Appl Environ Microbiol 66:3603–3607

Gonzalez-Toril E, Llobet-Brossa E, Casamayor EO, Amann R, Amils R (2003) Microbial ecology of an extreme acidic environment, the Tinto River. Appl Environ Microbiol 69:4853–4865

Hallberg KB, Coupland K, Kimura S, Johnson DB (2006) Macroscopic acid streamer growths in acidic, metal-rich mine waters in north Wales consist of novel and remarkably simple bacterial communities. Appl Environ Microbiol 72:2022–2030

Hallberg KB, Johnson DB (2003) Novel acidophiles isolated from moderately acidic mine drainage waters. Hydrometallurgy 71:139–148

Hallberg KB, Lindström EB (1996) Multiple serotypes of the moderate thermophile *Thiobacillus caldus*, a limitation of immunological assays for biomining microorganisms. Appl Environ Microbiol 62:4243–4246

Harneit K, Göksel A, Kock D, Klock JH, Gehrke T, Sand W (2005) Adhesion to metal sulphide surfaces by cells of *Acidithiobacillus ferrooxidans, Acidithiobacillus thiooxidans* and *Leptospirillum ferrooxidans*. In: Harrison STL, Rawlings DE, Petersen J (eds) Proceedings of the 16th international biohydrometallurgy symposium. 16th International Biohydrometallurgy Symposium, Cape Town, pp 635–646

Head IM, Saunders JR, Pickup RW (1998) Microbial evolution, diversity, and ecology: a decade of ribosomal RNA analysis of uncultivated microorganisms. Microb Ecol 35:1–21

Hugenholtz P, Goebel BM, Pace NR (1998) Impact of culture-independent studies on the emerging phylogenetic view of bacterial diversity. J Bacteriol 180:4765–4774

Jerez CA, Arredondo R (1991) A sensitive immunological method to enumerate *Leptospirillum ferrooxidans* in the presence of *Thiobacillus ferrooxidans*. FEMS Microbiol Lett 78:99–102

Johnson DB (1995) Selective solid media for isolating and enumerating acidophilic bacteria. J Microbiol Methods 23:205–218

Johnson DB, McGinness S (1991) A highly efficient and universal solid medium for growing mesophilic and moderately thermophilic iron-oxidizing acidophilic bacteria. J Microbiol Methods 13:113–122

Johnson DB, Okibe N, Hallberg KB (2005) Differentiation and identification of iron-oxidizing acidophilic bacteria using cultivation techniques and amplified ribosomal DNA restriction enzyme analysis (ARDREA). J Microbiol Methods 60:299–313

Karan G, Natarajan KA, Modak JM (1996) Estimation of mineral adhered biomass of Thiobacillus ferrooxidans by protein assay – some problems and remedies. Hydrometallurgy 42:169–175

Kolmert Å, Wikström P, Hallberg KB (2000) A fast and simple turbidimetric method for the determination of sulfate in sulfate-reducing bacterial cultures. J Microbiol Methods 41:179–184

Koppe B, Harms H (1994) Antigenic determinants and specificity of antisera against acidophilic bacteria. World J Microbiol Biotechnol 10:154–158

Lane DJ (1991) 16S/23S rRNA sequencing. In: Stackebrandt E, Goodfellow M (eds) Nucleic acid techniques in bacterial systematics. Wiley, New York, pp 115–175

Leclerc M, Delbes C, Moletta R, Godon J-J (2001) Single strand conformation polymorphism monitoring of 16S rDNA archaea during start-up of an anaerobic digester. FEMS Microbiol Ecol 34:213–220

Lindström EB, Sehlin HM (1989) High efficiency plating of the thermophilic sulfur-dependent archaebacterium *Sulfolobus acidocaldarius*. Appl Environ Microbiol 55:3020–3021

Lovley DR, Phillips EJP (1987) Rapid assay for microbially reduced ferric iron in aquatic sediments. Appl Environ Microbiol 53:1536–1540

Marchesi JR, Sato T, Weightman AJ, Martin TA, Fry JC, Hiom SJ, Wade WG (1998) Design and evaluation of useful bacterium-specific PCR primers that amplify genes coding for bacterial 16S rRNA. Appl Environ Microbiol 64:795–799

Marsh TL (1999) Terminal restriction fragment length polymorphism (T-RFLP): an emerging method for characterizing diversity among homologous populations of amplification products. Curr Opin Microbiol 2:323–327

Muyzer G, Smalla K (1998) Application of denaturing gradient gel electrophoresis (DGGE) and temperature gradient gel electrophoresis (TGGE) in microbial ecology. Antonie van Leeuwenhoek J Microbiol 73:127–41

Muyzer G, Hottenträger S, Teske A, Wawer C (1996) Denaturing gradient gel electrophoresis of PCR-amplified 16S rDNA – a new molecular approach to analyse the genetic diversity of mixed microbial communities. In: Akkermans ADL, van Elsas JD, de Bruijn FJ (eds) Molecular microbial ecology manual. Kluwer, Dordrecht, pp 1–23

Orita M, Suzuki Y, Sekiya T, Hayashi K (1989) Rapid and sensitive detection of point mutations and DNA polymorphisms using the polymerase chain reaction. Genomics 5:874–879

Orphan VJ, Taylor, LT, Hafenbradl D, DeLong EF (2000) Culture-dependent and culture-independent characterization of microbial assemblages associated with high-temperature petroleum reservoirs. Appl Environ Microbiol 66:700–711

Page RDM (1996) TREEVIEW: an application to display phylogenetic trees on personal computers. Comput Appl Biosci 12:357–358

Parro V, Moreno-Paz M (2003) Gene function analysis in environmental isolates: The nif regulon of the strict iron oxidizing bacterium *Leptospirillum ferrooxidans*. Proc Natl Acad Sci USA 100:7883–7888

Peccia J, Marchand EA, Silverstein J, Hernandez M (2000) Development and application of small-subunit rRNA probes for assessment of selected *Thiobacillus* species and members of the genus *Acidiphilium*. Appl Environ Microbiol 66:3065–40

Pernthaler A, Pernthaler J, Amann R (2002) Fluorescence *in situ* hybridization and catalyzed reporter deposition for the identification of marine bacteria. Appl Environ Microbiol 68:3094–3101

Pizarro J, Jedlicki E, Orellana O, Romero J, Espejo RT (1996) Bacterial populations in samples of bioleached copper ore as revealed by analysis of DNA obtained before and after cultivation. Appl Environ Microbiol 62:1323–1328

Ramsay B, Ramsay J, de Tremblay M, Chavarie C (1988) A method for quantification of bacterial protein in the presence of jarosite. Geomicrobiol J 6:171–177

Raskin L, Stromley JM, Rittmann BE, Stahl DA (1994) Group-specific 16S rRNA hybridization probes to describe natural communities of methanogens. Appl Environ Microbiol 60:1232–1240

Rawlings DE (1995) Restriction enzyme analysis of 16S rDNA genes for the rapid identification of *Thiobacillus ferrooxidans*, *Thiobacillus thiooxidans* and *Leptospirillum ferrooxidans* strains in leaching environments. In: Vargas T, Jerez CA, Wiertz JV, Toledo H (eds) Biohydrometallurgical processing. University of Chile, Santiago, pp 9–18

Rawlings DE, Coram NJ, Gardner MN, Deane SM (1999) *Thiobacillus caldus* and *Leptospirillum ferrooxidans* are widely distributed in continuous flow biooxidation tanks used to treat a variety of metal containing ores and concentrates. In: Amils R, Ballester A (eds) Biohydrometallurgy and the environment toward the mining of the 21st century. Elsevier, Amsterdam, pp 777–786

Rohwerder T, Schippers A, Sand W (1998) Determination of reaction energy values for biological pyrite oxidation by calorimetry. Thermochim Acta 309:79–85

Rohwerder T, Gehrke T, Kinzler K, Sand W (2003) Bioleaching review part A: Progress in bioleaching: fundamentals and mechanisms of bacterial metal sulfide oxidation. Appl Microbiol Biotechnol 63:239–248

Sasaki K, Ida C, Ando A, Matsumoto N, Saiki H, Ohmura N (2003) Respiratory isozyme, two types of rusticyanin of *Acidithiobacillus ferrooxidans*. Biosci Biotechnol Biochem 67:1039–1047

Schippers A, Bosecker K (2005) Bioleaching: analysis of microbial communities dissolving metal sulfides. In: Barredo JL (ed) Methods in biotechnology: microbial processes and products. Humana, Totowa, pp 405–412

Schrenk MO, Edwards KJ, Goodman RM, Hamers RJ, Banfield JF (1998) Distribution of *Thiobacillus ferrooxidans* and *Leptospirillum ferrooxidans*: implications for generation of acid mine drainage. Science 279:1519–1522

Silverman MP, Lundgren DG (1959) Studies on the chemoautotrophic iron bacterium *Ferrobacillus ferrooxidans*. I. An improved medium and harvesting procedure for securing high yields. J Bacteriol 77:642–647

Stackebrandt E, Goebel BM (1994) A place for DNA-DNA reassociation and 16S ribosomal-RNA sequence analysis in the present species definition in bacteriology. Int J Syst Bacteriol 44:846–849

Stahl DA, Amann R (1991) Development and application of nucleic acid probes in bacterial systematics. In: Stackebrandt E, Goodfellow M (eds) Nucleic acid techniques in bacterial systematics. Wiley, New York, pp 205–248

Stoner DL, Browning CK, Bulmer DK, Ward TE, MacDonell MT (1996) Direct 5S rRNA assay for monitoring mixed-culture bioprocesses. Appl Environ Microbiol 62:1969–1976

Thompson JD, Gibson TJ, Plewniak F, Jeanmougin F, Higgins DG (1997) The CLUSTAL_X windows interface: flexible strategies for multiple sequence alignment aided by quality analysis tools. Nucleic Acids Res 25:4876–4882

Tributsch H, Rojas-Chapana J (2004) Comparative study on pit formation and interfacial chemistry induced by *Leptospirillum* and *Acidithiobacillus ferrooxidans* during FeS^2 leaching. In: Tsezos M, Hatzikioseyian A, Remoudaki E (eds) Biohydrometallurgy; a sustainable technology in evolution. National Technical University of Athens, Zografou, pp 1047–1055

Tuovinen OH, Kelly DP (1973) Studies on growth of *Thiobacillus ferrooxidans*.1. Use of membrane filters and ferrous iron agar to determine viable numbers, and comparison with 14CO^2 fixation and iron oxidation as measures of growth. Arch Microbiol 88:285–298

Tyson GW, Chapman J, Hugenholtz P, Allen EE, Ram RJ, Richardson PM, Solovyev VV, Rubin EM, Rokhsar DS, Banfield JF (2004) Community structure and metabolism through reconstruction of microbial genomes from the environment. Nature 428:37–43

Tyson GW, Lo I, Baker BJ, Allen EE, Hugenholtz P, Banfield JF (2005) Genome-directed isolation of the key nitrogen fixer *Leptospirillum ferrodiazotrophum* sp. nov. from an acidophilic microbial community. Appl Environ Microbiol 71:6319–6324

von Wintzingerode F, Gobel UB, Stackebrandt E (1997) Determination of microbial diversity in environmental samples: pitfalls of PCR-based rRNA analysis. FEMS Microbiol Rev 21:213–229

Wagner M, Roger AJ, Flax JL, Brusseau GA, Stahl DA (1998) Phylogeny of dissimilatory sulfite reductases supports an early origin of sulfate respiration. J Bacteriol 180:2975–2982

White DC, Pinkhart HC, Ringelberg DB (1997) Biomass measurements: biochemical approaches. In: Hurst CJ, Knudsen GR, McInerney MJ, Stetzenbach LD, Walter MV (eds) Manual of environmental microbiology. American Society for Microbiology, Washington, D.C, pp 91–101

Zhou J (2003) Microarrays for bacterial detection and microbial community analysis. Curr Opin Microbiol 6:288–294

Zumstein E, Moletta R, Godon J-J (2000) Examination of two years of community dynamics in an anaerobic bioreactor using fluorescent-PCR single strand conformation polymorphism analysis. Environ Microbiol 2:69–78

13 Bacterial Strategies for Obtaining Chemical Energy by Degrading Sulfide Minerals

HELMUT TRIBUTSCH, JOSÉ ROJAS-CHAPANA

13.1 Introduction

Reduced inorganic sulfur compounds are oxidized by members of two domains of living beings namely Archaea and *Bacteria*. The sulfur compounds are mostly oxidized to sulfate and energy is, thereby, obtained for carbon dioxide fixation and cell growth. A number of genera have been shown to be involved in the sulfur cycle and in biomineralization. They include *Acidothiobacillus*, *Leptospirillum*, *Acidiphilum*, *Sulfobacillus*, *Ferroplasma*, *Sulfolobus*, *Metallosphaera* and *Acidianus*. Remarkably, these microorganisms survive in extremely inhospitable environments. They are usually highly acid tolerant and many can grow in environments of pH 1 or even lower. They can be subdivided into mesophilic (20–40°C temperature optima), moderately thermophillic (40–60°C), and (extremely) thermophilic (above 60°C). In some cases, both reduced sulfur and iron (Fe^{2+}) are used as energy sources, while some prokaryotes can oxidize only one of these species. In many ore deposits the iron- and the sulfur species are readily available in the form of the mineral iron pyrite (FeS_2). The reaction of pyrite with oxygen and water occurs exergonically with production of heat but at a very slow rate. The enthalpy of formation of pyrite is $\Delta H^f = -37.4$ kcal mol^{-1} and that of formation of $FeSO_4$ is $\Delta H^f = -222$ kcal mol^{-1}. The $T\Delta S^f$ values at 300 K are 3.79 and 8.6 kcal mol^{-1} respectively. The conversion of pyrite into two iron sulfate molecules thus involves a free-energy turnover of $\Delta G^f = 402.3$ kcal mol^{-1} for pyrite, to which approximately 10% of solvation energy has to be added. The process of reaction with oxygen is kinetically inhibited at ambient temperature. For this reason pyrite crystals are found to be quite stable in natural environments. The slow chemical oxidation process (Eq. 13.1) is greatly enhanced by bacteria that harvest part of the chemical energy, which would otherwise be dissipated as heat:

$$4FeS_2 + 15O_2 + 2H_2O \rightarrow 4Fe^{3+} + 8SO_4^{2-} + 4H^+. \tag{13.1}$$

Sulfuric acid is generated when ferric iron reacts with additional sulfide. An acid environment is thereby generated, which provides a favorable ecosystem for acid-loving bacteria. The acidity of the environment supports the leaching process by allowing protons to break additional chemical sulfur bonds during formation of interfacial $-SH^{\delta-}$ groups and by keeping Fe^{3+} complexes and iron hydroxide/iron oxides in solution, thus avoiding sedimentation and

Biomining
(ed. by Douglas E. Rawlings and D. Barrie Johnson)
© Springer-Verlag Berlin Heidelberg 2007

obstruction of the sulfide interface. Sulfide-oxidizing bacteria interfere in the natural sulfide oxidation process with oxygen by recovering part of the free energy of reaction in the form of chemical energy.

During recent decades, significant progress has been made in the understanding of sulfide chemistry and electrochemistry as well as bacterial activity and bacterial interaction with sulfides (d'Hugues et al. 2002; Suzuki 2001). Sophisticated techniques have been used to characterize bacterial attack and mineral disintegration. A new discipline has been suggested combining sulfide semiconductor electrochemistry and bacterial leaching mechanisms (Tributsch and Rojas-Chapana 2000). It combines semiconductor electronic knowledge about sulfides and bacterial leaching mechanisms aimed at dissolving sulfides. While some essential mechanisms have been elucidated, more work is required to penetrate the complex interactions of solid-state physics, interfacial electrochemistry and bacterial enzymatic mechanisms.

13.2 Pyrite As a Model System for Understanding Bacterial Sulfide Leaching Activities

Iron and sulfur are very common elements on earth and during the early evolution of life a reducing atmosphere facilitated the presence and turnover of large quantities of pyrite in the environment. It is, therefore, not surprising that a much discussed theory locates evolution of early life at pyrite interfaces (Wächtershäuser 1988, 2000). Interesting support for this idea is the fact that Fe–S complexes (ferredoxins) are essential components of catalytic centers in biological systems. Wächtershäuser's hypothesis describes the reduction of carbon dioxide during iron monosulfide reduction to pyrite in the presence of hydrogen disulfide. While the conditions for carbon dioxide fixation have been shown to be more complex (Tributsch et al. 2003), the role of pyrite interfaces for evolution of early life is intriguing and provides a very attractive model system. It suggests that primitive bacteria evolved by adapting to this energy-supplying sulfide mineral. It is, therefore, reasonable to use this abundant chemical energy source as a model system for understanding bacterial interaction. If bacteria evolved their energy-converting metabolism while interacting with pyrite, it may be possible to understand better the essential mechanistic details. This is especially true since the electronic properties of pyrite differ from those of other, much less abundant sulfides, for example, zinc sulfide or copper sulfide.

13.3 Electronic Structure and Thermodynamic Properties of Pyrite

The electrochemical, electronic and interfacial properties of pyrite have been discussed in numerous publications (Abd El-Halim et al. 1995; Ennaoui et al. 1993; Ennaoui and Tributsch 1984, 1986). It is convenient to analyze the properties of pyrite from a thermodynamic point of view by considering the potential–pH diagram for the iron–water–sulfur system (Ennaoui et al. 1993). From such a diagram it is seen that, at low pH, pyrite is thermodynamically unstable with respect to Fe^{2+} and H_2S or, at a more positive potential, with respect to Fe^{2+}, S^0 and SO_4^{2-}. However, at neutral and alkaline pH values pyrite may oxidize to FeOH and SO_4^{2-}. Thermodynamically speaking, and under conditions where most pyrite is present in natural environments, pyrite has, in principle, a relatively narrow stability domain. Therefore, the chemical energy stored in pyrite can, in principle, be quite easily released to form oxidized or reduced species and thermal energy can be released. However, the process of pyrite dissolution is kinetically hindered owing to the peculiar electronic structure of this semiconducting material. Pyrite is a semiconductor with an energy gap of approximately 0.95 eV and its valence and conduction bands are mostly derived from iron d states. Reliable calculations of the density of state in this material exist (Eyert et al. 1998). Figure 13.1 shows the energy diagram of both pyrite and RuS_2, which have identical electronic structures. It explains why the generation of a hole in the valence band does not directly lead to chemical disintegration of the mineral at the interface since no chemical bonds are broken, but a stepwise oxidation of the interfacial metal (Fe^{2+} or Ru^{2+}) occurs. In the case of pyrite, electron extraction from the valence band by Fe^{3+} does not directly break bonds but supports a stepwise oxidation of surface iron with water species. If iron is replaced by ruthenium as found in the mineral RuS_2, which has exactly the same crystal and electronic structure (Fig. 13.2) as pyrite, then eventually

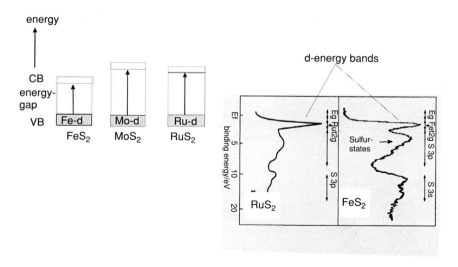

Fig. 13.1. Electronic band structure of FeS_2 (pyrite), RuS_2 and MoS_2 showing the valence band (*VB*) and the conduction band (*CB*) derived from metal d states (*left*). Comparative measured photoelectron emission spectra for FeS_2 and isocrystalline RuS_2 showing the d bands forming the valence bands of FeS_2 and RuS_2 (*right*)

oxygen will evolve from the water after an interfacial peroxo complex has formed. However, iron complexes cannot reach such high oxidation states and apparently the oxygen species is transferred to and reacts with the sulfur, the electronic state of which is situated at least 1 eV below the edge of the valence band. Owing to this interfacial reaction, the pyrite surface is destabilized and a stepwise oxidation of the sulfur species to sulfate may proceed. From this simplified electronic picture, it can be seen how the

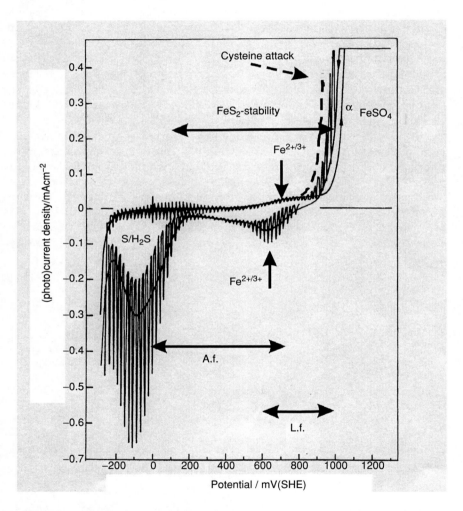

Fig. 13.2. Simplified scheme explaining the current–voltage behavior of pyrite in contact with an acidic electrolyte (0.5 mol L^{-1} H$_2$SO$_4$). The current–voltage curve is shown for the dark and for chopped light. The different reactions which are proceeding are indicated. The *dotted line* shows how the presence of cysteine enhances the anodic dissolution process. The preferred activity ranges of *Acidithiobacillus ferrooxidans* and *Leptospirillum ferrooxidans* are indicated

oxidative reaction of pyrite is inhibited, which is the reason for the quite high stability of pyrite in natural environments. However, bacteria have found a way to disintegrate this sulfide mineral effectively to get access to the chemical energy involved in Fe^{2+} and sulfide-to-sulfate oxidation. The very similar mineral RuS_2 (Table 13.1) cannot, on the other hand, be utilized by bacteria as a source of energy.

Many surface analytical and photoelectrochemical studies of pyrite have contributed to a good understanding of the interfacial behavior of this mineral (Ennaoui et al. 1985; Li et al. 1986; Ennaoui and Tributsch 1986). The fact that the electrons are transferred via energy bands derived from the iron implies that, when the electric potential changes, the energy bands shift positively or negatively depending on whether a positive or a negative potential is applied. The reason for this is simply that the exchange of charge between the transition metal d energy bands of pyrite, which are derived from iron, and the electrolyte occurs via coordination chemical mechanisms. That means, when holes are extracted from surface iron via d energy bands, the iron is oxidized by attaching water species. Ligands may be attached or released depending on the potential applied. The consequence is a high density of surface states, which change their state of charge depending on the potential applied. Electroreflection studies have underlined this interfacial property of pyrite (Salvador et al. 1991). When a negative potential is applied, the energy bands of pyrite shift negatively and when a positive potential is applied, they shift positively. The same happens when the electrical potential

Table 13.1. Comparison of nearly identical mineral properties of pyrite and laurite. The first is an excellent source of chemical energy for bacteria, the second is not a source of energy. This pinpoints the role of the metal oxidation state reached for the dissolution process and the bacterial leaching strategies

	Mineral Pyrite	Laurite
Chemical formula	FeS_2	RuS_2
Crystal structure	Face-centered cubic	Face-centered cubic
Electronic structure	$d{\rightarrow}d$ energy gap	$d{\rightarrow}d$ energy gap
Metal oxidation state reached	III–IV	VI–VIII
Heat of formation (kcal mol^{-1})	37.4	47.7
Chemical stability (aqua regia)	Dissolution	Stabile (kinetic)
Anodic oxidation product with water	$Fe^{2+/3+}/SO_4^{2-}$	O_2 (from water)
Energy gap (eV)	0.95	1.3
Density (g cm^{-3})	5.5	6.99
Acidithiobacillus ferrooxidans	Good activity	No activity
Leptospirillum ferrooxidans	Good activity	No activity

is applied via different redox properties of the electrolyte. These unusual properties of the pyrite mineral when interacting with an iron-containing electrolyte have to be considered for an understanding of the mechanisms which occur when bacteria interact with the surface of this mineral. Figure 13.2 shows schematically the electrochemical behavior of pyrite in the presence of an acid (H_2SO_4-containing) solution. The current–voltage diagram shows that, at negative potential (starting from slightly positive of the hydrogen potential), a reaction with protons occurs, leading to H_2S formation and a cathodic disintegration of the mineral. At positive potentials (up to 900 mV, standard hydrogen electrode) pyrite essentially shows oxidation of interfacial Fe^{2+} to Fe^{3+} and, during the reverse scan, the reduction of Fe^{3+} (near 700 mV). In this potential region, pyrite is kinetically stable since oxidation of interfacial iron does not dramatically interfere with the crystal structure of the sulfide. At still more positive potentials (approximately 900 mV), pyrite starts to react with water to form iron sulfate. This occurs via an iron oxide–iron hydroxide interfacial complex, which reacts with the sulfur species via a thiosulfate intermediate. Only via this reaction does the sulfur, having electronic states at approximately 1 eV below the edge of the valence band, get involved in the interfacial reaction and dissolution of the pyrite is possible. In this way iron is released from the crystal structure for electrochemical processes to proceed. Interestingly, it was found that addition of the thiol compound cysteine can shift anodic dissolution towards smaller positive potentials (Abd El-Halim et al. 1995). This indicates that a FeS_2–cysteine interaction as in ferredoxins (Österberg 1995) might disrupt chemical bonds in pyrite.

This electrochemical behavior of pyrite has been cross-checked with the electrochemical behavior of other transition metal disulfides which have exactly the same electronic structure but different chemical properties of the transition metals. These compounds were molybdenum disulfide and ruthenium disulfide (Table 13.1), which are both known minerals. The differences noted with molybdenum disulfide were (1) no H_2S evolution but only hydrogen evolution was observed at negative potentials (as measured using differential electrochemical mass spectroscopy) and (2) cysteine did not enhance electrochemical dissolution of MoS_2 at positive potentials. Ruthenium disulfide, which has the same cubic structure as iron disulfide and the same electronic structure of valence and energy bands, which are derived from the transition metal, behaves very differently from pyrite. Interestingly it does not dissolve anodically to yield ruthenium sulfate. Rather this sulfide reacts with water to liberate oxygen. It was found that bacteria that can leach FeS_2 (pyrite) could not leach RuS_2. This is especially remarkable since both sulfides, RuS_2 and FeS_2, are chemical energy storing minerals and should exergonically react to yield sulfate (Table 13.1). This shows how critical kinetic inhibition is and what role the chemistry of the transition metal plays. In the case of ruthenium, electron transfer processes via the ruthenium energy bands lead to a stepwise oxidation of interfacial ruthenium in contact with water to a peroxo complex, which releases oxygen. Iron in pyrite cannot reach such a high oxidation state and the oxidized complex, reacting interfacially, involves sulfur and causes pyrite to decompose to iron sulfate. In contrast, molybdenum in MoS_2 has still different properties of complex formation, and also a different crystal structure that shows a 2D S–Mo–S layer structures, held together via van der Waals interactions. MoS_2 also reacts with water via a thiosulfate intermediate to sulfate (Sand et al. 1999) as for pyrite; however, the different transition metal chemistry of Mo results in somewhat different interfacial chemistry. This reaction essentially occurs via step sites, perpendicular to the van der Waal's surfaces. MoS_2 does not favor significant bacterial leaching rates.

From these comparative considerations on the electrochemistry and interfacial chemistry of different transition metal disulfides, it is clear that the bacteria must have evolved very specific strategies towards their aim of gaining chemical energy from the oxidation of sulfides. If, as sometimes claimed, bacterial regeneration of Fe^{3+} ions is sufficient for the leaching of metal sulfides, then sulfides with comparable heats of formation should behave similarly. This is not the case. In the following we will analyze some of the common bacteria employed in mining activities and discuss their evolutionary adaptation with respect to the leaching of the most abundant and most ancient energy source, pyrite.

13.4   The Energy Strategy of *Leptospirillum ferrooxidans*

Leptospirillum ferrooxidans is a bacterium that is known to be able to oxidize only Fe^{2+} to Fe^{3+}. No ability to use reduced sulfur as an energy source has been demonstrated. When this bacterium is grown on a pure Fe^{2+} substrate, it grows well and no peculiarities are observed with its cellular morphology, which is shown in Fig. 13.3a in its typical spiral form. The question arises of how a bacterium with only Fe^{2+}-oxidizing properties attacks and dissolves pyrite. Taken together, the electrochemical characteristics of pyrite (Fig. 13.2) and the energy-band diagram (Fig. 13.1) suggest that the extraction of electrons via Fe^{3+} will only yield moderate changes in the pyrite interface. Extraction of electrons does not break essential bonds, and Fe^{2+} oxidation and Fe^{3+} reduction occur in a potential region where pyrite is kinetically stable. *L. ferrooxidans* has, however, learned how to dissolve pyrite. When this bacterium is cultivated on pyrite and bacterial cultures are observed, it is found that the bacteria may develop a capsule around their cells, which is dotted with small particles (Rojas-Chapana and Tributsch 2004; Fig. 13.3b). The ultrastructure of these particles reveals a typical pyrite crystallinity as confirmed by electron diffraction. *L. ferrooxidans* has, accordingly, learned to dissolve the microcrystalline structure of pyrite crystals to yield colloidal pyrite particles in its capsule, where they are apparently oxidized further to allow access to the chemical energy contained. This energy is derived from

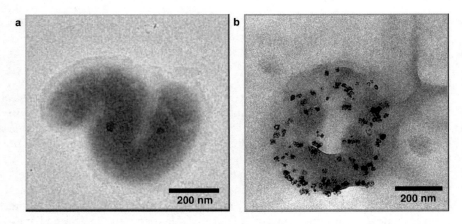

Fig. 13.3. Transmission electron microscopy (*TEM*) images of *L. ferrooxidans*. a *L. ferrooxidans* cells grown on pure Fe^{2+} (no capsule). **b** *L. ferrooxidans* cell grown on pyrite showing a characteristic ring shape and pyrite nanoparticles within the capsule

Fe^{2+}, the only species, which *Leptospirillum* can oxidize. As shown by Rawlings et al. (1999) and as deducible from Fig. 13.2, *L. ferrooxidans* uses its ability to work with high ferric ion concentrations (see the arrows indicating the activity region of *L. ferrooxidans*). It appears to concentrate Fe^{3+} in its capsule, thereby generating a highly positive oxidation potential. By contacting pyrite, it moves its electrochemical potential towards the region where electrochemical oxidation of iron sulfide to iron sulfate occurs. While in an electrochemical cell the liberated electrons are conducted into the external circuit, bacterial activity generates separated anodic and cathodic regions on the pyrite: anodic regions where the dissolution of pyrite occurs and cathodic regions where electrons reduce Fe^{3+} to Fe^{2+}. These Fe^{2+} species are then used for chemical energy gain by the bacteria. Topological evidence for such activity has been provided as shown in Fig. 13.4a. As the localized leaching proceeds, it can be seen that *L. ferrooxidans* accumulate in large numbers in deep holes formed through this type of "electrochemical machining" (Fig. 13.4b), which is also applied as a technical procedure to produce perforations and profiles in very hard materials (Ting et al. 2000; Li et al. 2000). Such hard materials are used as anodes, and an approaching cathode can induce local oxidation and dissolution. *L. ferrooxidans* apparently initiates similar processes and the deep craters produced are the consequence of their local-

Fig. 13.4. Scanning electron microscopy images depicting "electrochemical machining" patterns on pyrite. **a** Topological evidence in favor of localized pitting of pyrite by *L. ferrooxidans*. **b** Pyrite substrate showing subsurface pitting structures highly colonized by *L. ferrooxidans* populations. This is evidence for the ability of *L. ferrooxidans* to collectively attack solid pyrite

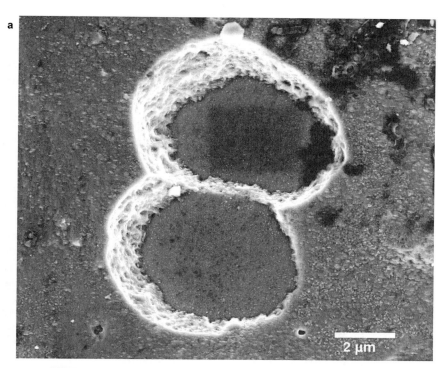

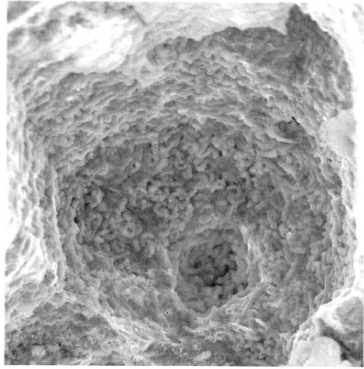

ized "electromachining" activity (Rojas-Chapana and Tributsch 2004). When pyrite is pushed into the electrochemical dissolution region, the comparatively weak bonds forming the links between different grains of the materials will be broken first, so pyrite particles disintegrate into colloids, which are then taken up into the capsule of *Leptospirillum*. The reason why this bacterium does not leach RuS_2 is simply that anodically RuS_2 will not react to form $RuSO_4$ but will liberate oxygen from water. No chemical energy carrier is released.

13.5   The Energy Strategy of *Acidothiobacillus ferrooxidans*

Acidothiobacillus ferrooxidans cannot harvest energy from Fe^{3+} at very positive potentials but rather gains energy by oxidizing Fe^{2+} at redox potentials approximately 200 mV more negative than *L. ferrooxidans*, where the energy content of Fe^{2+} is larger compared with the potential of the oxygen/water redox couple (oxygen being the final electron acceptor). *At. ferrooxidans* can oxidize both Fe^{2+} and reduced sulfur. It has apparently evolved a different strategy for dealing with pyrite. By contacting pyrite interfaces via its capsule it is able to disintegrate and perforate the pyrite to harvest the sulfur species, which is the main energy source. Previous studies have shown that during this process *At. ferrooxidans* accumulates sulfur colloids in its capsule, which serves as a temporary energy storage reserve (Rojas-Chapana et al. 1995, 1996; Tributsch and Rojas-Chapana 1998; Fig. 13.5). The only way in which pyrite could be electrochemically attacked at moderate potentials where it is kinetically stable is by breaking bonds via the formation of interfacial chemical complexes (Fig. 13.2). Evidence was found that thiol groups are involved in such an interfacial breakup and cysteine was identified as the probable candidate (Rojas-Chapana and Tributsch 2000). This amino acid, which plays a significant role in biological electron transfer and as a ligand for catalytic metal centers, was found to interact with pyrite (Abd El-Halim et al. 1995), making electrochemical dissolution possible at lower electrochemical potential (Fig. 13.2). It was demonstrated that, in the presence of cysteine, dissolution of pyrite was enhanced, both in the absence and in the presence of bacteria. On the basis of this finding it was suggested that the amino acid cysteine, which can now be produced very economically (Maier 2003), could be added to enhance bacterial leaching in mineral operations (Rojas-Chapana and Tributsch 2000, 2004). The acidic environment in leaching operations should favor the stability of cysteine by avoiding its oxidation to cystine. It could thus act as a leaching agent liberating Fe–S clusters and as a shuttle, transporting sulfide out from oxygen-deficient areas where bacterial activity is reduced.

   At. ferrooxidans has apparently developed a strategy to disrupt interfacial chemical bonds of pyrite by using cysteine or small thiol-containing peptides (e.g., gluthatione) as a catalyst for interaction. Evidence of the suggested ferredoxin-type complexes between cysteine and iron sulfides clusters has

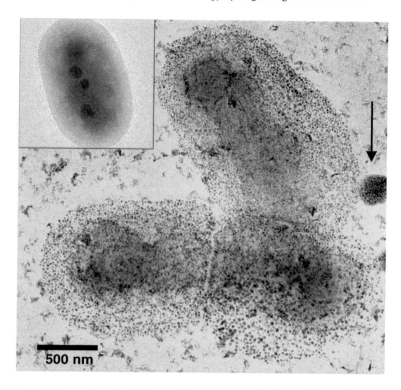

Fig. 13.5. TEM images showing *At. ferrooxidans* structure. *At. ferrooxidans* embedded in sulfur dotted capsules. Sulfur nanoparticles formed at the bacteria–pyrite interface have a narrow particle size and are uniformly dispersed within the bacterial capsule. Notable is that colloidal particles are intimately associated with the capsule, whose surface may be acting as a nucleation site for colloidal globule formation (*arrow*) *Insert*: *At. ferrooxidans* grown on Fe^{2+} without a capsule

also been obtained in other studies (Österberg 1995). They could also serve as vehicles for the transport of chemical sulfide energy into the bacterial cell. Again, in this case, a capsule or envelope is needed by bacteria for logistic reasons and has apparently evolved to mediate both this bond-breaking chemical interaction with pyrite and the storage of energy-rich intermediates.

The reason why *At. ferrooxidans* cannot access the chemical energy contained in RuS_2 is simply the high kinetic stability of RuS_2 against oxidation to thiosulfate and sulfate. When chemical bonds are broken, they capture electrons from the Ru complex which is oxidized, stepwise, to a high oxidation state, which then releases oxygen. The reason why MoS_2 is not a favorable energy source for *At. ferrooxidans* is the difficulty that thiol groups (of cysteine) encounter in breaking chemical sulfur bonds compared with the bonding energy to be gained in the formation of Mo_xS_y *complexes. Indeed, no electrochemical enhancement of MoS_2 dissolution is found in the presence of cysteine, in contrast to pyrite (Fig. 13.2).*

13.6   Surface Chemistry, Colloids and Bacterial Activity

High-resolution morphological studies have revealed that nanoparticles are involved in the bacterial leaching process. In order to elucidate the role played by colloids during mineral degradation, and whether they take part in a bacterial strategy for finding energy, this subject has to be considered in some detail. Although the importance of interfacial processes in bacterial-induced leaching is well recognized, the fundamental aspects are not yet understood. Therefore, it is important to consider biofilm formation, the interfacial leaching mechanism and biologically derived colloid formation (biomineralization). As mentioned earlier, during microbial leaching the generation of colloids dispersed within bacterial capsules and exopolymeric substances (EPS; Sand et al. 1995; Kinzler et al. 2003) may be governed by the corrosive properties of some thiol by-products that react with sulfides (Rojas-Chapana and Tributsch 2000) such as glutathione (Rohwerder and Sand 2003) and polysulfides (Schippers and Sand 1999). On the other hand, pyrite leached both by *At. ferrooxidans* and particularly by *L. ferrooxidans* shows massive pitting or perforation. This type of leaching is associated with microbially influenced corrosion. Though the rate of pitting attack should not be used to estimate general leaching attack as these are completely different leaching mechanisms, the subsurface bacterial activity and growth are higher than those observed as a consequence of simple Fe^{3+}-mediated chemical attack from the surrounding solution. The pitting patterns observed with *L. ferrooxidans* demonstrate clearly that this is due to a localized anodic dissolution of pyrite accelerated by a structured community of bacterial cells within a biofilm (Rojas-Chapana and Tributsch 2004). In addition, colloidal pyrite material is removed from the solid phase during interfacial activity of bacteria via localized pitting. Further, the leaching experiments showed that colloid formation takes advantage of the organic capsule, which in turn mediates the turnover of mineral colloids. Pyrite is hydrophobic; therefore, organic molecules should easily be adsorbed on its surface, as is commonly observed during flotation procedures. Thus, the formation of a bacterial capsule and EPS layers by leaching bacteria, as a result of enhanced adaptive capacity, plays a paramount role not only in "conditioning" the pyrite surface for bacterial attack, but also for the physical chemistry of colloid formation in microbial environments.

13.7   Mechanism of Colloidal Particle Uptake into the Capsule and Exopolymeric Substances

Sulfur is essential for all life, but it plays a particularly central role in the metabolism of sulfur-oxidizing prokaryotes that oxidize reduced forms of this element, such as sulfide (S^{2-}), to sulfate (SO_4^{2-}). Some of these microorganisms, including *At. ferrooxidans*, can store elemental sulfur in colloidal form for use when energy is in short supply (Tributsch and Rojas-Chapana 1998). *At. ferrooxidans* and *L. ferrooxidans* are both able to contact the surface of pyrite minerals and seem to be able to disaggregate it, allowing both coarse particle and fine colloidal material to flow into the bioleaching environment (biofilm) (Rojas-Chapana and Tributsch 2004; Rojas-Chapana et al. 1998). As a result, leaching bacteria create an interface with particles and colloids in order to use them as energy sources. Such interfaces provide an enormous specific surface in comparison to bulk pyrite. In the case of sulfur colloid production by *At. ferrooxidans*, it is apparent that there are physicochemical forces at work promoting the formation of the colloids. These forces can best be summarized as interfacial attractive and repulsive stabilizing forces. Attractive forces can be broadly divided into two categories – electrostatic and van der Waal's attraction.

Electrostatic forces are simply controlled by pH. Solid surfaces in water have a point of zero charge; at a pH above that point, surfaces will have a negative charge, and a positive charge below that point. Pyrite has a point of zero charge, a minimum charge, at a pH of approximately 2.4 (Widler and Seward 2002). The ideal condition for colloidal mobility would be to establish a pH at which the colloids have the same charge as the mineral matrix and are therefore electrostatically repelled from mineral surfaces. The surface charging of a pyrite interface will depend on its oxidation state as well as on the composition of the contacting electrolyte. Unfortunately we do not know what pH values occur at the pyrite–capsule interface, neither we do know well the composition of the electrochemically reacting capsule.

13.7.1 Sulfur Colloid Formation

To elucidate the mechanism by which *At. ferrooxidans* promotes sulfur colloid formation, a more fundamental understanding of the problem of localized leaching initiation that leads to pitting of the pyrite surface is necessary. This involves oxidation of the metal sulfide moiety and "biogenic" formation of stable nano-sized sulfur colloids embedded in a bioorganic matrix (Tributsch and Rojas-Chapana 1998). Sulfur atoms, as by-products of pyrite leaching, tend to be attracted to each other and agglomerate, leading to sulfur seeds. If the sulfur seed colloids formed in the interfacial water film can get close enough to a bioorganic phase, they will stick. It can be assumed that a thin water film at the pyrite–bacteria interface will assist in the enhanced uptake of sulfur colloids into the bacterial environment (capsule/EPS). The colloidal patterns formed in the *At. ferrooxidans* capsule (Fig. 13.5) show a pronounced influence of bioorganic components on the particle distribution and stabilization (Rojas-Chapana et al. 1996).

Since the sulfur nanoparticles are strictly confined to the area surrounding the bacteria, and since they are quite uniformly distributed within the bacterial capsule with an average particle size of 10 nm, it may be assumed that patterning occurs during uptake of the sulfur seeds, taking place at the pyrite–capsule interface. In addition, both the bacterial capsule and the nanoparticles form a reactive surface characterized by a chemical turnover associated with a gradual decrease in particle size during aging of the bacteria. Further, it has been observed that the sulfur nanoparticles are an integral part of the bacterial metabolism of At. ferrooxidans (Fig. 13.5).

The chemical nature of colloidal sulfur present in At. ferrooxidans was first described by Steudel et al. (1987). These authors proposed the so-called simple sulfur globule model which suggested the presence of polythionates, and mainly cyclooctasulfur. This model was more recently revised by Prange et al. (2002) and extended to other bacteria. Prange et al. concluded that sulfur in the globules of At. ferrooxidans mainly consists of polythionates. The same authors mentioned that no significant contribution of cyclooctasulfur (which can be formed in the medium by chemical processes) was found, though its presence could not be ruled out completely.

13.7.2   Pyrite Colloid Formation

Pyrite colloids have been observed within the capsule of L. ferrooxidans (Rojas-Chapana and Tributsch 2004). Edwards et al. (2001) described At. ferrooxidans bacteria covered with nanoparticulated mineral deposits, and Sand et al. (1999) proposed that At. ferrooxidans bacteria utilize nanoclusters of ferric iron compounds embedded within their exopolymer layers to afford a high rate of leaching at the cell–mineral interface. The evidence appears to support the theory that "colloid-bearing bacteria" are involved in the interfacial leaching of pyrite.

Despite the apparent limitation of L. ferrooxidans to use only iron from pyrite as the sole energy source, this acidophile shares several features in common with At. ferrooxidans. These include interfacial processes, bacterial adaptation to the pyrite surface through capsule/EPS formation, and bacteria-mediated uptake and storage of colloidal nanoparticles. However, unlike At. ferrooxidans, L. ferrooxidans activity on pyrite reveals striking differences in leaching patterns and colloid chemistry. Specifically, a pronounced pitting of the pyrite surface closely associated with determined crystal orientations (Fig. 13.4), and nanoparticles that are not the result of a by-product formed during the leaching reaction, but that rather consist of pyrite. Furthermore, in contrast to "biogenic" sulfur colloids in At. ferrooxidans, pyrite colloids neither show a narrow particle size distribution nor are they uniformly dispersed within the capsule. These three major differences suggest that the formation of colloidal nanoparticles at the pyrite interface involves weak material structures, including grain boundaries and dislocations, which leads to the exposure of underlying morphologies, allowing further removal of suspended solids and colloids. Further studies have suggested that L. ferrooxidans uses pyrite colloids as a transportable and storable form of energy. Energy-dispersive X-ray measurements made on areas of the *L. ferrooxidans* capsules where monodispersed colloids are not visible indicate that the capsule contains both Fe and S atoms corresponding to the stoichiometry of pyrite. This can be explained to be the consequence of a massive uptake of pyrite colloids into the capsule with their subsequent dissolution. Thus, they undergo a progressive decrease of grain size owing to their high surface reactivity. Under certain conditions, for example, when the pyrite particles have become very small, they start forming a cell-sized shell composed of pyrite mineral. At present, it is not possible to ascertain whether this is a side phenomenon or is a case of active biomineralization – the formation of a mineral protection that could assist bacteria in surviving especially harsh environmental conditions.

13.8 Energy Turnover at the Nanoscale, a Strategic Skill Evolved by Bacteria

As stated previously, *L. ferrooxidans* can utilize only ferrous iron as an energy source. One problem is that the Fe^{2+}/Fe^{3+} couple yields very little energy. In other words, ferrous iron has a quite positive redox potential. The other problem is that ferric iron, when extracting electrons from pyrite, cannot break essential bonds. Only when pyrite is polarized very positively can it be made to disintegrate electrochemically (Fig. 13.2). *L. ferrooxidans* has adapted its Fe^{3+}-mediated attack to such positive potentials, where not much energy is gained when electrons are transferred from ferrous iron to oxygen. To extract this energy *L. ferrooxidans* is obliged to oxidize ferrous iron in large quantities. To handle this problem, *L. ferrooxidans* has evolved a special colloid-based strategy. It involves a capsule structure comparable to that of *At. ferrooxidans*. This sulfide leaching bacterial capsule is made up in part of EPS (Kinzle et al. 2003). This is a gelatinous material that encloses either one cell or a group of cells. The capsule plays an important role in the attachment of bacterial cells to the pyrite surface and in the formation of a biofilm. Aside from the surface-active properties of the capsule, colloidal pyrite nanoparticles can be adsorbed and utilized by bacteria as a source of temporary energy. The beneficial effects of pyrite nanoparticles on *L. ferrooxidans* are due to the high reactivity of small particles and large-surface-area interfaces, so they act as discrete nanoelectrochemical units providing large reactive interfaces. In addition, thermodynamic quantities change with decreasing particle size, so, for example, dissolution processes become much more efficient (Ostwald ripening). The number of atoms located at the surface or in the interfacial region near the bacterial membrane increases as particle size decreases: they involve 1–2% of the atoms in a 100-nm particle, 10–15% at 10 nm, and 20–30% at 5 nm. When bacteria turnover nanoparticles within their capsule, they utilize more readily accessible sources of energy.

13.9   Summary

Pyrite is the one of the most important energy sources for sulfide-oxidizing bacteria and has shaped evolution of bacterial leaching strategies during a period of more than two billion years. Typical leaching strategies, as implemented by *At. ferrooxidans* and *L. ferrooxidans*, have been outlined. Physicochemical and morphological evidence reveals that the bacteria use nanoscale approaches to sulfide processing. A precondition is the presence of a suitable, chemically optimized reaction medium, which for the bacteria is an organic capsule. It acts as a contact medium to the sulfide as well as a reaction environment for the turnover of nanoparticles. This strategy associated with colloids and small particles is inherently related to a contact leaching mechanism of bacterial leaching, which guarantees a locally confined and efficient handling of sulfide energy. Owing to the long evolutionary period that bacteria have experienced, it can be expected that chemical mechanisms for interaction with solid mineral particles have been optimized. Interesting additional discoveries can therefore be expected in the future. It is reasonable to conclude that bacterial leaching mechanisms of much less abundant minerals will have evolved and diversified from originally pyrite designed processes. When being faced with alternative minerals as an energy source, bacteria may have attempted to adapt their strategies. Chalcopyrite, which is presently an important ore of copper, is not readily bioleached because intermediate oxidation products of this compound and their peculiar electronic structure slow down the reaction rate. Bacteria may only adapt well to such an energy source if evolution finds a way out of such energetic hindrance. Elevated-temperature leaching processes and thermophilic bacteria are therefore presently being exploited for commercial leaching of chalcopyrite, as described in Chap. 3.

When involving bacterial leaching mechanisms in mining operations, biomining engineers need to consider the evolved abilities of microorganisms to deal with solid-state sulfide sources and to evaluate carefully the new boundary conditions provided in the technical leaching operation. Questions have to be asked with respect to the mineral composition, the chemical environment, the coexistence and symbiotic interaction of different bacterial species as well as the physical conditions, such as temperature and pressure. Bacteria can only handle leaching problems within their genetically defined abilities. If leaching problems which bacteria cannot handle alone become apparent, the process has to be complemented by additional treatments including, but not limited to, thiol-active, molecules that may support the leaching abilities of bacteria and archaea.

References

Abd El-Halim AM, Alonso-Vante N, Tributsch H (1995) Iron/sulphur centre mediated photoinduced charge transfer at (100) oriented pyrite surfaces. J Electroanal Chem 399:29–39
d'Hugues P, Foucher S, Gallé-Cavalloni D, Morin D (2002) Continuous bioleaching of chalcopyrite using a novel extremely thermophilic mixed culture. Int J Miner Process 66:107–119

Edwards KJ, Hu B, Hamers RJ, Banfield JF (2001) A new look at microbial leaching patterns on sulfide minerals. FEMS Microbiol Ecol 34:197–206

Ennaoui A, Tributsch H (1984) Iron sulfide solar cells. Solar Cells 13:197–200

Ennaoui A, Tributsch H (1986) Energetic characterization of the photoactive FeS$_2$ (Pyrite) interface. Sol Energy Mater 14:461–474

Ennaoui A, Fiechter S, Goslowsky H, Tributsch H (1985) Photoactive synthetic polycrystalline pyrite (FeS$_2$). J Electrochem Soc 132:1579–1582

Ennaoui A, Fiechter S, Pettenkofer C, Alonso-Vante N, Büker K, Bronold M, Höpfner C, Tributsch H (1993) Iron disulfide for solar energy conversion. Sol Energy Mater Sol Cells 29:289–370

Eyert V, Höck KH, Fiechter S, Tributsch H (1998) Electronic structure of FeS$_2$: the crucial role of electron-lattice interaction. Phys Rev B 57:6350–6359

Kinzler K, Gehrke T, Telegdi J, Sand W (2003) Bioleaching – a result of interfacial processes caused by extracellular polymeric substances (EPS). Hydrometallurgy 71:83–88

Li XP, Alonso Vante N, Tributsch H (1986) Involvement of coordination chemistry electron transfer in the stabilization of the pyrite (FeS$_2$) photoanode. J Electroanal Chem Interface Electrochem 242:255–264

Li Y, Zhang D, Wu Y (2000) Biomachining of metal copper by *Thiobacillus* ferrooxidans. Wei Sheng Wu Xue Bao 40:327–330

Maier TH(2003) Semisynthetic production of unnatural L-α-amino acids by metabolic engineering of the cysteine-biosynthetic pathway. Nat Biotechnol 21:422–427

Österberg R (1995) The origin of metal ions occurring in living systems. In Berthon G (ed) Handbook of metal-ligand interactions in biological fluids. Dekke New York, pp 10–28

Prange A, Chauvistré R, Modrow H, Hormes J, Trüper HG, Dahl C (2002) Quantitative speciation of sulfur in bacterial sulfur globules: X-ray absorption spectroscopy reveals at least three different species of sulfur. Microbiology 148:267–276

Rawlings DE, Tributsch H, Hansford GS (1999) Reasons why 'Leptospirillum'-like species rather than Thiobacillus ferrooxidans are the dominant iron-oxidizing bacteria in many commercial processes for the biooxidation of pyrite and related ores. Microbiology 145:5–13

Rohwerder T, Sand W (2003) The sulfane sulfur of persulfides is the actual substrate of the sulfur-oxidizing enzymes from *Acidithiobacillus* and *Acidiphilium* spp. Microbiology 149:1699–1710

Rojas-Chapana JA, Tributsch H (2000) Bioleaching of pyrite accelerated by cysteine. Process Biochem 35:815–824

Rojas-Chapana JA, Tributsch H (2004) Interfacial activity and leaching patterns of *Leptospirillum ferrooxidans* on pyrite. FEMS Microbiol Ecol 47:19–29

Rojas-Chapana JA, Giersig M, Tributsch H (1986) The path of sulfur during the bio-oxidation of pyrite by *Thiobacillus ferrooxidans*. Fuel 75:923–93

Rojas-Chapana JA, Giersig M, Tributsch H (1995) Sulfur colloids as temporary energy reservoirs for *Thiobacillus ferrooxidans* during pyrite oxidation. Arch Microbiol 163:352–354

Rojas-Chapana JA, Giersig M, Tributsch H (1996) The path of sulfur during the bio-oxidation of pyrite by *Thiobacillus ferrooxidans*. Fuel 75:923–930

Rojas-Chapana JA, Bärtels CC, Pohlmann L, Tributsch H (1998) Co-operative leaching and chemotaxis of *Thiobacillus* studied with spherical sulfur/sulfide substrates. Process Biochem 33:239–248

Salvador P, Tafalla D, Tributsch H, Wetzel H (1991) Reaction mechanisms at the n-FeS$_2$/ I–interface/an electrolyte electroreflectance study. J Electrochem Soc 138:3361–3369

Sand W, Gehrke T, Josza P, Schippers A (1999) In: Proceedings of the international biohydrometallurgy symposium IBS-99. Elsevier, Amsterdam, pp 27–30

Sand W, Gerke T, Hallmann R, Schippers A (1995) Sulfur chemistry, biofilm, and the (in)direct attack mechanism – a critical evaluation of bacterial leaching. Appl Environ Microbiol 43:961–966

Schippers A, Sand W (1999) Bacterial leaching of metal sulfides proceeds by two indirect mechanisms via thiosulfate or via polysulfides and sulfur. Appl Environ Microbiol 65:319–321

Suzuki I (2001) Microbial leaching of metals from sulfide minerals. Biotechnol Adv 19:119–132

Steudel R, Holdt G, Göbel T, Hazeu W (1987) Chromatographic separation of higher poly-thionates $S_nO_6^{2-}$ ($n=3...22$) and their detection in cultures of *Thiobacillus ferrooxidans*: molecular composition of bacterial sulfur secretions. Angew Chem Int Ed Engl 26:151–153

Ting YP, Senthil Kumar A, Rahman M, Chia BK (2000) Innovative use of *Thiobacillus ferrooxidans* for the biological machining of metals. Acta Biotechnol 20:87–96

Tributsch H, Rojas-Chapana JA (1998) The role of transient iron disulfide films in microbial corrosion of steel. Corrosion 54:216–227

Tributsch H, Rojas-Chapana JA (2000) Metal sulfide semiconductor electrochemical mechanisms induced by bacterial activity. Electrochim Acta 45:4705–4716

Tributsch H, Fiechter S, Jokisch D, Rojas-Chapana JA, Ellmer K (2003) Photoelectrochemical power, chemical energy and catalytic activity for organic evolution on natural pyrite interfaces. Origins Life Evol Biosphere 33:129–162

Wächtershäuser G (1988) Pyrite formation, the first energy source for life: a hypothesis. Syst Appl Microbiol 10:207–210

Wächtershäuser G (2000) Perspectives: origin of life: life as we don't know it. Science 298:1307–1308

Widler AM, Seward TM (2002) The adsorption of gold(I) hydrosulphide complexes by iron sulphide surfaces. Geochim Cosmochim Acta 66:383–402

14 Genetic and Bioinformatic Insights into Iron and Sulfur Oxidation Mechanisms of Bioleaching Organisms

DAVID S. HOLMES, VIOLAINE BONNEFOY

14.1 Introduction

Arguably, the most important role played by microorganisms in the solubilization of metals in bioleaching operations is their ability to oxidize iron- and sulfur-containing minerals. Most progress has been made in understanding these fundamental processes in the mesophilic bacterium *Acidithiobacillus ferrooxidans* because its role in bioleaching was recognized earliest and, therefore, it has had the longest history of investigation. In addition, the complete genome sequence of *At. errooxidans* was recently released by The Institute for Genome Research (TIGR), which has allowed the bioinformatic prediction of several important biochemical pathways and has provided insight into the biochemistry and physiology of iron and sulfur oxidation. This chapter will describe some of the recent progress in understanding iron and sulfur oxidation reactions in this microorganism using molecular genetics and bioinformatics tools.

In the last decade it has become increasingly clear that several other bacteria and archaea play crucial roles in mineral solubilization, especially at the elevated temperatures that occur in tank reactors for gold recovery and in the later stages of copper heap leaching when exothermic biooxidation reactions have driven temperatures up to the 45–80 °C range. Several of these bacteria and archaea have been identified and some initial information regarding their iron and sulfur metabolisms is described in this chapter.

It has also become increasingly obvious that metal solubilization is promoted by the concerted effort of a consortium of microorganisms and that it is imperative to understand how these microorganisms cooperate in this endeavor. The identification of these microorganisms and the investigation of their interactions in bioleaching operations is a study in microbial ecology. To date, investigations into the microbial ecology of bioleaching operations has relied primarily on the use of standard techniques of molecular genetics, especially in the description of the spatial and temporal distribution of microorganisms in heap bioleaching operations and these advances are discussed in Chaps. 10–12. However, the emerging fields of comparative genomics and metagenomics are beginning to impact on our understanding of the microbial ecology and metabolic processes of bioleaching operations

Biomining
(ed. by Douglas E. Rawlings and D. Barrie Johnson)
© Springer-Verlag Berlin Heidelberg 2007

and some of the recent advances in these areas related to our understanding of microbial iron and sulfur oxidation will be described.

14.2 Relevant Biochemical and Chemical Reactions

The important biochemical and chemical reactions involved in iron and sulfur oxidation in bioleaching processes have been described elsewhere in this book (Chaps. 1, 8, 9). However, for the purposes of this chapter it is important to highlight the following two points:

1. The initial substrates for iron and sulfur oxidation enzymes in bioleaching microorganisms are insoluble metal sulfides. For example, pyrite (FeS_2) is the most common sulfide mineral present in sulfide ore deposits and its oxidation can be written

$$FeS_2 + 6Fe^{3+} + 3H_2O \rightarrow S_2O_3^{2-} + 7Fe^{2+} + 6H^+. \quad (14.1)$$

The Fe^{2+} generated in Eq. 14.1 can be biologically oxidized to Fe^{3+}:

$$4Fe^{2+} + O_2 + 4H^+ \rightarrow 4Fe^{3+} + 2H_2O. \quad (14.2)$$

The Fe^{3+} product of Eq.14.2 is a strong oxidant that can oxidize metals present in the ore, aiding in their solubilization and regenerating Fe^{2+} in the process. However, in the absence of this regeneration, Fe^{3+} may subsequently be precipitated as ferric oxyhydroxide:

$$Fe^{3+} + 2H_2O \rightarrow FeOOH + 3H^+. \quad (14.3)$$

The generation of a precipitate precludes the possibility that the oxidation of Fe^{2+} for energy generation occurs inside the cell and, therefore, it must be localized to the exterior of the outer membrane. This places an emphasis on where we must look for the key initial enzymes involved in iron oxidation.

2. The E'_0 of the redox couple Fe^{2+}/Fe^{3+} is +0.74 V at pH 2, whereas that of the $NA(P)D^+/NAD(P)H$ couple is –0.32 V at pH 7. This means that the electrons biologically extracted during the oxidation of Fe^{2+} must be pushed "uphill" against a thermodynamically favorable gradient to reduce $NAD(P)^+$ to $NAD(P)H$. Therefore, a source of energy must be found for the process and any proposed intermediate electron carriers must have values of E'_0 between those of Fe^{2+}/Fe^{3+} and $NA(P)D^+/NAD(P)H$.

14.3 Genetics of Bioleaching Microorganisms

14.3.1 Introduction

Molecular genetics refers to a collection of techniques and knowledge that can be used to investigate gene and protein function. For example, in the case of bioleaching organisms, it is relatively straightforward to isolate genes from bioleaching microorganisms and clone them into a well-characterized surrogate host such as *Escherichia coli* where their function can be analyzed by complementing mutants or where sufficient gene product (protein) can be isolated to carry out standard biochemical assays to investigate function. One problem is that these strategies seem to work best for proteins that are expressed in the cytoplasm of bioleaching organisms where the pH is close to neutral, as is the cytoplasm of *E. coli*. Analyzing the function of periplasmic or outer-membrane proteins of bioleaching microorganisms in *E. coli* results in proteins that may exhibit either no or poor function because they probably require an acid pH to fold and work correctly (Brown et al. 1994; Bengrine et al. 1998; Appia-Ayme 1998; Bruscella et al. 2005). This is unfortunate because many of the most relevant enzymes and electron carriers involved in iron and sulfur oxidation are outside the cytoplasm or are embedded within the membrane with loops exposed to the acid pH of the periplasm.

A significant challenge has been the difficulty of returning genes to bioleaching microorganisms once they have been studied in a surrogate host. This means that their function cannot be proved directly in the bioleaching organism in question, nor can genes that have been modified by genetic engineering in surrogate hosts be reintroduced back into the microorganisms from which they were derived. This has seriously impeded progress in understanding the physiology of bioleaching organisms and has prevented the development of genetically modified strains. Some promising progress has, however, been made in this direction recently and will be discussed in this chapter.

Genetic analysis of bioleaching microorganisms is also a challenge because, in general, they are strict or moderate acidophiles and obligatory or facultative chemoautolithotrophs. They grow slowly in laboratory culture with very low cell yields making them difficult to culture and posing severe problems for the isolation of sufficient quantities of enzymes for biochemical analysis.

A word of caution is in order regarding bioleaching microorganism strain classification and its implication for the identification of genes, enzymes and biochemical pathways. Even though two organisms may be classified as belonging to the same species, usually by ribosomal RNA spacer similarity, it does not necessarily mean that they share an identical repertoire of genes. This means that biological interpretations, extrapolated from one strain to another, should be treated with caution. This potential problem is exacerbated when the identification of a particular strain is not adequately described. The literature regarding the genetics and biochemistry of *At. ferrooxidans* is particularly rampant with examples of the use of "private" strains, i.e.. that are not deposited in public microbial banks and are often, lamentably, not adequately characterized. Such strains are often not publicly available and consequently results of investigations cannot be independently confirmed.

14.3.2   Gene Cloning

The majority of genes that encode proteins involved in the oxidation of sulfur (S^0), reduced inorganic sulfur compounds, or ferrous iron [Fe(II)] that have been cloned and studied are from *At. ferrooxidans*, and are described later. However, advances are being made in the cloning and analysis of Fe(II)- and S^0-related genes in other microorganisms. For example, genes from *Leptospirillum ferrooxidans* (Delgado et al. 1998) and *Thermoplasma acidophilum* have been investigated. Recently, a tetrathionate hydrolase gene from *At. caldus* has been sequenced, and preliminary evidence suggests that it may belong to an operon that also contains genes encoding a terminal quinol oxidase (Rzhepishevska et al. 2005, personal communication).

14.3.3   Gene Transfer Systems

The real challenge is to introduce DNA into bioleaching microorganisms. Without a reliable and efficient gene transfer system, expression of heterologous genes in these bacteria and the construction of mutants, precisely defined at the molecular level, are not possible. The classic approaches followed are (1) transduction, i.e., transfer of genetic information via a bacteriophage (virus) particle, (2) conjugation, i.e., the transfer of conjugative or mobilizable plasmids from one bacterium to another by cell-to-cell contact, or (3) electrotransformation, by exposing the cells in the presence of free DNA to a pulsed electric field which destabilizes transiently the bacterial membrane and permits the entry of the DNA into the cell.

14.3.3.1   *Acidiphilium* spp.

Acidiphilium spp. are acidophilic and facultative heterotrophic Gram-negative proteobacteria (Hiraishi et al. 1998). *Acidiphilium acidophilum* (formerly *Thiobacillus acidophilus*) can grow autotrophically with reduced inorganic sulfur and is capable of mixotrophic ferric iron [Fe(III)] reduction (Johnson and McGinness 1991). A bacteriophage, ϕAC1, that infects *Acidiphilium* and can integrate into its genome has been described (Ward et al. 1993) but no transduction system with this phage has been reported so far. However, genetic transfer by conjugation between *E. coli* and different *Acidiphilium* species (Roberto et al. 1991; Glenn et al. 1992; Quentmeier and Friedrich 1994; Bruhn and Roberto 1993) and by electrotransformation (Glenn et al. 1992; Inagaki et al. 1993) has succeeded with frequencies from 10^{-2} to 10^{-9} transconjugants per recipient cell and 10^3–10^4 transformants per milligram of plasmid DNA, respectively, depending on the plasmid and on the *Acidiphilium* species.

14.3.3.2 Acidithiobacillus thiooxidans

This Gram-negative proteobacterium is a strict acidophile and an obligate autotroph. However, in spite of these differences with the neutrophilic and heterotrophic E. coli, direct mating is possible, and self-transmissible broad-host-range plasmids were transferred between these two microorganisms. Transconjugants were obtained at a low but workable frequency (10^{-5}–10^{-7} per recipient) (Jin et al. 1992). With this approach, the E. coli phosphofructokinase was cloned and expressed successfully in At. thiooxidans (Tian et al. 2003).

14.3.3.3 Acidithiobacillus ferrooxidans

For many years, considerable but unsuccessful efforts were made by several groups to find conditions that would permit conjugation between E. coli and At. ferrooxidans. The main problem was to find a compatible medium in which both bacteria could survive and produce the energy for the conjugation process to take place. Nevertheless, Peng et al. (1994b, c) succeeded in designing a mating medium with which plasmids from different incompatibility groups were conjugated in several private strains of At. ferrooxidans. Low but workable frequencies of transconjugants (10^{-5}–10^{-7} transconjugants per recipient) were obtained, depending on the incompatibility group of the plasmid and on the strains (Peng et al. 1994c). In spite of this low frequency, the transposon Tn5 was successfully introduced with a mobilizable suicide vector and was shown to be able to transpose into the chromosome of At. ferrooxidans. This result is encouraging and opens the way to random transposon insertion mutagenesis (Peng et al. 1994c). In addition, the arsenic-resistance genes from a narrow-host-range nonmobilizable plasmid were cloned into a mobilizable plasmid and transferred by conjugation from E. coli to At. ferrooxidans and expressed (Peng et al. 1994a). The transfer of plasmids from different incompatibility groups from E. coli to different At. ferrooxidans strains, including strains ATCC33020 and ATCC19859, was later improved to get a frequency as high as 2.5×10^{-3} transconjugants per recipient (Liu et al. 2000, 2001a). More recently, transconjugants with the type strain ATCC23270 were obtained with even higher frequencies (Ratouchniak and Bonnefoy, unpublished data).

Electroporation of native and exogenous plasmids into At. ferrooxidans has been reported (Kusano et al. 1992a); however, only one out of the 30 private strains tested was transformed and 50% of colonies obtained were false positives and not transformants. Furthermore, in spite of all these efforts, the frequency was very low (120–200 colonies per milligram of DNA) (Kusano et al. 1992a; Rawlings and Kusano 1994). By following the efflux of ATP out of the cells, strain ATCC33020 was shown to be electropermeabilizable

(Guiliani et al. 1995); however, under these conditions, only false positives were obtained, even with a plasmid which had been purified from *At. ferrooxidans* ATCC33020 transconjugants, suggesting that the restriction barrier of this strain is very stringent (Liu, Ratouchniak, and Bonnefoy, unpublished data).

14.3.4   Mutant Construction

Only two *At. ferrooxidans* mutants, affected in their Fe(II) oxidation capacities, have been described. The first was obtained by nitrosoguanidine mutagenesis and was reported to affect the rusticyanin gene, but no proof was provided (Cox and Boxer 1986). The second was due to the insertion of IS*Afe1* (an insertion sequence element) within the *resB* gene, resulting in mutant cells that could grow on S^0 but not Fe(II) medium (Zhao and Holmes 1993; Cabrejos et al. 1999; Holmes et al. 2001). Interestingly, *resB* encodes a protein involved in the maturation of cytochrome *c*, suggesting that this electron carrier is required for Fe(II) oxidation but is not necessary for S^0 oxidation, a hypothesis which was later confirmed (Brasseur et al. 2004; Bruscella 2004). Unfortunately, this mutant has been lost and the attempts to construct a new one by marker exchange mutagenesis (MEM) have failed.

MEM, in contrast to the techniques described earlier, permits the construction of specifically designed mutants and is therefore called a "targeted" mutagenic approach. In MEM, a "suicide plasmid," carrying a selectable marker inserted inside the gene to be mutated, is introduced into *At. ferrooxidans* cells by conjugation. This plasmid is unable to replicate in this bacterium, and the only way to maintain the selection is by plasmid insertion into the chromosome (one crossover), or by exchange between the "targeted" normal (wild-type) gene in the chromosome and the mutated gene in the plasmid (two crossovers). A *recA* mutant was successfully constructed in this way (Liu et al. 2000, 2001b). The same strategy was later used to construct a *hip* mutant, the *hip* gene encoding a high redox potential iron–sulfur protein whose function remains controversial (Bruscella 2004; and see later). Unfortunately, even after 1 year of subculturing in Fe(II) medium without selection, only one crossover could be detected, and it was demonstrated that both the mutated and the wild-type alleles were present in the transconjugant (Bruscella 2004). Since the wild-type allele was still expressed, this strain behaves phenotypically as a wild type, suggesting that the *hip* gene is essential in the conditions tested and cannot be inactivated by MEM. In conclusion, the MEM approach, while feasible, is fastidious and can be only applied to nonessential genes.

14.4 Iron and Sulfur Oxidation and Reduction in *Acidithiobacillus ferrooxidans*

14.4.1 Ferrous Iron Oxidation

14.4.1.1 Introduction

The oxidation of Fe(II) by *At. ferrooxidans* can be considered to proceed in a "downhill" and an "uphill" reaction, although both pathways are interconnected in nature. In the "downhill" pathway, electrons removed biologically from Fe(II) proceed, through a series of electron carriers, from the outer membrane to the cytoplasm, where they are used to reduce O_2 to water, consuming protons in the process. The majority of the protons consumed have entered the cell via the ATP synthetase complex embedded in the inner membrane. *At. ferrooxidans* generates ATP using this proton motive force (PMF) generated by the tremendous proton gradient across the inner membrane and the electrons from the oxidation of iron are used only to neutralize the incoming protons.

However, in addition, *At. ferrooxidans* has to regenerate the reduced pyridine nucleotides NADPH and NADH necessary for CO_2 and N_2 fixation and other anabolic processes. Because the standard reduction half-potential of the Fe(II)/Fe(III) couple (+0.77 V at pH 2, the pH of the medium) is much more positive than that of the NAD(P)/NAD(P)H couple (–0.32 V at the cytoplasmic pH 7), the electrons have to be "pushed uphill" from Fe(II) to NAD(P) against the redox potential gradient. This "uphill" electron transfer requires energy which is probably provided by the PMF and the "uphill" flow of electrons can be considered somewhat similar to a mitochondrion working in reverse, a hypothesis originally proposed by Ingledew (Ingledew and Cobley 1980; Ingledew 1982) over 20 years ago. In addition, ATP hydrolysis via the ATP synthetase working in reverse may be used to generate an electrochemical proton gradient that may provide some of the force to push electrons "uphill" (Elbehti et al. 2000). This invokes an interesting possible regulatory mechanism for adjusting the balance between the production of NAD(P)H and ATP. When ATP concentrations are high, the ATP synthetase functions like an ATPase and the ATP hydrolysis may provide additional PMF to push electrons "uphill" to reduce NAD(P) to NAD(P)H that can then be used to fix carbon. Conversely, when the concentration of fixed carbon is high, ATP synthetase, working in the normal respiratory direction, will produce ATP (Elbehti et al. 2000).

14.4.1.2 The "Downhill" Electron Pathway

Because of the relatively high redox potential of the Fe(II)/Fe(III) couple, a considerable amount of Fe(II) has to be oxidized to sustain the growth of

Fe(II)-oxidizing acidophiles and, as a result, many electrons have to be transferred from this compound to oxygen. On the basis of the assumption that the proteins involved in the "downhill" electron transfer chain have to be synthesized in large quantities, the redox proteins present in relative abundance in Fe(II)-grown cells were proposed to be involved in this pathway and were characterized. From these studies, a number of models have been designed, which differ slightly in the proteins involved and in their order in the respiratory chain (Ingledew et al. 1977; Yamanaka et al. 1991; Yamanaka and Fukumori 1995; Blake and Shute 1994; Bruschi et al. 1996; Giudici-Orticoni et al. 2001). However, these models contained several questionable points, the main one being that the first electron acceptor proposed in all these models is located in the periplasm, while it is known that (1) an important natural substrate is pyrite, an insoluble sulfide mineral, and (2) soluble Fe(II) cannot enter the cell because of its rapid auto-oxidization and the highly insoluble nature of the ferric oxy–hydroxide product at the pH of the cell. Following a completely different approach, another model has been proposed based on the studies on genetic organization, gene regulation, and subcellular localization of the different partners (Appia-Ayme 1998; Appia-Ayme et al. 1999; Yarzabal et al. 2002b, 2004).

The first gene encoding a redox protein was the *iro* gene from Fe-1 strain (Kusano et al. 1992b), encoding a high-potential redox iron–sulfur protein (HiPIP) which was proposed by several authors to be the first electron acceptor from Fe(II) (Fukumori et al. 1988; Cavazza et al. 1995; Yamanaka and Fukumori 1995; Bruschi et al. 1996; Yamanaka et al. 1991). This gene is located between the *purA* and a transfer RNA encoding genes but is transcribed independently (Kusano et al. 1992b). It should be noted that in three collection strains including the type strain (ATCC23270, ATCC33020 and ATCC19859), only one gene encoding a HiPIP was identified and this gene, referred to as the *hip* gene, encodes a protein presenting only about 51% similarity to Iro (Bruscella et al. 2005). This *hip* is in a completely different genetic context from *iro* and its expression is higher in S^0-grown cells than in Fe(II)-grown cells (Bruscella 2004; Bruscella et al. 2005; Quatrini et al. 2005a). These data, and the periplasmic location of the corresponding Hip protein, raise doubts about its proposed role at least in these three strains (see later). For several years, a number of laboratories worldwide have tried to isolate the gene encoding rusticyanin, whose concentration is very high in Fe(II)-grown cells (up to 5% of the cell protein; Cox and Boxer 1978) but without success. The sequence of a 260-bp internal fragment of the *rus* gene was determined by Pulgar et al. (1993), and a synthetic *rus* gene was constructed, as well as site-specific mutants, and was overexpressed in *E. coli* to study the biochemical and biophysical properties of the protein (Casimiro et al. 1995). Also, in 1995, the sequence of the entire *rus* gene and of the flanking regions from the ATCC33020 strain was determined (Guiliani et al. 1995; Bengrine et al. 1995, 1998). One year later, the DNA sequence encoding the mature rusticyanin from the type strain ATCC23270 was published (Hall et al. 1996). From the analysis of the genomic organization of the *rus* locus, a model for the iron respiratory chain was proposed (Appia-Ayme 1998; Appia-Ayme et al. 1999) and was later confirmed by subcellular localization of the different components (Yarzabal et al. 2002a, b) and by genetic regulation (Yarzabal et al. 2003, 2004; Quatrini et al. 2005a). The *rus* gene belongs to an operon which is more highly expressed in Fe(II)-grown cells than in S^0-grown cells, supporting the involvement of the *rus* operon encoded products in the oxidation of Fe(II) (Yarzabal et al. 2003, 2004; Quatrini et al. 2005a). This operon encodes, in addition to rusticyanin, three other electron transfer proteins, two cytochromes *c* (Cyc1 and Cyc2) and an aa_3-type cytochrome oxidase (CoxABCD) (Appia-Ayme 1998; Appia-Ayme et al. 1999). These redox proteins constitute likely a "respiratory supercomplex" involved in Fe(II) respiration, as proposed by Appia-Ayme et al. (1999). This operon encodes also a protein (ORF) which presents some similarities (32%) to the *Halobacterium* sp. NRC-1 Pan1 protein of unknown function. ORF has been located in the outer membrane (Yarzabal and Bonnefoy, unpublished data) but its function remains unknown.

As discussed earlier, it was assumed that the electrons have to pass from the outside medium, where pyrite (FeS_2) is oxidized, to the cytoplasm, where H_2O reduction takes place. The order of the different partners encoded by the *rus* operon in the respiratory chain is dependent on their subcellular localization. The cytochrome *c* Cyc2 has been shown to be located in the outer membrane, with a domain facing the external environment where it may interact with insoluble substrates (Yarzabal et al. 2002b), while the cytochrome *c* Cyc1 is bound to the inner membrane (Yarzabal et al. 2002a). Rusticyanin was confirmed to be periplasmic and the cytochrome oxidase to be an integral inner-membrane complex (Yarzabal et al. 2002b). Therefore, the electron transporters appear to constitute an "electron wire spanning both the outer and the inner membranes to conduct electrons from pyrite to oxygen," as suggested by Yarzabal et al. (2002b), through Cyc2→rusticyanin→Cyc1→CoxABCD (Fig. 14.1).

14.4.1.3   The "Uphill" Electron Pathway

Ingledew (1982) suggested a role for the bc_1 complex in the pathway to regenerate NAD(P)H in Fe(II)-grown cells. The existence of a reverse electron flow from a cytochrome *c* through the cytochrome bc_1 complex to quinone and the NAD(P)H dehydrogenase was later clearly demonstrated in *At. ferrooxidans* (Elbehti et al. 2000; Brasseur et al. 2002). An operon encoding a bc_1 complex has also been characterized in the *At. ferrooxidans* ATCC19859 (Levican et al. 2002) and ATCC33020 (Bruscella 2004) strains. This operon has been shown to be more highly expressed in Fe(II)-grown cells than in S^0-grown cells (Bruscella 2004; Quatrini et al. 2005a), in agreement with its role proposed in the "uphill" pathway between Fe(II) and NAD(P). This operon encodes, in addition to the three subunits of the bc_1 complex (PetA1B1C1), a cytochrome *c*

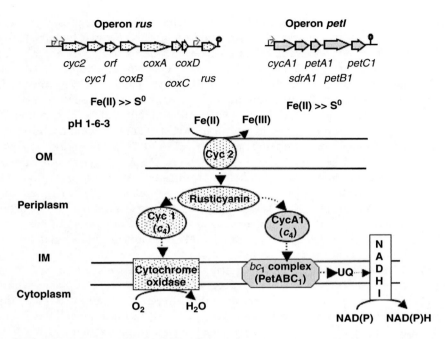

Fig. 14.1. Proposed model for Fe(II) energetic metabolism in *Acidithiobacillu ferrooxidans*. The transcriptional units and the corresponding redox proteins are presented with the same background pattern. *Dotted lines* represent the electron transport. The *small arrows* and *black dots* in the transcriptional units represent predicted promoters and *rho*-independent stop sites, respectively. *IM* inner membrane, *OM* outer membrane

(CycA1) and a ribitol/glucose dehydrogenase (SdrA1), the latter of unknown function (Levican et al. 2002). The cytochrome c_4 encoded by the *cycA1* gene has been characterized and proposed to belong to the electron transfer chain between Fe(II) and oxygen, and more precisely to receive the electrons directly from Fe(II) and to transfer them to rusticyanin (Giudici-Orticoni et al. 2000). Because *cycA1* belongs to the *petI*, and not to the *rus* operon, this cytochrome is more likely to be involved in the same electron transfer chain than the bc_1 complex, i.e., the reverse electron pathway between Fe(II) and NAD(P) (Bruscella 2004; Quatrini et al. 2005a).

In the reverse electron pathway, the bc_1 complex receives electrons from a cytochrome *c* and transfers them to the quinol pool (Griesbeck et al. 2000). This cytochrome *c* has been suggested to be the cytochrome c_4 encoded by the *cycA1* gene (Bruscella 2004; Quatrini et al. 2005a; Fig. 14.1). Fe(II) oxidation and NAD(P) reduction have been proposed to be coupled to explain the balance of the reducing equivalent from Fe(II) between the two pathways: the exergonic one, through the aa_3-type oxidase toward oxygen, and the endergonic one, through a bc_1 complex toward NAD(P) (Elbehti et al. 2000). As previously reported (Bruscella 2004; Brasseur et al. 2004; Quatrini et al., 2005a), the bifurcation is likely at the level of rusticyanin, which gives electrons to two different cytochromes c_4: CycA1 encoded by the *petI* operon, or Cyc1 encoded by the *rus* operon. In the former case, electrons are transferred to the endergonic pathway, while in the later case, they are transferred to the exergonic pathway (Fig. 14.1). Recent evidence from a combination of inhibitors and electron uncouplers supports this model (Chen and Suzuki 2005).

14.4.2   Sulfur Oxidation

Unraveling the biochemistry of sulfur oxidation has proved particularly challenging. Sulfur can exist in various oxidation states from -2 to $+6$, complicating the resolution of the enzymatic steps involved. Also, some steps can proceed spontaneously, exacerbating the search for reactions catalyzed by enzymes. Sulfur oxidation is widespread in prokaryotes and can proceed by a number of distinct routes. Recent genomic and biochemical data suggest that the sulfur oxidation pathways in archaea and bacteria differ substantially (Friedrich et al. 2005). Furthermore, acidophilic bacteria, including *At. ferrooxidans* and *At. thiooxidans*, oxidize sulfur by a system that is different from that for the majority of other bacteria (Friedrich et al. 2005). For example, neither the *sox* genes that encode the sulfur-oxidizing system widely distributed in a number of bacteria nor the *sor* gene encoding the archaeal-type sulfur oxygenase reductase were identified in the partial genome sequence of the type strain ATCC23270 (Urich et al. 2004; Friedrich et al. 2005), an observation that has now been confirmed by inspection of the complete genome sequence of this bacterium (Appia-Ayme, Bonnefoy, Quatrini, and Holmes unpublished data). It is clear that the sulfur oxidation pathways of *At. ferrooxidans* are different from those of most archaea and bacteria, which makes their study an exciting challenge.

Several enzymes of *At. ferrooxidans* have been suggested to be involved in the oxidation of sulfur, sulfide, and reduced inorganic sulfur compounds (Fig. 14.2): a thiol-bearing outer-membrane protein, which mobilizes elemental sulfur and transports it into the periplasmic space as persulfide sulfur, a sulfur dioxygenase which oxidizes persulfide sulfur to sulfite (Silver and Lundgren 1968a; Rohwerder and Sand 2003; Rohwerder et al. 2003), a sulfite: oxidoreductase which oxidizes sulfite to sulfate (Vestal and Lundgren 1971), a sulfide:quinone oxidoreductase which oxidizes sulfide to sulfur (Wakai et al. 2004), a thiosulfate oxidase which catalyzes the oxidation of thiosulfate to tetrathionate (Silver and Lundgren 1968b), a rhodanase which splits the thiosulfate to sulfur and sulfite (Tabita et al. 1969), and a tetrathionate hydrolase which hydrolyzes tetrathionate to thiosulfate, sulfur, and sulfate (De Jong et al. 1997). While the enzymatic steps to oxidize polythionates are still not very clear, it is now generally agreed that S^0 is transported into the periplasmic space as persulfide-sulfur where it is oxidized by sulfur dioxygenase to sulfite, which is further oxidized to sulfate by a sulfite oxidoreductase

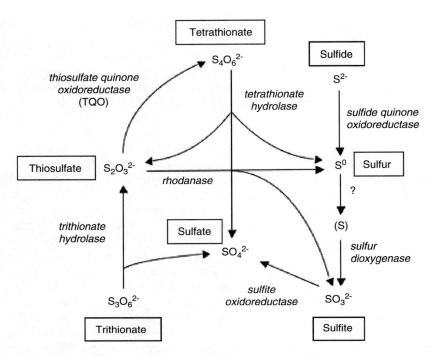

Fig. 14.2. Proposed enzymatic steps involved in the oxidation of inorganic sulfur compounds by *At. ferrooxidans* (after Kuenen et al. 1993; Rohwerder and Sand 2003)

(Pronk et al. 1990; Kuenen et al. 1993; Lorbach et al. 1993; Rohwerder and Sand 2003).

Some of the electrons that are transferred in the reactions described previously can be fed into the respiratory chain for energy conservation. Corbett and Ingledew (1987) proposed that the electrons from S^0 oxidation are passed to oxygen through a bc_1 complex and a terminal oxidase. However, lower levels of total cytochromes c were observed in S^0-grown cells compared with Fe(II)-grown cells, suggesting that these electron transporters are not absolutely required for S^0 oxidation (Yarzabal et al. 2002a). The same conclusion was inferred from the analysis of a *resB* mutant in which cytochromes c cannot be matured but in which growth on S^0 is not affected (Cabrejos et al. 1999). In addition, oxygen reduction in Fe(II)-grown cells and S^0-grown cells was suggested to occur via two separate terminal oxidases (Pronk et al. 1991; Harahuc et al. 2000). Together, these data suggested that electrons from S^0 are transferred to oxygen by at least two respiratory chains, one of which is devoid of cytochrome c. Indeed, a branching point at the level of the quinol pool has been demonstrated and the electrons can either be transferred to a bd-type oxidase or to a ba_3-type (or bo_3-type, depending on the strain) oxidase via a bc_1 complex (Brasseur et al. 2004). This bc_1 complex has been shown to function in direct (downhill) mode (Brasseur et al. 2004), in contrast to the one detected in Fe(II)-grown cells which can function only in reverse ("uphill") mode (Brasseur et al. 2002). The most likely explanation is therefore that *At. ferrooxidans* has two bc_1 complexes, one functioning in direct mode in S^0-grown cells, and one functioning in reverse mode in Fe(II)-grown cells. Indeed, two operons encoding bc_1 complexes have been detected in the genome sequence of the ATCC23270 strain (Brasseur et al. 2002; Bruscella 2004). These two operons are also present in strains ATCC33020 and ATCC19859, indicating that the presence of two bc_1 complexes is a general property of bacteria categorized as "*At. ferrooxidans*," a feature that is unique so far to this species (Bruscella 2004). Surprisingly, not only *petA*, *petB* and *petC* genes are duplicated but also the cytochrome *c* gene, *cycA*, and the ribitol/glucose dehydrogenase gene, *sdrA*. In addition, in the second operon, referred to as *petII*, a gene encoding a HiPIP, *hip*, is located downstream from *petC2*. All these genes, including the *hip* gene, have been shown to be cotranscribed (Bruscella 2004). Furthermore, they are expressed in S^0-grown cells (Bruscella 2004; Quatrini et al. 2005a), suggesting that the proteins encoded by this operon, including the HiPIP, are involved in S^0 oxidation. Therefore, the bc_1 complex encoded by *petI* is the one functioning in reverse and transfers the electrons from Fe(II) to NAD(P), while the bc_1 complex encoded by *petII* is the one functioning in direct mode and transfers electrons from S^0 to oxygen. When functioning in direct mode, the bc_1 complex receives electrons from the quinol pool and transfers them to a membrane-bound cytochrome *c*, and/or to a soluble redox protein such a cytochrome *c*, or to a HiPIP which then transfers the electron to the terminal oxidase where oxygen reduction takes place (Trumpower 1990; Bonora et al. 1999; Pereira et al. 1999). Since a membrane-bound cytochrome *c* (CycA2) and a periplasmic HiPIP are encoded by the *petII* operon, these two redox proteins are good candidates for transferring electrons between the two integral membrane complexes: the bc_1 and the terminal oxidase (Bruscella 2004; Quatrini et al. 2005a; Fig. 14.3).

The model is far from being complete. For example, how are the electrons transferred to the quinol pool? Which is the terminal oxidase involved, since it is known from the genome sequence that there are at least three oxidases, an aa_3 type, encoded by the *rus* operon and involved in the downhill pathway between Fe(II) and oxygen, and a *bd* and a *bo*3 types (Brasseur et al. 2004). Furthermore, *At. ferrooxidans* is able to oxidize not only S^0, but also sulfide and reduced inorganic sulfur compounds, such as thiosulfate, tetrathionate, and sulfite, which are oxidized to sulfate. Where are the electrons produced by the oxidation reactions fed in the respiratory chain? To answer all these questions, a global approach using a genome-wide microarray transcript profiling analysis was undertaken to facilitate an overall view of the genes involved in S^0 or Fe(II) oxidation. Oligonucleotides corresponding to each of the genes of the *At. ferrooxidans* type strain ATCC23270 were spotted onto glass slides and hybridized with complementary DNA retrotranscribed from

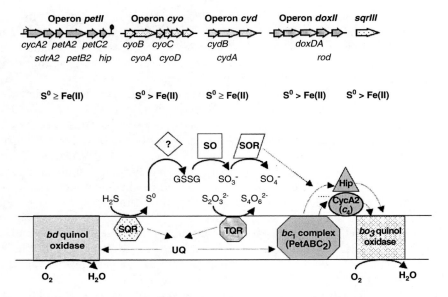

Fig. 14.3. Model proposed for sulfur energetic metabolism in *At. ferrooxidans*. The transcriptional units and the corresponding redox proteins are presented with the same background pattern. *Dotted lines* represent the electron transport. *GSSG* oxidized glutathione, *TQR* thiosulfate quinol reductase, *SQR* sulfide quinone reductase, *SO* sulfur oxygenase, *SOR* sulfite oxidoreductase, *Hip* high-potential iron–sulfur protein, *UQ* ubiquinone

RNA extracted from Fe(II)-grown cells and S^0-grown cells (Quatrini et al. 2005a). The expression of the redox protein encoding genes which are preferentially transcribed in Fe(II) or S^0 conditions has been validated by real-time PCR, Northern blot hybridization, and/or immunodetection analysis (Yarzabal et al. 2004; Bruscella 2004; Quatrini et al. 2005a). These transcriptome analyses have shown that the operons encoding the *bd* and the *bo$_3$* terminal oxidases, the operon encoding thiosulfate quinone reductase, and the gene encoding sulfide quinone oxidoreductase are more highly expressed in S^0 than in Fe(II) conditions (Quatrini et al. 2005a), in agreement with previous results obtained on the ATCC19859 strain (Brasseur et al. 2004; Ramirez et al. 2004; Acosta et al. 2005). These data suggest that the electrons from S^0 and reduced sulfur compounds enter the respiratory chain through the sulfide quinone oxidoreductase or the thiosulfate quinone reductase at the level of the quinol pool and through the sulfite oxidoreductase at the level of the cytochrome c_4 or of the HiPIP. These electrons can then either be transferred to a *bd* terminal oxidase or, through a *bc$_1$* complex, to a *bo$_3$* terminal oxidase (Fig. 14.3).

An alternative mechanism has been proposed in which S^0 oxidation is coupled to the reduction of Fe(III) (Sugio et al. 1985, 1987, 1988a, b, 1989, 1992a, b; Fig. 14.4), but this model has been contested mainly on the basis of bioenergetic considerations (Corbett and Ingledew 1987; Pronk et al. 1991; Kuenen et al. 1993).

14.4.3 Ferric Iron and Sulfur Reduction in *Acidithiobacillus ferrooxidans*

In large-scale leaching operations, where dissolved oxygen concentrations are low and Fe(III) concentrations high, anaerobic Fe(III) reduction may be an important process. It has been known for a number of years that *At. ferrooxidans* is able to reduce Fe(III) under anaerobic conditions coupled to the oxidation of S^0 (Brock and Gustafson 1976; Kino and Usami 1982; Sugio et al. 1985; Das et al. 1992; Pronk et al. 1992) or formate (Pronk et al. 1991).

Anaerobic S^0 oxidation coupled to Fe(III) reduction appears to be catalyzed by a series of enzymes apparently different from those involved in the aerobic oxidation of S^0 (Sugio et al. 1985): after chemical reduction of S^0 by glutathione to hydrogen sulfide (Sugio et al. 1989), a periplasmic hydrogen sulfide:Fe(III) oxidoreductase reduces hydrogen sulfide to sulfite (Sugio et al. 1987), and a membrane-bound sulfite:Fe(III) oxidoreductase reduces sulfite to sulfate (Sugio et al. 1988a, 1992a; Fig. 14.4).

Anaerobic S^0 oxidation coupled to Fe(III) reduction was shown to be an energy-transducing process (Pronk et al. 1991, 1992). The pathway of electrons in S^0 oxidation seems to be the same whether Fe(III) (anaerobiosis) or oxygen (aerobiosis) is the terminal electron acceptor (Corbett and Ingledew 1987). A pathway for aerobic and anaerobic oxidation of formate and S^0 was proposed (Corbett and Ingledew 1987; Pronk et al. 1991; Fig. 14.5). In this model, the electrons from formate and S^0 are transported to oxygen, or Fe(III), via a bc_1 complex. The oxidoreductase involved in the reduction of Fe(III) is proposed to be that which catalyzes the aerobic oxidation of Fe(II). Finally, different terminal oxidases are proposed to be involved in the reduction of oxygen to water.

Ohmura et al. (2002) have shown that some *At. ferrooxidans* strains are able to grow under anaerobic conditions using Fe(III) or S^0 as the electron acceptor and hydrogen as the electron donor. Furthermore, a 28-kDa cytochrome c has been shown to be synthesized in the cells growing by respiring Fe(III), but

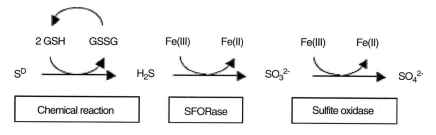

Fig. 14.4. Model proposed by Sugio et al. (1989) for the oxidation of S^0 coupled to the reduction of Fe(III). *SFORase* hydrogen sulfide:Fe(III) oxidoreductase; *GSH* glutathione

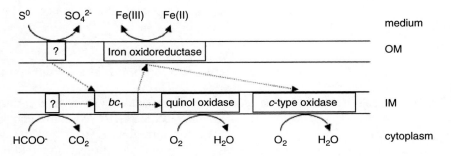

Fig. 14.5. Proposed model for the aerobic and anaerobic electron pathway in S^0 and formate oxidation. *Dotted lines* represent the electron transport. (Adapted from Pronk et al. 1991)

not S^0, with either hydrogen or S^0 as the electron donor. However, the type strain ATCC23270 was shown to be unable to grow on hydrogen by the anaerobic reduction of Fe(III).

14.5 Iron Oxidation in Other Bioleaching Microorganisms

14.5.1 Introduction

Although *At. ferrooxidans* remains the paradigm for understanding acidophilic Fe(II) and S^0 oxidation in acidic conditions, evidence for electron transport proteins involved in these two processes is beginning to be accumulated for other microorganisms involved in bioleaching or close relatives of these microorganisms (Table 14.1).

Table 14.1. Some of the microorganisms involved in bioleaching that can oxidize iron and/or sulfur

Fe(II) → Fe(III)	$S^0 \rightarrow SO_4$
Acidithiobacillus ferrooxidans	*Acidithiobacillus. ferrooxidans*
Leptospirillum spp.	*Acidithiobacillus thiooxidans*
Ferroplasma acidarmanus	*Acidithiobacillus caldus*
Acidimicrobium ferrooxidans	*Alicydobacillus tolerans*
Ferromicrobium acidophilus	*Sulfolobus sp.*
Alicydobacillus tolerans	*Thermoplasma acidophilum*
Sulfolobus metallicus	*Picrophilus oshimae*
Metallosphaera sp.	*Sulfobacillus acidophilus*
Acidianus sp.	*Metallosphaera sp.*
	Acidianus sp.

More than a decade ago, the acidophilic bacteria *At. ferrooxidans*, *L. ferrooxidans*, an unidentified iron-oxidizing bacterium (m1), and two archaea (*Sulfobacillus metallicus* and *Metallosphaera sedula*) were shown to contain spectrally distinct redox-active proteins during autotrophic growth on soluble Fe(II) (Barr et al. 1990; Blake et al. 1992, 1993). This suggested that different pathways or variations of pathways for Fe(II) oxidation have evolved, an idea that is largely supported by recent bioinformatic analysis of sequenced or partially sequenced genomes of several acidophilic bacteria and archaea, coupled with experimental evidence in a few cases.

14.5.2   *Ferroplasma* spp.

An electron transport chain involved in iron oxidation has been proposed for *Ferroplasma* type II and "*F. acidarmanus*" Fer1 based on an analysis of sequence information derived from an environmental genome shotgun library (Tyson et al. 2004; and reviewed in Golyshina and Timmis 2005). Genes coding for putative heme-copper terminal oxidases, cytochrome *b*, an associated Rieske iron–sulfur protein, and blue copper proteins were predicted. The blue copper protein had sequence similarity with rusticyanin of *At. ferrooxidans* and sulfocyanin (SoxE) of *Sulfolobus acidocaldarius*. It was suggested that in *Ferroplasma* type II, these proteins formed a terminal oxidase supercomplex similar to the SoxM supercomplex of *Sulfolobus acidocaldarius*. However, this view is not entirely supported from spectral studies of membrane proteins and proteomic analysis of *Ferroplasma* strains (Dopson et al. 2005). In addition, some genes of a SoxM complex are missing from the nearly complete genome of "*Fp. acidarmanus*" Fer1 and the complete genome of *Ferroplasma* type II.

A preliminary model for electron transport for "*Fp. acidarmanus*" Fer1 has been proposed by Dopson et al. (2005) (Fig. 14.6). This model was based on inhibitor studies, spectral information, and differential protein expression. Several key differences between this scheme and the one proposed for *At. ferrooxidans* (Fig. 14.1) can be detected. Firstly, the absence in *Ferroplasma* of an "uphill" electron pathway from Fe(II) to the NADH ubiquinone oxidoreductase complex. Presumably, this pathway is not required because low-potential electrons can be derived from the oxidation of organic compounds in *Ferroplama* and fed directly into the NADH ubiquinone oxidoreductase complex to generate reducing power. Secondly, *Ferroplama* lacks the outer-membrane cytochrome Cyc2, proposed to be the primary catalyst of Fe(II) oxidation in *At. ferrooxidans*. The role of Cyc2 is suggested to be assumed by the blue copper protein sulfocyanin. *Ferroplasma* is an archaeon and lacks the outer membrane found in Gram-negative bacteria such as *At. ferrooxidans*. Therefore, direct contact with insoluble substrates such as pyrite or soluble Fe(II) could presumably be made by the blue copper protein embedded in the cell envelope. Also, in *At. ferrooxidans* the cytochrome c_4 (Cyc1) is proposed to connect rusticyanin with the terminal oxidase, whereas it is absent in

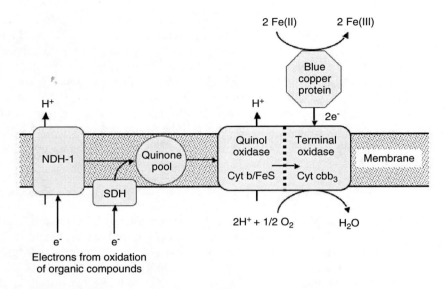

Fig. 14.6. Model for electron transport during chemotrophic or chemomixotrophic growth of "*Ferroplasma acidarmanus*" Fer1. *NDH-1* NADH ubiquinone oxidoreductase, *SDH* succinate dehydrogenase. (Redrawn after Dopson et al. 2005)

Ferroplasma and the blue copper protein connects directly with the terminal oxidase. Thus, the "downhill" electron components connecting Fe(II) oxidation to the terminal oxidase are fewer in *Ferroplasma* (one protein) than in *At. ferrooxidans* (three proteins). Finally, the cytochrome bc_1 complex of *At. ferrooxidans* postulated to be involved in "downhill" electron flow of electrons during S^0 oxidation has been replaced by a cytochrome *b*/FeS complex with similarity to the cytochrome *b* and Rieske protein of *At. ferrooxidans* but lacking the cytochrome c_1 of the complex.

14.5.3   *Leptospirillum* spp.

Leptospirillum spp. belong to the deep-branching *Nitrospirae* phylum, and consist of three main groups represented by *L. ferrooxidans* (group I), *L. ferriphilum* (group II), and *L. ferrodiazotrophum* (group III) (Tyson et al. 2005). On the basis of the almost complete genome sequence of a *Leptospirillum* group II like bacterium derived from an environmental genome shotgun library, Tyson et al. (2004) reported the discovery of putative genes potentially encoding a red cytochrome, a cytochrome cbb_3-type heme-copper terminal oxidase, a cytochrome *b* and an associated FeS-containing protein and a cytochrome *bd*-type quinol oxidase. A preliminary electron transport chain for *Leptospirillum* group II was proposed for Fe(II) oxidation, including both a "downhill" respiration and an "uphill" NADH synthesis electron flow

(Fig. 14.7). Most notably there is an absence of a blue copper protein corresponding to rusticyanin of *At. ferrooxidans* or sulfocyanin of *Ferroplasma*. The initial oxidation of Fe(II) is suggested to occur via the red cytochrome which is positioned, according to the scheme of Tyson et al. (2004), within the periplasmic space. Interestingly, a metaproteomic analysis of the community revealed the presence of copious quantities of this red cytochrome in the biofilm and it was suggested to play an important role in Fe(II) oxidation in the community (Ram et al. 2005).

Uphill electron flow is postulated to occur via a cytochrome b/FeS complex, similar to the bc_1 complex of *At. ferrooxidans* but lacking the cytochrome c_1 component. This then feeds electrons to a quinone pool and subsequently to a NADH ubiquinone oxidoreductase as has been proposed for *At. ferrooxidans*. Although not discussed by Tyson et al. (2004), implicit in this model is that the energy for "uphill" electron flow comes from the PMF as has been postulated for *At. ferrooxidans*. Clearly, additional bioinformatic analysis of potential genes and pathways involved in Fe(II) oxidation coupled with experimental validation is now required.

The dissimilarity of the components of the Fe(II) oxidation electron transfer pathways between *At. ferrooxidans* and *Leptospirillum* group II could account for the observed differences in their Fe(II) oxidation capabilities in bioleaching operations. Optimum bioleaching efficiency was obtained at lower substrate concentrations with *L. ferrooxidans* than with *At. ferrooxidans*

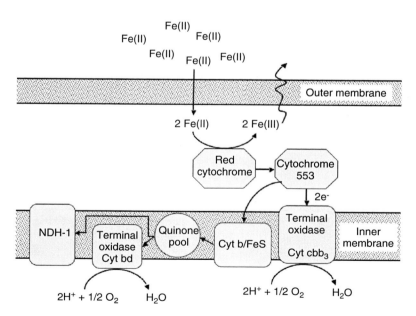

Fig. 14.7. Preliminary model of iron oxidation in *Leptospirillum* group II (redrawn according to Tyson et al. 2004)

(Sand et al. 1992). This may be explained by the greater affinity for Fe(II) of *L. ferrooxidans* (K_m=0.25 mM) compared with *At. ferrooxidans* (K_m=1.34 mM), implying that its Fe(II)-oxidizing system needs less substrate for saturation than the system of *At. ferrooxidans* (Norris et al. 1988). Also, the tolerance of *L. ferrooxidans* to Fe(III) is significantly greater than that of *At. ferrooxidans* (Norris et al. 1988).

Furthermore, while oxidation of Fe^{+2} by *At. ferrooxidans* was possible only at redox potentials of up to +850 mV, Fe(II) oxidation by *Leptospirillum* was able to occur at redox potentials of up to +950 mV (Boon et al. 1999) This accounts for the observation that *At. ferrooxidans* can outgrow *L. ferrooxidans* at high ratios of Fe(II) to Fe(III), which occurs during the earlier stages of Fe(II) oxidation in bioleaching, but that *L. ferrooxidans* outcompetes *At. ferrooxidans* once the Fe(III) concentration becomes high (Rawlings et al. 1999). An additional explanation that could play a role in accounting for the reduced Fe(II) oxidation capabilities of *At. ferrooxidans* at high Fe(III) concentrations is that this microorganism possesses more predicted Fe(II) and Fe(III) uptake complexes than *Leptospirillum* spp., perhaps rendering it more susceptible to higher Fe(III) concentrations (Quatrini et al. 2005b).

14.5.4 *Metallosphaera sedula*

A recent paper provides an initial glimpse at the proteins involved in electron transport in the thermoacidophilic crenarchaeon *M. sedula* (Kappler et al. 2005). Respiratory complexes were investigated when grown heterotrophically or chemolithotrophically on either S^0 or pyrite. Gene clusters, encoding two terminal oxidase complexes, a quinol oxidase SoxABCD and a SoxM oxidase supercomplex, were detected; the former was upregulated in cells grown on S^0 and the latter was upregulated when cells were grown on yeast extract. Both terminal oxidase complexes were downregulated when the cells were grown on pyrite, but there appeared to be oxidase-associated hemes in these conditions, suggesting the presence of additional, as yet uncharacterized, genes encoding terminal oxidases perhaps involved in Fe(II) oxidation. A gene cluster encoding a high-redox-potential membrane-bound cytochrome *b* and components of a bc_1 complex system was also detected. The cytochrome *b* was strongly upregulated when cells were grown on pyrite compared with yeast extract, suggesting a role for this protein in Fe(II) oxidation. This cytochrome *b* was not cotranscribed with the bc_1 complex genes and it was suggested that it was unlikely to be part of the bc_1 complex. No mention of the possible presence of a blue copper protein was made in the report. Further work is required to firmly establish its presence or absence.

14.5.5 Sulfur Oxidation in Other Bioleaching Microorganisms

Substantial progress has been made in understanding sulfur oxidation in a wide range of bacteria and archaea, including some known to be involved in bioleaching such as "*Fp. acidarmanus*" or close relatives of known bioleaching microorganisms such as *Acidianus ambivalens* and the reader is directed to an excellent recent review that covers current knowledge of prokaryotic sulfur oxidation (Friedrich et al. 2005).

14.6 Outstanding Questions and Future Directions

Of all the major energetic pathways in nature, Fe(II) oxidation is perhaps the least understood. One of the reasons for this lacuna in our knowledge is the apparent diversity of proteins that can extract electrons from iron and the multiplicity of ways to subsequently feed them into energy-yielding pathways. This would suggest that biological Fe(II) oxidation has evolved separately many times. However, future work might reveal common underlying mechanisms such as the use of multiheme cytochromes and small copper proteins that could be homologous members of multifamily proteins. If this proves to be the case, then Fe(II) oxidation might have evolved just a few times and the apparent diversity of pathways results from variations on a limited number of themes. This issue will become clearer as more genomes are sequenced.

More research is needed to understand Fe(III) reduction processes and how these might impact dump and heap bioleaching where at times, or in specific locations, there might be an inadequate supply of air to support biooxidation. Nor is enough known about S^0 oxidation and S^0 reduction to suggest ways that might prevent passivation of mineral surfaces by S^0 deposits resulting from mesophilic bacterial activity.

Cross-species genome analysis is already beginning to impact our understanding of bioleaching as shown, for example, by the discovery of possible reasons why *Leptospirillum* is outcompeted by *At. ferrooxidans* in early stages of a bioleaching operation but how, at later stages, it is able to continue biooxidation at higher Fe(III) loads that inactivate *At. ferrooxidans*.

An exciting potential of metagenomics is to provide community-wide assessment of metabolic and biogeochemical function. Analysis of specific functions across all members of a community can generate integrated models about how organisms share the workload of maintaining the nutrient and energy budgets of the community. The models can then be tested with genetic and biochemical approaches. The best example of such an analysis is the nearly complete sequencing of the metagenome of a community in acid drainage of the Richmond Iron Mountain mine (Tyson et al. 2005). The metagenomic sequence challenged a number of significant hypotheses. First, it appears that *Leptospirillum* group III contains genes with similarity to those known to be involved in nitrogen fixation, suggesting that it provides the community with fixed nitrogen. This was a surprise because the previous supposition was that a numerically dominant member of the community, such as *Leptospirillum* group II, would be responsible for nitrogen fixation.

However, no genes for nitrogen fixation were found in the *Leptospirillum* group II genome, leading the authors to suggest that the group III organism is a keystone species that has a low numerical representation but provides a service that is essential to community function.

Furthermore, the prevailing idea that *Ferroplasma* strains, including those found at Iron Mountain, can fix CO_2 has been challenged (Dopson et al. 2005). If it turns out that they are organomixotrophs incapable of fixing CO_2, then some other member such as *Leptospirillum* must be providing them with fixed carbon. Lessons learnt from the Iron Mountain metagenomic project can be applied to further our understanding of bioleaching. For example, it is already known that tank biooxidation (reviewed in Rawlings 2005) and heap bioleaching (Demergasso et al. 2005) proceed in three stages, resulting from temperature increases due to exothermic biological oxidation of Fe(II) and S^0: an early stage favoring mesophilic microorganisms (30–40°C) such as *At. ferrooxidans*, *At. thiooxidans* like bacteria, and *Sulfurisphaera*-like archaea; a second stage when the temperature begins to rise (40–55°C) when *At. caldus*, *Leptospirillum*, and *Ferroplasma* groups become dominant; and a final stage (55–65°C or higher) where *Sulfobacillus*-like and *Alicyclobacillus*-like bacteria (Karavaiko et al. 2005) become dominant and archaea such as *Ferroplasma* thrive. This means that the development and interaction of each of these microbial communities, including possible community biofilm formation in the case of heap bioleaching, must be considered in order to comprehend bioleaching processes and suggest ways by which bioleaching can be improved.

Acknowledgements. This work was supported by Fondecyt 1050063, Ecos/Conicyt, CNRS/Conicyt, "Geomex" and "Puces à ADN" from the Centre National de la Recherche Scientifique and "BIOMINE" European project (sixth PCRD no, NM2-Ct-2005-500329).

References

Acosta M, Beard S et al (2005) Identification of putative sulfurtransferase genes in the extremophilic *Acidithiobacillus ferrooxidans* ATCC23270 genome: structural and functional characterization of the proteins. OMICS 9:13–29

Appia-Ayme C (1998) Caractérisation d'un opéron codant pour sept proteins transporteurs d'électrons chez *Thiobacillus ferrooxidans*. Université de la Méditerranée, Aix-Marseille II

Appia-Ayme C, Guiliani N et al (1999) Characterization of an operon encoding two *c*-type cytochromes, an aa_3-type cytochrome oxidase, and rusticyanin in *Thiobacillus ferrooxidans* ATCC33020. Appl Environ Microbiol 65:4781–4787

Barr DW, Ingledew WJ, Norris PR (1990) Respiratory chain components of iron-oxidizing, acidophilic bacteria. FEMS Microbiol Lett 70:85–90

Bengrine A, Guiliani N et al (1995) Studies of the rusticyanin encoding gene of *Thiobacillus ferrooxidans* ATCC33020. In: Jerez CA, Vargas T, Toledo H, Wiertz J (eds) Biohydrometallurgical processing, vol 2. University of Chile, Santiago, pp 75–83

Bengrine A, Guiliani N et al (1998) Sequence and expression of the rusticyanin structural gene from *Thiobacillus ferrooxidans* ATCC33020 strain. Biochim Biophys Acta 1443:99–112

Blake RC 2nd, Shute EA (1994) Respiratory enzymes of *Thiobacillus ferrooxidans*. Kinetic properties of an acid-stable iron:rusticyanin oxidoreductase. Biochemistry 33:9220–9228

Blake RC 2nd, Shute EA, Waskovsky J, Harrison AP Jr (1992) Respiratory components in acidophilic bacteria that respire on iron. Geomicrobiol J 10:173–192

Blake RC 2nd, Shute EA et al (1993) Enzymes of aerobic respiration on iron. FEMS Microbiol Rev 11:9–18

Bonora P, Principi II et al (1999) On the role of high-potential iron-sulfur proteins and cytochromes in the respiratory chain of two facultative phototrophs. Biochim Biophys Acta 1410:51–60

Boon M, Brasser HJ, Hansford GS, Heijnen JJ (1999) Comparison of the oxidation kinetics of different pyrites in the presence of *Thiobacillus ferrooxidans* or *Leptospirillum ferroxidans*. Hydrometallurgy 53:57–72

Brasseur G, Bruscella P et al (2002) The bc_1 complex of the iron-grown acidophilic chemolithotrophic bacterium *Acidithiobacillus ferrooxidans* functions in the reverse but not in the forward direction. Is there a second bc_1 complex? Biochim Biophys Acta 1555:37–43

Brasseur G, Levican G et al (2004) Apparent redundancy of electron transfer pathways via bc_1 complexes and terminal oxidases in the extremophilic chemolithoautotrophic *Acidithiobacillus ferrooxidans*. Biochim Biophys Acta 1656:114–126

Brock TD, Gustafson J (1976) Ferric iron reduction by sulfur- and iron-oxidizing bacteria. Appl Environ Microbiol 32:567–571

Brown LD, Dennehy ME, Rawlings DE (1994) The F_1 genes of the F_1F_0 ATP synthase from the acidophilic bacterium *Acidithiobacillus ferrooxidans* complement *Escherichia coli* F_1 *unc* mutants. FEMS Microbiol Lett 122:19–26

Bruhn DF, Roberto FF (1993) Maintenance and expression of enteric arsenic resistance genes in *Acidiphilium*. In: Torma AE, Wey JE, Lakshmanan VI (eds) Biohydrometallurgical technologies, vol 2. The Minerals, Metals and Materials Society, Warrendale, pp 745–754

Bruscella P (2004) Etude des opérons *petI* et *petII* codant pour deux complexes bc_1 chez la bactérie acidophile chimioautotrophe stricte *Acidithiobacillus ferrooxidans*. Université de la Méditerranée, Aix-Marseille II

Bruscella P, Cassagnaud L et al (2005) The HiPIP from the acidophilic *Acidithiobacillus ferrooxidans* is correctly processed and translocated in *Escherichia coli*, in spite of the periplasm pH difference between these two microorganisms. Microbiology 151:1421–1431

Bruschi M, Cavazza C et al (1996) Biooxidation des minéraux sulfurés et dissolution de métaux par la bactérie acidophile: *Thiobacillus ferrooxidans*. Déchets 4:27–30

Cabrejos ME, Zhao HL et al (1999) IST1 insertional inactivation of the *resB* gene: implications for phenotypic switching in *Thiobacillus ferrooxidans*. FEMS Microbiol Lett 175:223–229

Casimiro DR, Toy-Palmer A et al (1995) Gene synthesis, high-level expression, and mutagenesis of *Thiobacillus ferrooxidans* rusticyanin: His 85 is a ligand to the blue copper center. Biochemistry 34:6640–6648

Cavazza C, Guigliarelli B et al (1995) Biochemical and EPR characterization of a high potential iron-sulfur protein in *Thiobacillus ferrooxidans*. FEMS Microbiol Lett 130:193–200

Chen Y, Suzuki I (2005) Effects of electron transport uncouplers on the oxidation compounds interacting *Acidithiobacillus ferrooxidans* Can J Microbiol 51:695–703

Corbett CM, Ingledew WJ (1987) Is $Fe^{3+/2+}$ cycling an intermediate in sulphur oxidation by Fe^{2+}-grown *Thiobacillus ferrooxidans*. FEMS Microbiol Lett 41:1–6

Cox JC, Boxer DH (1978) The purification and some properties of rusticyanin, a blue copper protein involved in iron(II) oxidation from *Thiobacillus ferrooxidans*. Biochem J 174:497–502

Cox JC, Boxer DH (1986) The role of rusticyanin, a blue copper protein, in the electron transport chain of *Thiobacillus ferrooxidans* grown on iron or thiosulfate. Biotechnol Appl Biochem 8:269–275

Das A, Mishra AK et al (1992) Anaerobic growth on elemental sulfur using dissimilar iron reduction by autotrophic *Thiobacillus ferrooxidans*. FEMS Microbiol Lett 97:167–172

De Jong GA, Hazeu W et al (1997) Polythionate degradation by tetrathionate hydrolase of *Thiobacillus ferrooxidans*. Microbiology 143:499–504

Delgado M, Toledo H et al (1998) Molecular cloning, sequencing, and expression of a chemoreceptor gene from *Leptospirillum ferrooxidans*. Appl Environ Microbiol 64:2380–2385

Demergasso CS, Galleguillos PA, Escudero LV, Zepeda VJ et al (2005) Molecular characterization of microbial populations in a low-grade copper ore bioleaching test heap. Hydrometallurgy 80:241–253

Dopson M, Baker-Austin C, Bond PL (2005) Analysis of differential protein expression during growth states of *Ferroplasma* strains and insights into electron transport for iron oxidation. Microbiology 151;4127–4137

Elbehti A, Brasseur G et al (2000) First evidence for existence of an uphill electron transfer through the bc_1 and NADH-Q oxidoreductase complexes of the acidophilic obligate chemolithotrophic ferrous ion-oxidizing bacterium *Thiobacillus ferrooxidans*. J Bacteriol 182:3602–3606

Friedrich CG, Bardischewsky F et al (2005) Prokaryotic sulfur oxidation. Curr Opin Microbiol 8:253–259

Fukumori Y, Yano T et al (1988) FeII oxidizing enzyme purified from *Thiobacillus ferrooxidans*. FEMS Microbiol Lett 20:169–172

Giudici-Orticoni MT, Leroy G et al (2000) Characterization of a new dihemic c_4-type cytochrome isolated from *Thiobacillus ferrooxidans*. Biochemistry 39:7205–7211

Giudici-Orticoni MT, Leroy G et al (2001) Two c-type cytochromes in *Thiobacillus ferrooxidans*: structure, comparison and functional role. In: Ciminelli VST, Garcia O Jr (eds) Biohydrometallurgy: fundamentals, technology and sustainable development, vol A. Elsevier, Amsterdam, pp 291–298

Glenn AW, Roberto FF et al (1992) Transformation of *Acidiphilium* by electroporation and conjugation. Can J Microbiol 38:387–393

Golyshina OV, Timmis KN (2005) *Ferroplasma* and relatives, recently discovered cell wall-lacking archaea making a living in extremely acid, heavy metal-rich environments. Environ Microbiol 7:1277–1288

Griesbeck C, Hauska G et al (2000) Biological sulfide oxidation: sulfide-quinone reductase (SQR), the primary reaction. In: Pandalai SG (ed) Recent research developments in microbiology, vol 4. Research Signpost, Trivadrum, pp 179–203

Guiliani N, Bengrine A et al (1995) Genetics of *Thiobacillus ferrooxidans*: advancement and projects: sequence and analysis of the *rus* gene encoding rusticyanin and *alaS* encoding alanyl-tRNA-synthetase. In: Holmes D, Smith RW (eds.) Mineral bioprocessing, vol 2. The Minerals, Metals and Materials Society, Warrendale, pp 95–110

Hall JF, Hasnain SS et al (1996) The structural gene for rusticyanin from *Thiobacillus ferrooxidans*: cloning and sequencing of the rusticyanin gene. FEMS Microbiol Lett 137:85–89

Harahuc L, Lizama HM et al (2000) Selective inhibition of the oxidation of ferrous iron or sulfur in *Thiobacillus ferrooxidans*. Appl Environ Microbiol 66:1031–1037

Hiraishi A, Nagashima KV et al (1998) Phylogeny and photosynthetic features of *Thiobacillus acidophilus* and related acidophilic bacteria: its transfer to the genus *Acidiphilium* as *Acidiphilium acidophilum* comb. nov. Int J Syst Bacteriol 48:1389–1398

Holmes DS, Zhao HL et al (2001) ISAfe1, an ISL3 family insertion sequence from *Acidithiobacillus ferrooxidans* ATCC19859. J Bacteriol 183:4323–4329

Inagaki K, Tomono J et al (1993) Transformation of the acidophilic heterotroph *Acidiphilium facilis* by electroporation. Biosci Biotechnol Biochem 57:1770–1771

Ingledew WJ (1982) *Thiobacillus ferrooxidans*. The bioenergetics of an acidophilic chemoautotroph. Biochim Biophys Acta 683:89–117

Ingledew WJ, Cobley JG (1980) A potentiometric and kinetic study on the respiratory chain of ferrous-iron-grown *Thiobacillus ferrooxidans*. Biochim Biophys Acta 590:141–158

Ingledew WJ, Cox JC et al (1977) A proposed mechanism for energy conservation during Fe^{2+} oxidation by *Thiobacillus ferrooxidans*: chemiosmotic coupling to net H+ influx. FEMS Microbiol Lett 2:193–197

Jin SM, Yan WM et al (1992) Transfer of IncP plasmids to extremely acidophilic *Thiobacillus thiooxidans*. Appl Environ Microbiol 58:429–430

Johnson DB, McGinness S (1991) Ferric iron reduction by acidophilic heterotrophic bacteria. Appl Environ Microbiol 57:207–211

Kappler U, Sly LI, McEwan AG (2005) Respiratory gene clusters of *Metallosphaera sedula* – differential expression and transcriptional organization. Microbiology 151:35–43

Karavaiko GI, Bogdanova TI, Tourova TP, Kondrat'eva TF, Tsaplina IA, Egorova MA, Krasil'nikova EN, Zakharchuk LM (2005) Reclassification of '*Sulfobacillus thermosulfidooxidans* subsp. *thermotolerans*' strain K1 as *Alicyclobacillus tolerans* sp. nov. and *Sulfobacillus disulfidooxidans* Dufresne *et al.* 1996 as *Alicyclobacillus disulfidooxidans* comb. nov., and emended description of the genus *Alicyclobacillus*. Int J Syst Evol Microbiol 55:941–947

Kino K, Usami S (1982) Biological reduction of ferric iron by iron- and sulfur-oxidizing bacteria. Agric Biol Chem 46:803–805

Kuenen JG, Pronk JT et al (1993) A review of bioenergetics and enzymology of sulfur compound oxidation by acidophilic thiobacilli. In: Torma AE, Apel ML, Brierley CL (eds) Biohydrometallurgical technologies, vol 2. The Minerals, Metals and Materials Society, Warrendale, pp 487–494

Kusano T, Sugawara K et al (1992a) Electrotransformation of *Thiobacillus ferrooxidans* with plasmids containing a *mer* determinant. J Bacteriol 174:6617–6623

Kusano T, Takeshima T et al (1992b) Molecular cloning of the gene encoding *Thiobacillus ferrooxidans* FeII oxidase. High homology of the gene product with HiPIP. J Biol Chem 267:11242–11247

Levican G, Bruscella P et al (2002) Characterization of the *petI* and *res* operons of *Acidithiobacillus ferrooxidans*. J Bacteriol 184:1498–1501

Liu Z, Guiliani N et al (2000) Construction and characterization of a *recA* mutant of *Thiobacillus ferrooxidans* by marker exchange mutagenesis. J Bacteriol 182:2269–2276

Liu Z, Borne F et al (2001a) Genetic transfer of IncP, IncQ and IncW plasmids to four *Thiobacillus ferrooxidans* strain by conjugation. Hydrometallurgy 59:339–345

Liu Z, Guiliani N et al (2001b) Mutagenesis by reverse genetics of the acidophilic chemolithoautotrophic *Acidithiobacillus ferrooxidans*: construction of a *recA* mutant. In: Ciminelli VST, Garcia OJ (eds) Biohydrometallurgy: fundamentals, technology and sustainable development, vol A. Elsevier, Amsterdam, pp 489–498

Lorbach SC, Buonfiglio V et al (1993) Oxidation of reduced sulfur compounds by *Thiobacillus ferrooxidans* 23270 and *Thiobacillus ferrooxidans* FC. In: Torma AE, Apel ML, Brierley CL (eds) Biohydrometallurgical technologies. The Minerals, Metals and Materials Society, Warrendale, pp 443–452

Norris PR, Barr DW, Hinson D (1988) Iron and mineral oxidation by acidophilic bacteria: affinities for iron and attachment to pyrite. In: Norris PR, Kelly DP (eds) Biohydrometallurgy. Proceedings of the international symposium. Science and Technology Letters, Kew, pp 43–59

Ohmura N, Sasaki K et al (2002) Anaerobic respiration using Fe^{3+}, S^0, and H^2 in the chemolithoautotrophic bacterium *Acidithiobacillus ferrooxidans*. J Bacteriol 184:2081–2087

Peng J, Yan W, Bao X (1994a) Expression of heterologous arsenic resistance genes in the obligately autotrophic biomining bacterium *Thiobacillus ferrooxidans*. Appl Environ Microbiol 60:2653–2656

Peng JB, Yan W et al (1994b) Solid medium for the genetic manipulation of *Thiobacillus ferrooxidans*. J Gen Appl Microbiol 40:243–253

Peng JB, Yan WM et al (1994c) Plasmid and transposon transfer to *Thiobacillus ferrooxidans*. J Bacteriol 176:2892–2897

Pereira MM, JN Carita et al (1999) Membrane-bound electron transfer chain of the thermohalophilic bacterium *Rhodothermus marinus*: characterization of the iron-sulfur centers from the dehydrogenases and investigation of the high-potential iron-sulfur protein function by in vitro reconstitution of the respiratory chain. Biochemistry 38:1276–1283

Pronk JT, Meulenberg R et al (1990) Oxidation of reduced inorganic sulfur compounds by acidophilic thiobacilli. FEMS Microbiol Rev 75:293–306

Pronk JT, Meijer WM et al (1991) Energy transduction by anaerobic ferric iron respiration in *Thiobacillus ferrooxidans*. Appl Environ Microbiol 57:2063–2068

Pronk JT, De Bruyn JC et al (1992) Anaerobic growth of *Thiobacillus ferrooxidans*. Appl Environ Microbiol 58:2227–2230

Pulgar V, Nunez L, Moreno F et al (1993) Expression of rusticyanin gene is regulated by growth condition in *Thiobacillus ferrooxidans*. In: Torma AE, Wey JE, Lakshmanan VI (eds) Biohydrometallurgical technologies, vol 2. The Minerals, Metals and Materials Society, Warrendale, pp 541–548

Quatrini R, Appia-Ayme C et al (2005a) Global analysis of the ferrous iron and sulfur energetic metabolism of *Acidithiobacillus ferrooxidans* by microarrays transcriptome profiling. In: Harrison STL, Rawlings DE, Petersen J (eds) 16th international biohydrometallurgy symposium proceedings, Cape Town, pp 761–771

Quatrini R, Jedlicki E, Holmes DS (2005b) Genomic insights into the iron uptake mechanisms of the biomining microorganism *Acidithiobacillus ferrooxidans*. J Ind Microbiol Biotechnol 32:606–614

Quentmeier A, Friedrich CG (1994) Transfer and expression of degradative and antibiotic resistance plasmids in acidophilic bacteria. Appl Environ Microbiol 60:973–978

Ram RJ, VerBerkmoes NC et al (2005) Community proteomics of a natural microbial biofilm. Science 308:1915–1920

Ramirez P, Guiliani N et al (2004) Differential protein expression during growth of *Acidithiobacillus ferrooxidans* on ferrous iron, sulfur compounds, or metal sulfides. Appl Environ Microbiol 70:4491–4498

Rawlings DE (2005) Characteristics and adaptability of iron- and sulfur-oxidizing microorganisms used for the recovery of metals from minerals and their concentrates. Microb Cell Fact 4:13

Rawlings DE, Kusano T (1994) Molecular genetics of *Thiobacillus ferrooxidans*. Microbiol Rev 58:39–55

Rawlings DE, Tributsch H, Hansford GS (1999) Reasons why '*Leptospirillum*'-like species rather than *Thiobacillus ferrooxidans* are the dominant iron-oxidizing bacteria in many commercial processes for the biooxidation of pyrite and related ores. Microbiology 145:5–13

Roberto FF, Glenn AW, Bulmer D, Ward TE (1991) Genetic transfer in acidophilic bacteria which are potentially applicable in coal beneficiation. Fuel 70:595–598

Rohwerder T, Sand W (2003) The sulfane sulfur of persulfides is the actual substrate of the sulfur-oxidizing enzymes from *Acidithiobacillus* and *Acidiphilium* spp. Microbiology 149:1699–1710

Rohwerder T, Gehrke T, Kinzler K, Sand W (2003) Bioleaching review part A: progress in bioleaching: fundamentals and mechanisms of bacterial metal sulfide oxidation. Appl Microbiol Biotechnol 63:239–248

Sand W, Rohde K, Sobotke B, Zenneck C (1992) Evaluation of *Leptospirillum ferrooxidans* for leaching. Appl Environ Microbiol 58:85–92

Silver M, Lundgren DG (1968a) Sulfur-oxidizing enzyme of *Ferrobacillus ferrooxidans* (*Thiobacillus ferrooxidans*). Can J Biochem 46:457–461

Silver M, Lundgren DG (1968b) The thiosulfate-oxidizing enzyme of *Ferrobacillus ferrooxidans* (*Thiobacillus ferrooxidans*). Can J Biochem 46:1215–1220

Sugio T, Domatsu C et al (1985) Role of a ferric ion-reducing system in sulfur oxidation of *Thiobacillus ferrooxidans*. Appl Environ Microbiol 49:1401–1406

Sugio T, Mizunashi W et al (1987) Purification and some properties of sulfur:ferric ion oxidoreductase from *Thiobacillus ferrooxidans*. J Bacteriol 169:4916–4922

Sugio T, Katagiri T et al (1988a) Existence of a new type of sulfite oxidase which utilizes ferric ions as an electron acceptor in *Thiobacillus ferrooxidans*. Appl Environ Microbiol 54:153–157

Sugio T, Wada K et al (1988b) Synthesis of an iron-oxidizing system during growth of *Thiobacillus ferrooxidans* on sulfur-basal salts medium. Appl Environ Microbiol 54:150–152

Sugio T, Katagiri T et al (1989) Actual substrate for elemental sulfur oxidation by sulfur:ferric ion oxidoreductase purified from *Thiobacillus ferrooxidans*. Biochim Biophys Acta 973:250–256

Sugio T, Hirose T et al (1992a) Purification and some properties of sulfite:ferric ion oxidore-ductase from *Thiobacillus ferrooxidans*. J Bacteriol 174:4189–4192

Sugio T, White KJ et al (1992b) Existence of a hydrogen sulfide:ferric ion oxidoreductase in iron-oxidizing bacteria. Appl Environ Microbiol 58:431–433

Tabita R, Silver M et al (1969) The rhodanese enzyme of *Ferrobacillus ferrooxidans* (*Thiobacillus ferrooxidans*). Can J Biochem 47:1141–1145

Tian KL, Lin JQ et al (2003) Conversion of an obligate autotrophic bacteria to heterotrophic growth: expression of a heterogeneous phosphofructokinase gene in the chemolithotroph *Acidithiobacillus thiooxidans*. Biotechnol Lett 25:749–54

Trumpower BL (1990) Cytochrome bc_1 complexes of microorganisms. Microbiol Rev 54:101–109

Tyson GW, Chapman J et al (2004) Community structure and metabolism through reconstruc-tion of microbial genomes from the environment. Nature 284:37–43

Tyson GW, Lo I et al (2005) Genome-directed isolation of the key nitrogen fixer *Leptospirillum ferrodiazotrophum* sp. nov. from an acidophilic microbial community. Appl Environ Microbiol 71:6319–6324

Urich T, Bandeiras TM et al (2004) The sulphur oxygenase reductase from *Acidianus ambivalens* is a multimeric protein containing a low-potential mononuclear non-haem iron centre. Biochem J 381:137–146

Vestal JR, Lundgren DG (1971) The sulfite oxidase of *Thiobacillus ferrooxidans* (*Ferrobacillus ferrooxidans*). Can J Biochem 49:1125–1130

Wakai S, Kikumoto M, Kanao T, Kamimura K (2004). Involvement of sulfide:quinone oxidore-ductase in sulfur oxidation of an acidophilic iron-oxidizing bacterium, *Acidithiobacillus ferrooxidans* NASF-1. Biosci Biotechnol Biochem 68:2519–2528

Ward TE, Bruhn DF et al (1993) Characterization of a new bacteriophage which infects bacteria of the genus *Acidiphilium*. J Gen Virol 74:2419–2425

Yamanaka T, Fukumori Y (1995) Molecular aspects of the electron transfer system which par-ticipates in the oxidation of ferrous ion by *Thiobacillus ferrooxidans*. FEMS Microbiol Rev 17:401–413

Yamanaka T, Yano T et al (1991) The electron transfer system in an acidophilic iron-oxidizing bacterium. In: Mukohata Y (ed) New era of bioenergetics. Academic, Tokyo, pp 223–246

Yarzabal A, Brasseur G et al (2002a) Cytochromes *c* of *Acidithiobacillus ferrooxidans*. FEMS Microbiol Lett 209:189–195

Yarzabal A, Brasseur G et al (2002b) The high-molecular-weight cytochrome *c* Cyc2 of *Acidithiobacillus ferrooxidans* is an outer membrane protein. J Bacteriol 184:313–317

Yarzabal A, Duquesne K, Bonnefoy V (2003) Rusticyanin gene expression of *Acidithiobacillus ferrooxidans* ATCC33020 in sulfur- and in ferrous iron-media. Hydrometallurgy 71:107–114

Yarzabal A, Appia-Ayme C et al (2004) Regulation of the expression of the *Acidithiobacillus fer-rooxidans rus* operon encoding two cytochromes *c*, a cytochrome oxidase and rusticyanin. Microbiology 150:2113–2123

Zhao HL, Holmes DS (1993) Insertion sequence IST1 and associated phenotypic switching in *Thiobacillus ferrooxidans*. In: Torma AE, Apel ML, Brierley CL (eds) Biohydrometallurgical technologies, vol 2. The Minerals, Metals and Materials Society, Warrendale, pp 667–671

Index